Trends in Sustainable Buildings and Infrastructure

Trends in Sustainable Buildings and Infrastructure

Editors

Víctor Yepes
Ignacio Navarro Martínez

MDPI • Basel • Beijing • Wuhan • Barcelona • Belgrade • Manchester • Tokyo • Cluj • Tianjin

Editors
Víctor Yepes
Universitat Politècnica de
València
Spain

Ignacio Navarro Martínez
Universitat Politècnica de
València
Spain

Editorial Office
MDPI
St. Alban-Anlage 66
4052 Basel, Switzerland

This is a reprint of articles from the Special Issue published online in the open access journal *International Journal of Environmental Research and Public Health* (ISSN 1660-4601) (available at: https://www.mdpi.com/journal/ijerph/special_issues/sustainable_infrastructure).

For citation purposes, cite each article independently as indicated on the article page online and as indicated below:

LastName, A.A.; LastName, B.B.; LastName, C.C. Article Title. *Journal Name* **Year**, *Volume Number*, Page Range.

ISBN 978-3-0365-0914-3 (Hbk)
ISBN 978-3-0365-0915-0 (PDF)

Contents

About the Editors

Víctor Yepes is a full professor of Construction Engineering; he holds a Ph.D. degree in civil engineering. He serves at the Department of Construction Engineering, Universitat Politècnica de València, Valencia, Spain. He has been the Academic Director of M.Sc. studies in concrete materials and structures since 2007 and a Member of the Concrete Science and Technology Institute (ICITECH). He is currently involved in several projects related to the optimization and life-cycle assessment of concrete structures, as well as optimization models for infrastructure asset management. He is currently teaching courses in construction methods, innovation, and quality management. He authored more than 250 journals and conference papers, including more than 100 papers published in journals quoted in *JCR*. He acted as an expert for project proposal evaluations for the Spanish Ministry of Technology and Science, and he is a main researcher for many projects. He currently serves as an Editor-in-Chief of the *International Journal of Construction Engineering and Management* and a member of the editorial board of 12 international journals (*Structure & Infrastructure Engineering, Structural Engineering and Mechanics, Mathematics, Sustainability, Revista de la Construcción, Advances in Civil Engineering,* and *Advances in Concrete Construction,* among others).

Ignacio Navarro Martínez holds a Ph.D. degree in civil engineering. He works at the Department of Construction Engineering, Universitat Politècnica de València, Valencia, Spain. He has published 11 articles and 9 conference papers in *JCR* journal. He combines his research activity with his professional career as a structural designer. During his professional experience, he has been dedicated to the calculation of steel and concrete structures related to renewable energies, especially in the field of wind energy, both onshore and offshore, as well as to the calculation of road and port structures. He has specialized in the numerical calculation of steel and concrete structures in onshore and offshore environments.

Preface to "Trends in Sustainable Buildings and Infrastructure"

The Sustainable Development Goals agreed by the United Nations in 2015 advocate for a profound paradigm shift in the way that infrastructures are designed. Actual practices usually fall short in assessing issues beyond the economic ones. Aspects such as the environmental impacts resulting from the life cycle of our structures, as well as the positive and negative effects that their construction and maintenance can have on society, are new criteria that need to be effectively included in our designs by 2030. To face such a challenging task, actual practices need to be reinvented, approaching the design of infrastructures from a holistic perspective that simultaneously integrates each of the three dimensions of sustainability, namely economy, environment and society.

This book comprises 11 chapters that explore the actual sustainability-related trends in the construction sector. The chapters collect the papers included in the Special Issue "Trends in Sustainable Buildings and Infrastructure" of the *International Journal of Environmental Research and Public Health*.

We would like to thank both the MDPI publishing and editorial staff for their excellent work, as well as the authors who have collaborated in its preparation. The papers included in this book cover a broad range of topics directly related to the sustainable design of infrastructures, addressing maintenance design criteria towards sustainability, life-cycle-oriented building and infrastructure design, design optimization based on sustainable criteria, inclusion of the social dimension in the design of infrastructures and the application of decision-making processes that effectively integrate the three dimensions of sustainability, resilience and the use of sustainable materials.

Víctor Yepes, Ignacio Navarro Martínez
Editors

International Journal of
Environmental Research and Public Health

MDPI

Article

Investigating the Dynamics of China's Green Building Policy Development from 1986 to 2019

Zezhou Wu [1], Qiufeng He [1], Kaijie Yang [1], Jinming Zhang [2,*] and Kexi Xu [3]

[1] Sino-Australia Joint Research Center in BIM and Smart Construction, Shenzhen University, Shenzhen 518060, China; wuzezhou@szu.edu.cn (Z.W.); 1900474011@email.szu.edu.cn (Q.H.); 2060474011@email.szu.edu.cn (K.Y.)
[2] School of Political Studies, Nanjing Agricultural University, Nanjing 210095, China
[3] School of Public Administration, Zhejiang University of Finance and Economics, Hangzhou 310018, China; xkxzj2017@zufe.edu.cn
*   Correspondence: zjming@njau.edu.cn

Abstract: China has enacted numerous green building policies (GBPs) to promote green building (GB) development in the past decades. Investigating the evolution characteristics of China's GBPs is significant for the future optimization of the GBP system. However, few studies on this topic have been conducted. To bridge this research gap, this paper adopted the methods of bibliometric analysis and text mining to probe the dynamic evolution of the GBPs in China. Firstly, a total 199 collected policies from 1986 to 2019 were grouped into five stages according to the Five-Year Plan. Then, the topics emphasized in different stages and the cooperative relationships among policymaking agencies were discovered by mapping and visualizing the co-word network and co-author network. Based on the derived results, an in-depth discussion was further conducted from five aspects: targets, objects, instruments, GB performance indicators, and the collaboration structure of policymaking agencies. It was revealed that the topics of GBPs evolved from macro to specific, and the types of policy targets, objects, instruments, and GB performance indicators evolved from few to multiple. Additionally, the collaboration structure of policymaking agencies went from dispersive to centralized. This study sheds lights on the dynamic evolution of China's GBPs and provides valuable references for other countries in need.

Keywords: green building; policy evolution; bibliometric analysis; text mining; China

Citation: Wu, Z.; He, Q.; Yang, K.; Zhang, J.; Xu, K. Investigating the Dynamics of China's Green Building Policy Development from 1986 to 2019. *Int. J. Environ. Res. Public Health* **2021**, *18*, 196. https://doi.org/10.3390/ijerph18010196

Received: 19 November 2020
Accepted: 25 December 2020
Published: 29 December 2020

Publisher's Note: MDPI stays neutral with regard to jurisdictional claims in published maps and institutional affiliations.

1. Introduction

Due to the overexploitation and uncontrolled use of resources, global resources are being consumed at an alarming rate and the global environment is being seriously damaged [1–3]. At present, the world is faced with increasing greenhouse gas emissions, reducing forest cover, diminishing biodiversity, and depleting freshwater systems and natural resources [4–10]. To some extent, the building sector should be responsible for such resource abuse and environmental damage because it has brought a lot of negative impacts to society, including waste of building materials, dust production, water pollution, high energy consumption, harmful gases, and so on [11–16]. Against this backdrop, there are more and more calls to promote the sustainable development of the building sector, and green building (GB) came into being due to its advantages of minimizing the negative impacts on the environment and improving the living conditions of occupants [17–19]. Given these advantages, GB has been advocated and promoted all over the world as a guiding paradigm for the sustainable development of the building sector [20,21].

Green building policies (GBPs) are regarded as playing an important role in promoting GB practice [22,23]. In China, the government has issued numerous policies to promote green building development. As early as 1986, the promulgation of the "Design Standard for Energy-Saving of Civil Buildings (Heating Residential Buildings)" marked the beginning of GBP formulation in China. Since then, to better guide and support green building practice, the Chinese government has been constantly improving the GBP system and shifting the topics of GBPs [24]. For instance, the focus of the topic "GB technology" shifted several times over the past decades, from the new wall material technology in 1992 to resource utilization of construction waste and prefabricated technology in 2013 [25]. In addition, in terms of the green building assessment standard, the focus turned from emphasizing resource saving and environmental protection technologies to people-oriented aspects [26].

As the topics of GBPs keep shifting, a clear understanding of the policy evolution dynamics is essential for stakeholders to better grasp the key points of green building development. For GB practitioners, understanding the dynamics of GBPs can help them to address the most innovative GB technologies and the most recent incentive measures. For policymakers, the analysis of the collaboration structures can facilitate better understanding the distribution of responsibility in different government departments with respect to policy design. In addition, since China owns one of the largest construction industries in the world [27], a study of the dynamic evolution of China's GBPs can offer a valuable reference for other countries in need.

In existing literature, however, few studies have attempted to systematically investigate the dynamics of GBPs in China. Some existing studies just investigated green building policy in a short historical period (e.g., the 11th Five-Year Plan period) or mainly focused on specific policies (e.g., green retrofit policies). For instance, Shi et al. [28] evaluated the effectiveness of green building policies during the 11th Five-Year Plan period. Liu et al. [29] reviewed China's green retrofit policies during 1996–2019 and identified the deficiencies of the current policy system. Ye et al. [30] investigated more than 70 green building standards in China and proposed development suggestions for green building standards. To bridge this research gap, this paper conducts a bibliometric analysis and text mining to investigate the dynamics of China's national-level GBPs from 1986 to 2019.

The data collection and analysis methods are described in Section 2. Then, the topics in different stages and the collaboration structures of government departments are analyzed in Section 3. Following the analysis results, China's GBPs are discussed from five aspects (e.g., policy targets, objects, instruments, GB performance indicators, and the collaboration structure of policymaking agencies) in Section 4. Finally, a conclusion is given in Section 5.

2. Materials and Methods

2.1. Data Collection

The policy documents investigated in this research were retrieved from three official websites and two well-known databases. The three official websites included the Central Government, the Ministry of Housing and Urban-Rural Development (MOHURD), and the National Development and Reform Commission, which are usually used to publish national-level policies related to the building sector [31]. The two databases, namely, the Magic Weapon of Peking University and the China's National Knowledge Infrastructure database [32], were selected to serve as supplementary retrieval databases to ensure the integrity of data collection. The two databases were chosen because they are the largest policy database and the largest academic database in China, respectively.

According to the release date of China's first GBP, the retrieval time range was determined as being from 1 January 1986 to 31 September 2019. The search keywords were retrieved from numerous review literature on GB [33–37] covering green buildings, ecological buildings, sustainable buildings, high-performance buildings, green building technology, and green construction. The keywords were applied to the content of policy and the document type was set at the national level. Initially, a total of 557 policy documents were collected from all websites and databases. A GBP document usually constitutes a title and contents. As the search keywords were applied to the policy content, it was possible that some of the retrieved documents just mentioned those keywords rather than explaining or introducing them in detail. Thus, a manual check was subsequently conducted to filter the initially collected policies according to the following criteria: (1) The policies that were duplicates or had no substantive information related to GB were eliminated. (2) The policies needed to be in the form of a law, regulation, measure, notice, opinion, or other document representing government policy, excluding news reports or government daily work report documents. (3) The policy needed to be a national-level policy issued by the central government or its directly affiliated agencies. Ultimately, 199 national GBPs were obtained for policy analysis. The number distribution of the identified GBPs is shown in Figure 1.

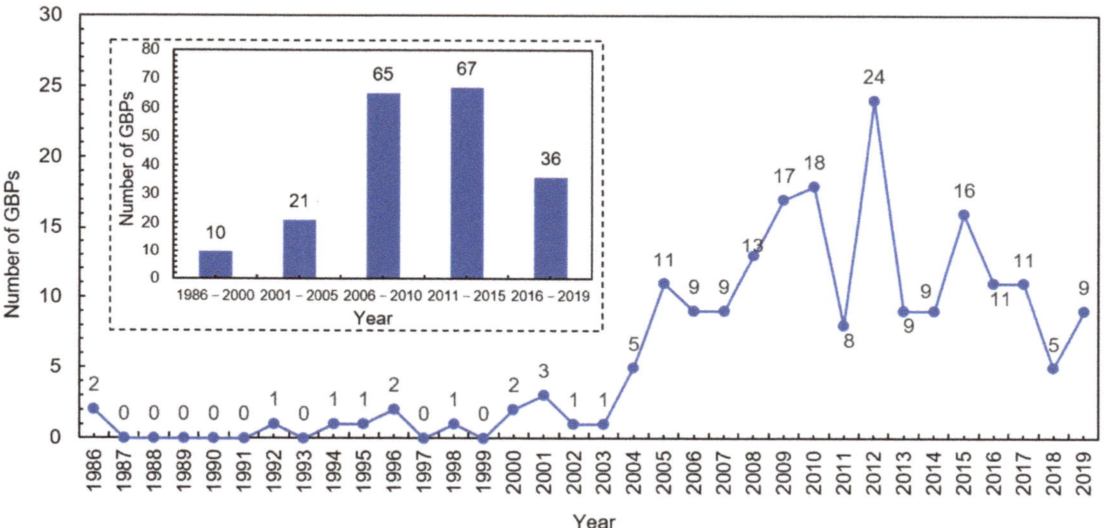

Figure 1. Distribution of China's green building policies (GBPs) from 1986 to 2019.

From Figure 1, it can be seen that before 2004, the number of GBPs published each year was less than four. However, since 2004, the number of GBPs has increased steadily, with five or more GBPs published each year. According to the number of policies published each year and the Five-Year Plan, five stages (i.e., 1986–2000, 2001–2005, 2006–2010, 2011–2015, and 2016–2019) were grouped for further analysis. The Five-Year Plan was selected because it is regarded one of the most important national policies in China and has been successfully applied in other policy studies [38,39].

2.2. Research Methods

2.2.1. Bibliometric Analysis

Bibliometric analysis employs a quantitative and visual processes approach for the description, evaluation, and monitoring of published research to measure scientific progress and production results in a specific field over a period of time [40]. This method has been applied in some previous studies related to policy analysis, and its effectiveness has been well confirmed [41–43]. Compared with qualitative research, bibliometric analysis reduces the dependence on researchers' knowledge and experience and makes the research results more repeatable and verifiable [31,41]. Bibliometric analysis mainly includes five specific methods: bibliographical coupling, co-citation analysis, citation analysis, co-author analysis, and co-word analysis [40,44]. The latter three methods are the most commonly employed in policy analysis. Citation analysis is usually applied to evaluate the impact of policies [43]. If a policy is heavily cited, it is considered to be important [40]. Co-author analysis uses co-authorship data to reveal the collaborative relationship between policymaking agencies [42]. Co-word analysis is usually utilized in conjunction with social network analysis to capture the historical dynamics of policy topics through the keywords in the policy [43]. In this study, co-word analysis and co-author analysis are adopted to analyze China's GBPs.

Co-word analysis is a content analysis technique that uses the keywords in documents to establish relationships and builds a conceptual structure of the domain [45,46]. This method has been used to search management information systems, analyze research trends [47], discover research hotspots [48], and identify the evolution of research topics [49]. Generally, co-word analysis of policy documents includes three steps [50]: (1) extracting 3–6 keywords from policy documents; (2) utilizing Bibexcel to establish a co-word matrix; (3) adopting the social network analysis to establish a modular matrix; and (4) employing Gephi software to generate a visualization network of the keywords by running its Layout module and Modularity module [51,52].

Co-author analysis explores the cooperative relationships among policymaking agencies in the release of policy documents, similar to the steps of co-word network analysis. Firstly, Bibexcel was utilized to establish the frequency statistics of policymaking agencies and to generate their co-occurrence matrix. Then, Gephi software was utilized to generate a visualization graph to clarify cooperative relationships among policymaking agencies.

2.2.2. Text Mining

Text mining is considered to be an effective solution to extract keywords from documents [43,53]. A three-sub-step approach was used in this step. The first step was to separate words. Based on the Jieba package in Python, a series of sentences was separated into individual words to reduce the dimension of the computer processing text. Nevertheless, some technical terms were not expected to be separated. It was expected that the term "green building" be presented in this way, instead of "green" and "building." Thus, the custom dictionary was implemented in this step. The second step was to remove stop words, which are meaningless and are frequently used in the document, such as "the first item" and "increase strength." Ultimately, term frequency–inverse document frequency (TF–IDF) was utilized to extract keywords for each document, which is an effective method to capture words that do not emerge frequently but are uniquely representative in different documents [54]. The specific calculation process of TF–IDF is shown in Equations (1)–(3) [55]:

$$tf_{ij} = \frac{n_{i,j}}{\sum_k n_{k,j}} \tag{1}$$

$$idf_i = log\frac{|D|}{|1 + \{j : t_i \in d_j\}|} \tag{2}$$

$$TF - IDF = tf_{i,j} \times idf_{i,j} \tag{3}$$

In Equation (1), i is a specific word, j is a document containing the word i, $n_{i,j}$ represents the number of times the word i appears in document j, and $\sum_k n_{k,j}$ is the sum of the occurrences of all the words in document j.

In Equation (2), $|D|$ is the total number of documents and $\{j : t_i \in d_j\}$ represents the number of documents containing the word i.

Equation (3) is the product of $tf_{i,j}$ with $idf_{i,j}$, where $tf_{i,j}$ is the frequency that word i appears in document j, and $idf_{i,j}$ is the frequency that word i appears in all documents.

3. Results

3.1. GBP Topics in Different Stages

The co-word network graphs of each stage visualized by Gephi are shown in Figures 2–6. In these figures, the node represents the keyword and its size implies the word frequency. The lines and their thicknesses represent the co-occurrence relationship and co-occurrence intensity of the keywords, respectively. All keywords in the same cluster are displayed as nodes with the same color and are used to explain policy topics [51].

3.1.1. Topics Discovered in Stage 1 (1986–2000)

The co-occurrence relationships of 35 keywords extracted from 13 GBPs of stage 1 are visualized in Figure 2. These keywords were clustered into five groups, representing the different GBP topics. It can be seen that "energy-saving," "technology," "wall material renovation," "standard," and "pilot demonstration" were the most important GBP topics in stage 1.

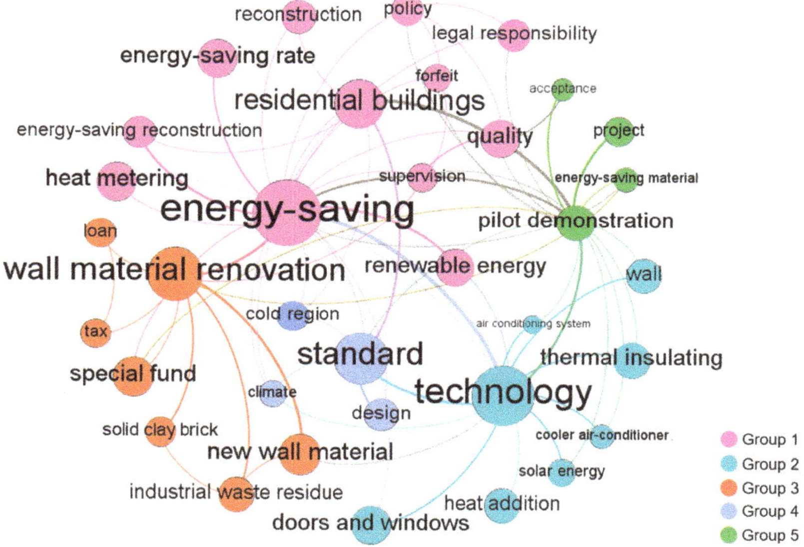

Figure 2. Co-word network of China's GBPs in stage 1.

"Energy-saving," the keyword in Group 1, had the highest frequency of occurrence in the co-word network. In 1994, MOHURD pointed out that "Building energy saving is the most direct and cheap fundamental measure to ease the energy shortage contradiction and reduce environmental pollution in China" and proposed that the energy-saving rate of new buildings should reach 50% by 2000 [56]. Thus, energy-saving was the main policy topic and GB performance indicator in stage 1. In this stage, the main object of GBPs was residential buildings, which was one of the high-frequency words in cluster 1. Technology was an important guarantee to achieve building energy-saving. In Group 2, "doors and windows," "wall," and "thermal insulating" were high-frequency keywords, indicating that energy-saving doors and windows and thermal insulation walls were the main energy-saving technology in stage 1. In Group 3, "wall material renovation" was identified as a high-frequency keyword. Due to new wall materials having potential advantages in the protection of cultivated land and the utilization of industrial waste residue, as early as 1992, the Chinese government issued the wall material innovation policy and implemented relevant financial incentive regulations such as special funds and tax reduction to promote the innovation of wall materials. In Group 4, "standard" was identified as a critical GBP topic. The Chinese government promulgated the first building energy-saving design standard in 1986 and revised it in 1995. This standard was only applicable to residential buildings in severe cold and cold regions of China—the keywords in Group 4 clearly reflect such a phenomenon. In Group 5, "pilot demonstration" had a lower frequency than other topic keywords, indicating that the pilot demonstration of green building has gradually attracted the attention of the government.

3.1.2. Topics Discovered in Stage 2 (2001–2005)

A total of 76 keywords were extracted to identify the policy topics in this stage. As shown in Figure 3, these keywords were clustered into seven groups. Generally, the five policy topics of the previous stage continued to play an important role. In addition, "supervision" and "innovation award" were added in this stage.

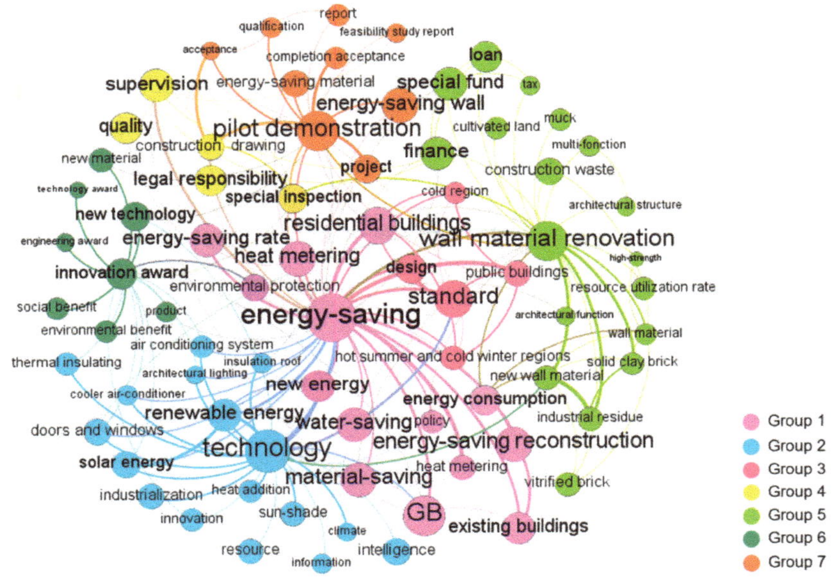

Figure 3. Co-word network of China's GBPs in stage 2.

"Energy-saving" was still the most frequent keyword in this stage, which confirmed that it remained a key policy target. Different from the previous stage, "energy-saving" had strong linkages with new keywords "public buildings" and "existing buildings" in Group 1. In 2002, MOHURD [57] announced that during the 10th Five-Year Plan period, the foci of building energy-saving work were to promote the energy-saving transformation of existing buildings and promote the energy-saving of public buildings. Consistent with this announcement, the main objects of GBPs during stage 2 expanded from residential buildings to public buildings and existing buildings. "Technology" was identified as a high-frequency keyword in Group 2. In 2005, the first "Technical Guidelines for Green Buildings" was issued. In these guidelines, MOHURD clarified the meaning of GBs, the key points technology, and the performance indicator system of green building for the first time. Among them, GB technology and GB performance indicators mainly included five aspects: water-saving, energy-saving, material-saving, land-saving, and indoor environmental quality. In Group 3, "standard" was identified as a high-frequency keyword. Since there were few standards related to GB in stage 1, MOHURD issued some energy-saving design standards suitable for different climatic regions and different building types in this stage, such as the Energy-Saving Design Standards for Public Buildings. In 2002, MOHURD [57] stated that during the 10th Five-Year Plan period, the Regulations on the Management of Energy-Saving of Civil Buildings must be fully implemented, which was the first document in China to incorporate quality supervision and management into the policy [58]. Against this backdrop, "supervision" in Group 4 was identified as a critical policy topic, and this keyword was closely related to "construction drawing," indicating that the government paid more attention to the supervision of the building design stage. In Group 5, "wall material renovation" was still a high-frequency keyword. Compared with the previous stage, "wall material renovation was closely related to the new keyword "architectural function." By reviewing the policy documents in this stage, it was found that the Chinese government proposed to develop new wall materials with different functions to meet the needs of different building structures and functions. "Innovation award" was identified as a high-frequency keyword in cluster 6 because MOHURD issued the National Green Building Innovation Award in this stage to promote the development of GB and its technology. In Group 7, the high-frequency keyword "pilot demonstration" was identified as a critical policy topic. In light of the fact that pilot demonstration projects play an important role in accumulating technical experience and influencing neighboring construction projects to adopt sustainable measures, the Chinese government proposed to carry out numerous pilot demonstration projects for GB in this stage so as to promote the popularization of green building concepts and the development of green building technology.

3.1.3. Topics Discovered in Stage 3 (2006–2010)

Regarding GBPs in stage 3, 138 keywords were extracted to identify the policy topics and were clustered into 10 groups. As shown in Figure 4, "energy-saving," "GB evaluation," "technology," "renewable energy building," "standard," "supervision," "special fund," "innovation award," "exposition," and "wall material renovation" were the most frequent keywords in each group.

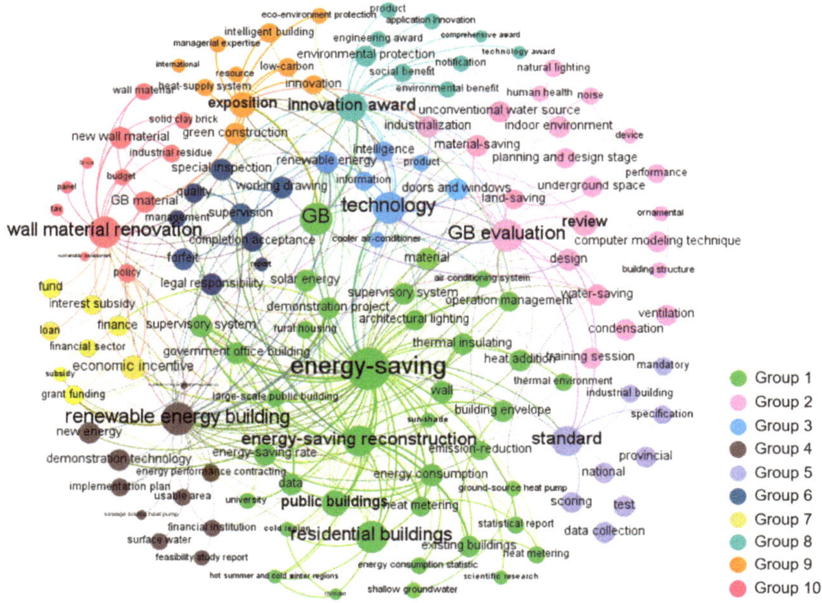

Figure 4. Co-word network of China's GBPs in stage 3.

In Group 1, "energy-saving" and "energy-saving reconstruction" increased compared with their frequency in the previous stage, which illustrated that the energy-saving reconstruction of existing buildings became the key policy target in this stage. For example, in the 11th Five-Year Plan Building Energy Conservation Task, the Chinese government clearly proposed that 150 million square meters of reconstruction area should be completed in 2010 [59]. In addition, an interesting finding is that "energy-saving" had strong linkages with the keywords "supervisory system" and "energy consumption," which showed that the Chinese government began to focus on the application of modern technology to better manage the using of energy in high-energy buildings [60,61]. "GB evaluation" was a new high-frequency keyword. Influenced by countries leading the way in GB development, MOHURD issued China's first green building evaluation standard in 2006. "Renewable energy building" was also a new high-frequency keyword in this stage. With the improvement of people's life quality, the contradiction between energy supply and demand in buildings was becoming more and more serious. Promoting the development of renewable energy buildings in the building sector became the most economical and reasonable choice [60,61]. In 2006, MOHURD and Ministry of Finance (MOF) [62] issued the "Implementation Opinions on Promoting the Application of Renewable Energy in the building sector," which clearly proposed that the policy and regulation system of renewable energy buildings should be basically formed in this stage. Against this backdrop, a large number of policies related to renewable energy building were issued. In Group 7, "special funds" was identified as a high-frequency keyword because the government began to attach importance to the financial incentive policies during this period and provide financial support for the building projects that met green building requirements. In 2010, the exposition initiated by MOHURD was held in China. According to the keywords in Group 9, it can be seen that the "exposition" mainly discussed the latest achievements, development trends, new technologies, and new products of GBs at home and abroad. This provided a learning platform for China to produce independent innovation technology.

3.1.4. Topics Discovered in Stage 4 (2011–2015)

In stage 4, 218 keywords were used to identify policy topics, and these keywords were divided into 13 groups, as shown in Figure 5. "GB material," "demonstration project," and "development planning" were new high-frequency keywords, reflecting three emerging topics of the GBPs in stage 4. The policy topics reflected by the remaining high-frequency keywords were similar to those in stage 3.

"Energy-saving" was still the highest frequency of occurrence in the co-word network. It is worth noting that "energy-saving" had a strong linkage with some new keywords in Group 1, including "industrialization," "prefabrication technology," and "prefabrication." Due to the advantages of saving material and protecting the environment, prefabrication technology has been identified as an effective technique to improve the environmental performance of buildings. In the "Implementation Opinions on Accelerating the Development of Green Buildings in China," MOHURD and MOF [63] mentioned promoting housing industrialization and promoting the use of prefabrication technology in buildings. In the "Green Building and Green Ecological Urban Development Planning" (hereinafter referred to as the "Planning"), MOHURD [64] proposed six suggestions to accelerate the development of the GB industry during the 12th Five-Year Plan period, including the use of prefabrication technology. In conclusion, similar to the study of Wang et al. [31], a strong linkage between "green building" and "prefabrication technology" appeared in stage 3. In addition, in the "Planning," MOHURD [64] proposed to formulate green building evaluation standards suitable for different climate regions and building types during the 12th Five-Year Plan period. Therefore, in Group 4, the keyword "standard" had a strong linkage with "hospitals," "data center buildings," and "existing buildings." In Group 5, the keyword "renewable energy building" was connected with some new keywords such as "energy management company" and "energy contract management," which reflected that the Chinese government was trying to promote energy-saving and emission reduction in the building sector through a market mechanism. In Group 11, "GB material" was identified as a high-frequency keyword that had a strong linkage with "informatization." In recent years, the Chinese government began to pay attention to the establishment of a GB material traceability information system by using two-dimensional code, cloud computing, and other technologies so as to improve the level of GB material logistics informatization and supply chain coordination.

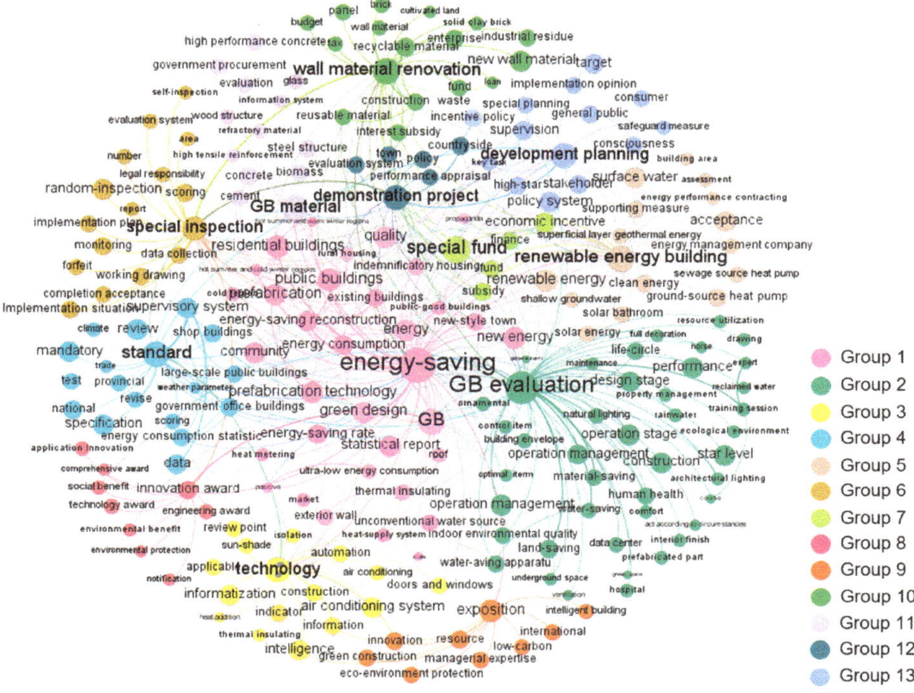

Figure 5. Co-word network of China's GBPs in stage 4.

3.1.5. Topics Discovered in Stage 5 (2016–2019)

In the last stage, 104 keywords were clustered into nine topics (see Figure 6): energy-saving, GB evaluation, technology, standard, supervision, GB material, construction waste, development planning, and laboratory.

"Energy-saving" remained the highest frequency unchanged. But different from the previous stage, "energy-saving" had a strong linkage with the new keywords "zero energy consumption building" and "mild region." In 2017, MOHURD [65] first mentioned zero energy consumption buildings [66], and released the Near Zero Energy Consumption Building Technical Standard, which indicated that a higher target for China's building energy-saving was established. In 2019, MOHURD issued the Design Standard for Energy Efficiency of Residential Buildings in the Mild Region. So far, China's design standards for building energy-saving covered all climatic regions of the country. In Group 2, the high-frequency keyword "GB evaluation" had a strong linkage with "green campus," "green hotel," and "post evaluation," which showed that the objects of GBPs were further expanded and that the Chinese government paid more attention to the quality supervision of GB evaluation projects in the operation stage. The keyword "technology" in Group 3 had a strong linkage with "operation and maintenance." In 2016, MOHURD issued the Technical Specification for Green Building Operation and Maintenance, which stipulated the key operation technologies and established the evaluation index system of GB operation and maintenance. The release of this policy once again confirmed that the Chinese government attached great importance to the operation stage of green building. The keyword "prefabricated building" and some high-frequency keywords increased compared with their frequency in the previous stage, such as "development planning" and "supervision." In light of the advantages of prefabricated buildings, the Chinese government gradually realized the importance of prefabricated technology in promoting GBD. Since 2016, the

Chinese government has repeatedly proposed to promote the integrated development of GBs and prefabricated buildings, and has formulated corresponding regulatory policies. "Construction waste" was a new high-frequency keyword, which had a strong linkage with "resource utilization" and "standard." With the rapid development of urban construction in China, more and more construction waste was produced in the process of urban demolition, housing construction, and decoration [67]. In the "Outline of the 13th Five-Year Plan," the State Council (SC) proposed to promote the resource utilization of construction waste. Subsequently, the Chinese government issued numerous policies related to construction waste to promote the healthy development of the whole construction waste treatment industry, such as the Industrial Standard for the Resource Utilization of Construction Waste. Against this backdrop, construction waste became a critical policy topic and a new performance indicator for GB in this stage. In Group 10, "laboratory" was a new emerging keyword. In 2018, the State Key Laboratory of GB in Western China was established in Shaanxi province, creating conditions for training high-level GB researchers.

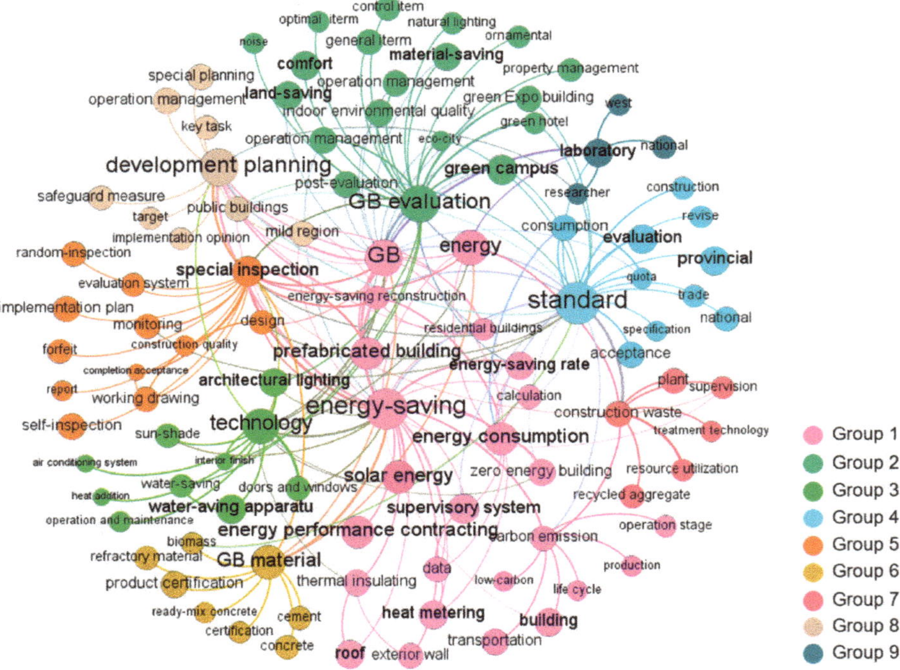

Figure 6. Co-word network of China's GBPs in stage 5.

3.2. GB Policymaking Agencies

A review of the GBPs issued from 1986 to 2019 showed that a total of 14 agencies participated, as shown in Table 1. Since the names of some government agencies have changed in the past 34 years, to avoid confusion, similar to the study by Wang et al. [31], this paper adopted the latest names of these agencies.

Table 1. Green building (GB) policymaking agencies in China from 1986 to 2019.

Government Department	Abbreviations	Description	No. of GBPs
Ministry of Housing and Urban-Rural Development	MOHURD	Government department	180
Ministry of Finance	MOF	Government department	25
Ministry of Industry and Information Technology	MIIT	Government department	16
Ministry of Science and Technology	MOST	Government department	9
National Development and Reform Commission	NDRC	Government department	8
Ministry of Ecology and Environment	MOEE	Government department	4
State Council	SC	The highest organ of state administration	4
State Administration of Market Regulation	SAMR	Government department	3
Ministry of Commerce	MOC	Government department	2
State Economic and Trade Commission	SETC	Government department	1
Chinese Society of Urban Science	CSUS	Social organization	1
State Administration of Taxation	SAT	Government department	1
National Government Offices Administration	NGOA	Government department	1
Ministry of Education	MOE	Government department	1

The number of different government agencies participating in the GBP release is also presented in Table 1. It can be found that the Ministry of Housing and Urban-Rural Development (MOHURD) was the core policymaking agency, and its participation in issuing policies accounted for 90.45% of the total policies. In China, one of the responsibilities of MOHURD is to promote building energy-saving and emission reduction. As GB is not only a kind of building type but also a key development object of building energy-saving and emission reduction, MOHURD plays an important role in the formulation of GBP. The Ministry of Finance (MOF), the Ministry of Industry and Information Technology (MIIT), and the Ministry of Science and Technology (MOST) each published at least nine policies. MOF is responsible for formulating policies on special funds and financial subsidies for GB. MIIT formulates industry planning, policies, and standards, and guides information construction. MOST is mainly responsible for the formulation of planning, policies, and measures related to innovation and advanced technology.

The cooperation networks of policymaking agencies in different stages are shown in Figure 7. In this figure, the nodes represent the policymaking agencies, and the size of the node represents the number of policies that the policymaking agency participates in releasing. The edges represent the cooperation relationships among these policymaking agencies, which are quantified by the number of collaborative GBP documents between each pair of nodes. The dynamics of these co-author network analyses reflect the historical variations of policymaking agencies as well as their evolving roles and powers in policymaking [43].

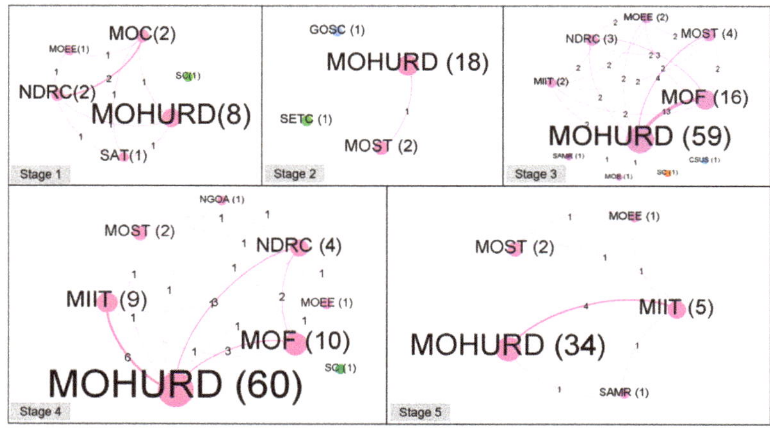

Figure 7. Co-author networks of GB policymaking agencies in five stages.

In stage 1, a total of six government agencies contributed to policy release, and four of them cooperated closely because they jointly released the "Design Standard for Energy-Saving of Civil Buildings (Heating Residential Buildings)."

In stage 2, the number of government agencies decreased to four, and their cooperative relationships were loose. In the above two stages, due to China's GB being still in its development infancy, the number of policy agencies involved in policymaking was relatively small and their partnerships were simple.

In stage 3, the number of government agencies increased significantly, and their cooperative relationships became complex and close, which showed that the Chinese government began to pay more attention to the GBD. In addition, it can be found that MOHURD and MOF had the closest cooperation, indicating that the Chinese government tried to stimulate the enthusiasm of the market to develop GBs through the promulgation of financial incentive policies in this stage.

In stage 4, the number of agencies decreased, whereas their collaborative relationships became closer. An interesting finding is that policies issued by MOHURD alone in this stage account for a higher proportion than those issued by MOHURD alone in the previous stage, which showed that MOHURD's centralized responsibility in GB policymaking began to return. In addition, MOHURD and MIIT had the closest cooperative relationships, as numerous industry planning and standards were issued in this stage.

In stage 5, the number of government agencies was further reduced, and their cooperative relationships became simpler. Among these connections, the alliance between MOHURD and MOF was eliminated and the interactions between MOHURD and MIIT became closer, indicating that the Chinese government mainly promulgated some standards and planning in this stage. In addition, 94.44% of the policies were released by MOHURD, implying the return of centralized responsibility in GB policymaking.

4. Discussion

Based on the above results and primary analyses, the historical dynamics of GBPs are further considered from five aspects: policy targets, policy objects, policy instruments, GB performance indicators, and the collaboration structure of policymaking agencies, as illustrated in Table 2. The policy targets were mainly extracted from the significant policy documents in each stage, such as the "Outline of the 10th Five-Year Plan for Building Energy Saving." The policy instruments, including direction-based policies (DP), regulation-based policies (RP), evaluation-based policies (EP), financial support policies (FP), supervision policies (SP), organization and professional training (OP), and knowledge and information (KI), were employed in this study. Referring to previous studies [37,68], 10 GB performance indicators were adopted such as quality, energy, CO_2 emissions, and water (see Table 2 for details).

Table 2. Comparison of GBPs in different stages.

	Stage 1 (1986–2000)	Stage 2 (2001–2005)	Stage 3 (2006–2010)	Stage 4 (2011–2015)	Stage 5 (2016–2019)
Policy targets		<ul><li>Building energy-saving rate is 50%</li><li>Wall material renovation</li></ul>	<ul><li>Building energy-saving rate is 65%</li><li>Wall material renovation</li></ul>	<ul><li>More than 95% of the newly built buildings in cities and towns implement the mandatory energy-saving standards</li><li>Promotion of the development of renewable energy buildings</li><li>The application area of solar energy and shallow ground energy accounts for more than 25% of the new building area</li></ul>	<ul><li>Implementation of demonstration construction of 100 green ecological urban areas</li><li>GB standards be implemented for government invested buildings, large public buildings, and commercial real estate projects</li><li>Carrying out of energy-saving reconstruction of existing buildings</li></ul>	<ul><li>Improvement of building energy-saving standards</li><li>Carrying out of energy-saving reconstruction of existing buildings</li><li>Expansion of the application scale of renewable energy buildings</li><li>Promotion of energy-saving of rural buildings</li></ul>
Policy objects	<ul><li>Residential buildings</li></ul>	<ul><li>Residential buildings</li><li>Public buildings</li></ul>	<ul><li>Public buildings</li><li>Residential buildings</li><li>Government office buildings</li><li>Large-scale public buildings</li><li>Colleges and universities</li><li>Industrial buildings</li></ul>	<ul><li>Residential buildings</li><li>Public buildings</li><li>Government office buildings</li><li>Large-scale public buildings</li><li>Public welfare buildings (schools, hospitals, etc.)</li><li>Industrial buildings</li><li>Indemnificatory housing</li><li>New town</li><li>Urban area</li><li>Rural housing</li></ul>	<ul><li>Residential buildings</li><li>Public buildings</li><li>Government office buildings</li><li>Large-scale public buildings</li><li>Public welfare buildings (schools, hospitals, etc.)</li><li>Industrial buildings</li><li>Indemnificatory housings</li><li>Rural housing</li><li>New urban buildings</li><li>Community (urban area)</li><li>Office and store buildings</li><li>Hotel buildings</li><li>Prefabricated buildings</li></ul>
Policy instruments	DP/RP	DP/RP/FP/SP/KI	DP/RP/EP/FP/SP/OP/KI	DP/RP/EP/FP/SP/OP/KI	DP/RP/EP/FP/SP/OP/KI

Table 2. *Cont.*

	Stage 1 (1986–2000)	Stage 2 (2001–2005)	Stage 3 (2006–2010)	Stage 4 (2011–2015)	Stage 5 (2016–2019)
GB performance indicators	• Energy • Employment of new wall materials • Quality	• Energy • Employment of new wall materials • Quality	• Energy • Quality • Water • GB materials • Lad • Employment of new wall materials • CO_2 emissions • Construction waste	• Energy • Quality • GB materials • Water • Lad • Employment of innovation and advanced technology • CO_2 emissions • Construction waste	• Energy • Quality • GB materials • Employment of innovation and advanced technology • Health and well-being • Water • Lad • Construction waste • CO_2 emissions
Collaboration structure	Some agencies with close collaboration	Few agencies with simple collaboration	More agencies with complex and close collaboration	More agencies with decreased and decentralized collaboration	Few agencies with centralized collaboration

DP: Direction-based policies, RP: Regulation-based policies, FP: Financial support policies, SP: Supervision policies, KI: Knowledge & information, EP: Evaluation-based policies, OP: Organization & professional training.

Based on the analysis results, it can be seen that China's GB originated from implementing energy-saving technologies. At the end of the 20th century, building energy-saving was considered to be the most direct and fundamental measure for alleviating the energy shortage contradiction and reducing environmental pollution [56]. Therefore, the main target of China's early GBP target was energy-saving. For example, in the "Design Standard for Energy-Saving of Civil Buildings (Heating Residential Buildings)," the Chinese government proposed that the energy-saving rate of buildings should reach 50%. With the rapid development of the economy and the acceleration of urbanization, the building energy demand was also increasing. Thus, the Chinese government first increased the building energy-saving target from 50% to 65%, and then raised the building energy-saving target to 75% [69]. In addition, in stage 4, MOHURD [25] had formulated four other detailed policy targets, including the implementation of the demonstration construction of 100 green ecological cities and the promotion of energy-saving reconstruction of existing buildings. In stage 5, the targets of GBPs included the improvement of the GB standard system and the further expansion of GB application scale [65].

The main object of China's GBP was residential buildings before 2000 and in the following year, MOHURD [57] proposed to strengthen the energy-saving work of residential buildings and public buildings as well as to promote the energy reconstruction of existing buildings. With the substantial increase in the numbers of GBPs in stage 3, the policy objects were further extended, including government office buildings, large-scale public buildings with high energy consumption, industrial buildings, universities, and rural housing. In stage 4, to explore the sustainable development model in the process of urbanization, the Chinese government issued the "National New Urbanization Plan (2014–2020)," which proposed to develop green towns and green urban areas. In stage 5, the policy objects had been continuously refined, including green hotel building, green exhibit building, etc.

From 1986 to 2019, China's GBPs always included an RP instrument, DP instrument, and FP instrument. With its expertise in promoting technological progress and standardizing market order, the RP instrument was the most widely employed GBP instrument and received increasing attention from the Chinese government. Second to the RP instrument, the DP instrument was also highly valued by the Chinese government because it played a critical role in summarizing the development experience of GB and planning the development direction of GB. The FP instrument was also usually employed by the Chinese government due to its expertise in mobilizing the market to actively implement GBs. Based on the existing policy instruments in stage 1, the SP instrument and KI instrument were added in stage 2, indicating that the Chinese government began to attach importance to the use of a mandatory supervision policy to regulate the implementation of GBs. In stage 3, the number of policies under the KI instrument increased, which showed that the Chinese government attached great importance to enhancing public awareness of GB. The policy instruments in stage 4 remained unchanged, whereas the FP instrument was absent in stage 5, which shows that China's green building is well on the way to becoming mandatory for all construction projects, rather than a socially conscious, idealistic option [20].

The early GBPs were devoted to building energy-saving, which was regarded as a means for addressing the energy shortage. Furthermore, building energy-saving mainly focused on the thermal insulation of the building envelope brought on by the use of new wall materials. Therefore, the early GB performance indicators mainly consisted of energy, quality, and employment of new wall materials. In stage 2, energy was still the most important performance indicator of GBs because the Chinese government proposed to ease the pressure on building energy demand by increasing the use of new and renewable energy. In addition, the performance indicators of this stage also included water, lad, material, and indoor environmental quality, which were first mentioned by MOHURD in "Technical Guidelines for Green Building." In stage 3, construction waste was mentioned by more and more policies, such as China's first GB evaluation standard and the "Notice on Supervision of Building Energy Saving," which has been regarded as an emerging indicator to measure the performance of GBs. In stage 4, an emerging indicator, the employment of innovation and advanced technology, was utilized to measure the performance of GBs. Specifically, these technologies included two-dimensional code, a building information model (BIM), prefabrication technology, and so on. In stage 5, China's GB evaluation standard was revised again in 2019, and the GB performance indicators changed from focusing on saving resources and environmental protection to being people-oriented. Therefore, health and well-being were recognized as additional emerging GB performance indicators.

From 1986 to 2019, there were several changes in the number of GB policymaking agencies and their cooperative relationships in China. From stage 1 to stage 2, the number of GB policymaking agencies in China decreased slightly, and their cooperative relationship tended to be simple. In stage 3, the number of policymaking agencies increased, and then they decreased again stage 4 and in stage 5. Meanwhile, the collaboration network of policymaking agencies went from dispersive to centralized, and MOHURD was the core institution with the largest number of policies issued in each stage.

5. Conclusions

A systematic and insightful investigation of the dynamic evolution of GBPs is of great significance for understanding of GBP background. This study investigated the development dynamics of GBPs in China from 1986 to 2019 by mapping and visualizing the co-word network and co-author network. The derived results revealed that China's GBPs have evolved through five stages: theoretical exploration, pilot demonstration, rapid growth, scope expansion, and deepening development. Throughout GBP development, energy-saving has always been the main policy topic and the target of most concern in the past decades. Meanwhile, the topics related to material saving and construction waste management emerged in recent years. Though energy was always the main GB performance indicator, the GBPs began to pay attention to the indicators concerning health and well-being. In addition, it was revealed that MOHURD played an important role in the development of GBPs in China.

The investigation of the dynamic evolution of China's GBPs can facilitate the stakeholders from both China and overseas in promoting GB practice. For the Chinese stakeholders, a clear understanding of the policy dynamics can help them to address the most innovative GB technologies and the most recent incentive measures. For the overseas stakeholders, the findings can provide valuable insights for countries in which the basic economic development and building sector conditions are similar to those in China. However, it should be mentioned that this study had specific limitations. For example, only national-level GBPs of China are considered in this study, whereas the GBPs of different cities in China are ignored. In fact, the historical dynamic of China's local policies varies with the level of the local economy and are recommended for future studies.

Author Contributions: Conceptualization, Z.W. and Q.H.; data curation, Q.H. and J.Z.; investigation, Q.H.; methodology, Q.H.; writing—original draft, Z.W., Q.H., and K.X.; writing—review and editing, J.Z., K.Y., and Z.W. All authors have read and agreed to the published version of the manuscript.

Funding: This research was conducted with the support of the National Natural Science Foundation of China (grant number 7200010881), the Foundation for Basic and Applied Basic Research of Guangdong Province (grant number 2019A1515110247), the Natural Science Foundation of Zhejiang Province of China (grant number LQ18G030011), and the Natural Science Foundation of SZU.

Institutional Review Board Statement: Not applicable.

Informed Consent Statement: Not applicable.

Conflicts of Interest: The authors declare no conflict of interest.

References

1. Li, Y.A.; Yang, L.; He, B.J.; Zhao, D.D. Green building in China: Needs great promotion. *Sustain. Cities Soc.* **2014**, *11*, 1–6. [CrossRef]
2. Zuo, J.; Zhao, Z. Green building research-current status and future agenda: A review. *Renew. Sustain. Energy Rev.* **2014**, *30*, 271–281. [CrossRef]

3. Wu, Z.; Yu, A.T.W.; Poon, C.S. An off-site snapshot methodology for estimating building construction waste composition—A case study of Hong Kong. *Environ. Impact Assess. Rev.* **2019**, *77*, 128–135. [CrossRef]
4. Tang, E.Z.; Peng, C.; Xu, Y.L. Changes of energy consumption with economic development when an economy becomes more productive. *J. Clean. Prod.* **2018**, *196*, 788–795. [CrossRef]
5. Sun, H.; Edziah, B.K.; Sun, C.; Kporsu, A.K. Institutional quality, green innovation and energy efficiency. *Energy Policy* **2019**, *135*, 111002. [CrossRef]
6. Scherer, A.G.; Palazzo, G.; Seidl, D. Managing legitimacy in complex and heterogeneous environments: Sustainable development in a globalized world. *J. Manag. Stud.* **2013**, *50*, 259–284. [CrossRef]
7. Varma, C.R.S.; Palaniappan, S. Comparision of green building rating schemes used in North America, Europe and Asia. *Habitat Int.* **2019**, *89*, 101989. [CrossRef]
8. Opoku, A. Biodiversity and the built environment: Implications for the sustainable development goals (SDGs). *Resour. Conserv. Recycl.* **2019**, *141*, 1–7. [CrossRef]
9. Singha, R.; Samanta, S.; Chatterjee, S.; Pariari, A.; Majumdar, D.; Satpati, B.; Wang, L.; Singha, A.; Mandal, P. Probing lattice dynamics and electron-phonon coupling in the topological nodal-line semimetal ZrSiS. *Phys. Rev. B* **2018**, *97*, 94112. [CrossRef]
10. Wu, Z.; Yu, A.T.W.; Wang, H.; Wei, Y.; Huo, X. Driving factors for construction waste minimization: Empirical studies in Hong Kong and Shenzhen. *J. Green Build.* **2019**, *14*, 155–167. [CrossRef]
11. Santamouris, M. Innovating to zero the building sector in Europe: Minimising the energy consumption, eradication of the energy poverty and mitigating the local climate change. *Sol. Energy* **2016**, *128*, 61–94. [CrossRef]
12. Wen, B.; Musa, S.N.; Onn, C.C.; Ramesh, S.; Liang, L.; Wang, W.; Ma, K. The role and contribution of green buildings on sustainable development goals. *Build. Environ.* **2020**, 107091. [CrossRef]
13. Vyas, G.S.; Jha, K.N. What does it cost to convert a non-rated building into a green building? *Sustain. Cities Soc.* **2018**, *36*, 107–115. [CrossRef]
14. Wu, Z.; Jiang, M.; Cai, Y.; Wang, H.; Li, S. What Hinders the Development of Green Building? An Investigation of China. *Int. J. Environ. Res. Public Health* **2019**, *16*, 3140. [CrossRef] [PubMed]
15. Wei, Y.; Zhu, X.; Li, Y.; Yao, T.; Tao, Y. Influential factors of national and regional CO_2 emission in China based on combined model of DPSIR and PLS-SEM. *J. Clean. Prod.* **2019**, *212*, 698–712. [CrossRef]
16. Bao, Z.; Lee, W.M.; Lu, W. Implementing on-site construction waste recycling in Hong Kong: Barriers and facilitators. *Sci. Total Environ.* **2020**, *747*, 141091. [CrossRef]
17. Darko, A.; Zhang, C.; Chan, A.P.C. Drivers for green building: A review of empirical studies. *Habitat Int.* **2017**, *60*, 34–49. [CrossRef]
18. Zhang, J.; Li, H.; Olanipekun, A.O.; Bai, L. A successful delivery process of green buildings: The project owners' view, motivation and commitment. *Renew. Energy* **2019**, *138*, 651–658. [CrossRef]
19. Balaban, O.; Puppim de Oliveira, J.A. Sustainable buildings for healthier cities: Assessing the co-benefits of green buildings in Japan. *J. Clean. Prod.* **2017**, *163*, S68–S78. [CrossRef]
20. Olubunmi, O.A.; Xia, P.B.; Skitmore, M. Green building incentives: A review. *Renew. Sustain. Energy Rev.* **2016**, *59*, 1611–1621. [CrossRef]
21. Shan, M.; Hwang, B.-g. Green building rating systems: Global reviews of practices and research efforts. *Sustain. Cities Soc.* **2018**, *39*, 172–180. [CrossRef]
22. Sharma, M. Development of a 'green building sustainability model' for green buildings in India. *J. Clean. Prod.* **2018**, *190*, 538–551. [CrossRef]
23. Liang, X.; Yu, T.; Guo, L. Understanding Stakeholders' Influence on Project Success with a New SNA Method: A Case Study of the Green Retrofit in China. *Sustainability* **2017**, *9*, 1927. [CrossRef]
24. Zhang, Y.; Kang, J.; Jin, H. A review of green building development in China from the perspective of energy saving. *Energies* **2018**, *11*, 334. [CrossRef]
25. MOHURD. 12th Five-Year Plan for Green Building Development Plan. Available online: http://www.gov.cn/gzdt/2013-04/18/content_2380994.htm (accessed on 7 October 2020).
26. Wu, Z.; Liu, L.; Li, S.; Wang, H. Investigating the Crucial Aspects of Developing a Healthy Dormitory based on Maslow's Hierarchy of Needs—A Case Study of Shenzhen. *Int. J. Environ. Res. Public Health* **2020**, *17*, 1565. [CrossRef] [PubMed]
27. Chang, R.d.; Soebarto, V.; Zhao, Z.y.; Zillante, G. Facilitating the transition to sustainable construction: China's policies. *J. Clean. Prod.* **2016**, *131*, 534–544. [CrossRef]
28. Shi, Q.; Lai, X.; Xie, X.; Zuo, J. Assessment of green building policies-A fuzzy impact matrix approach. *Renew. Sustain. Energy Rev.* **2014**, *36*, 203–211. [CrossRef]
29. Liu, G.; Tan, Y.; Li, X. China's policies of building green retrofit: A state-of-the-art overview. *Build. Environ.* **2020**, *169*, 106554. [CrossRef]
30. Ye, L.; Cheng, Z.; Wang, Q.; Lin, H.; Lin, C.; Liu, B. Developments of green building standards in China. *Renew. Energy* **2015**, *73*, 115–122. [CrossRef]
31. Wang, Y.N.; Xue, X.L.; Yu, T.; Wang, Y.W. Mapping the dynamics of China's prefabricated building policies from 1956 to 2019: A bibliometric analysis. *Build. Res. Inf.* **2020**. [CrossRef]
32. Dou, Y.; Xue, X.; Wang, Y.; Luo, X.; Shang, S. New media data-driven measurement for the development level of prefabricated construction in China. *J. Clean. Prod.* **2019**, *241*, 118353. [CrossRef]

33. Darko, A.; Chan, A.P.C. Critical analysis of green building research trend in construction journals. *Habitat Int.* **2016**, *57*, 53–63. [CrossRef]

34. Darko, A.; Chan, A.P.C.; Huo, X.; Owusu-Manu, D.-G. A scientometric analysis and visualization of global green building research. *Build. Environ.* **2019**, *149*, 501–511. [CrossRef]

35. Dat Tien, D.; Ghaffarianhoseini, A.; Naismith, N.; Zhang, T.; Ghaffarianhoseini, A.; Tookey, J. A critical comparison of green building rating systems. *Build. Environ.* **2017**, *123*, 243–260.

36. Liu, X.; Wang, M.; Fu, H. Visualized analysis of knowledge development in green building based on bibliographic data mining. *J. Supercomput.* **2020**, *76*, 3266–3282. [CrossRef]

37. Zhang, Y.; Wang, J.; Hu, F.; Wang, Y. Comparison of evaluation standards for green building in China, Britain, United States. *Renew. Sustain. Energy Rev.* **2017**, *68*, 262–271. [CrossRef]

38. Yuan, X.; Zuo, J. Transition to low carbon energy policies in China-from the Five-Year Plan perspective. *Energy Policy* **2011**, *39*, 3855–3859. [CrossRef]

39. Hu, A.-G. The Five-Year Plan: A new tool for energy saving and emissions reduction in China. *Adv. Clim. Chang. Res.* **2016**, *7*, 222–228. [CrossRef]

40. Zupic, I.; Cater, T. Bibliometric methods in management and organization. *Organ. Res. Methods* **2015**, *18*, 429–472. [CrossRef]

41. Yang, C.; Huang, C.; Su, J. A bibliometrics-based research framework for exploring policy evolution: A case study of China's information technology policies. *Technol. Forecast. Soc. Chang.* **2020**, *157*, 120116. [CrossRef]

42. Zhang, Q.; Lu, Q.; Zhong, D.; Ye, X. The pattern of policy change on disaster management in China: A bibliometric analysis of policy documents, 1949–2016. *Int. J. Disaster Risk Sci.* **2018**, *9*, 55–73. [CrossRef]

43. Huang, C.; Su, J.; Xie, X.; Ye, X.; Li, Z.; Porter, A.; Li, J. A bibliometric study of China's science and technology policies: 1949–2010. *Scientometrics* **2015**, *102*, 1521–1539. [CrossRef]

44. Khanra, S.; Dhir, A.; Mantymaki, M. Big data analytics and enterprises: A bibliometric synthesis of the literature. *Enterp. Inf. Syst.* **2020**, *14*, 737–768. [CrossRef]

45. Feng, J.; Zhang, Y.Q.; Zhang, H. Improving the co-word analysis method based on semantic distance. *Scientometrics* **2017**, *111*, 1521–1531. [CrossRef]

46. Feng, Y.; Zhu, Q.; Lai, K.-H. Corporate social responsibility for supply chain management: A literature review and bibliometric analysis. *J. Clean. Prod.* **2017**, *158*, 296–307. [CrossRef]

47. Du, H.; Li, N.; Brown, M.A.; Peng, Y.; Shuai, Y. A bibliographic analysis of recent solar energy literatures: The expansion and evolution of a research field. *Renew. Energy* **2014**, *66*, 696–706. [CrossRef]

48. Du, H.; Li, B.; Brown, M.A.; Mao, G.; Rameezdeen, R.; Chen, H. Expanding and shifting trends in carbon market research: A quantitative bibliometric study. *J. Clean. Prod.* **2015**, *103*, 104–111. [CrossRef]

49. Li, J.; Wang, M.; Ho, Y. Trends in research on global climate change: A science citation index expanded-based analysis. *Glob. Planet. Chang.* **2011**, *77*, 13–20. [CrossRef]

50. Nan, Y.; Feng, T.; Hu, Y.; Qi, X. Understanding aging policies in China: A bibliometric analysis of policy documents, 1978–2019. *Int. J. Environ. Res. Public Health* **2020**, *17*, 5956. [CrossRef]

51. Yao, H.; Zhang, C. A bibliometric study of China's resource recycling industry policies: 1978–2016. *Resour. Conserv. Recycl.* **2018**, *134*, 80–90. [CrossRef]

52. Kauffman, J.; Kittas, A.; Bennett, L.; Tsoka, S. DyCoNet: A Gephi plugin for community detection in dynamic complex networks. *PLoS ONE* **2014**, *9*, e101357. [CrossRef]

53. Wang, X.L.; Liu, Y.; Ju, Y.B. Sustainable public procurement policies on promoting scientific and technological innovation in China: Comparisons with the US, the UK, Japan, Germany, France, and South Korea. *Sustainability* **2018**, *10*, 2134. [CrossRef]

54. Ding, Z.; Liu, R.; Li, Z.; Fan, C. A thematic network-based methodology for the research trend identification in building energy management. *Energies* **2020**, *13*, 4621. [CrossRef]

55. Ding, Z.K.; Li, Z.J.; Fan, C. Building energy savings: Analysis of research trends based on text mining. *Autom. Constr.* **2018**, *96*, 398–410. [CrossRef]

56. MOHURD. Ninth Five-Year Plan and the 2010 Plan for Building Energy-Saving. Available online: http://www.cnki.com.cn/Article/CJFDTotal-SGJS608.000.htm (accessed on 18 September 2020).

57. MOHURD. Outline of the 10th Five-Year Plan for Building Energy Saving. Available online: http://www.mohurd.gov.cn/wjfb/200611/t20061101_158478.html (accessed on 18 September 2020).

58. Ma, M.; Cai, W.; Wu, Y. China act on the energy efficiency of civil buildings (2008): A decade review. *Sci. Total Environ.* **2019**, *651*, 42–60. [CrossRef] [PubMed]

59. MOHURD. 11th Five-Year Plan for Building Energy Conservation Task. Available online: http://www.gov.cn/gzdt/2010-05/18/content_1608438.htm (accessed on 18 September 2020).

60. Kong, X.; Lu, S.; Wu, Y. A Review of Building Energy Efficiency in China during "Eleventh Five-Year Plan" Period. *Energy Policy* **2012**, *41*, 624–635. [CrossRef]

61. Guo, Q.; Wu, Y.; Ding, Y.; Feng, W.; Zhu, N. Measures to enforce mandatory civil building energy efficiency codes in China. *J. Clean. Prod.* **2016**, *119*, 152–166. [CrossRef]

62. MOHURD; MOF. Implementation Opinions on Promoting the Application of Renewable Energy in the Building Sector. Available online: http://www.mohurd.gov.cn/wjfb/200611/t20061101_158510.html (accessed on 18 September 2020).

63.   MOHURD; MOF. Implementation Opinions on Accelerating the Development of Green Buildings in China. Available online: http://www.mohurd.gov.cn/fgjs/xgbwgz/201205/t20120510_209831.html (accessed on 18 September 2020).
64.   MOHURD. Green Building and Green Ecological Urban Development Planning. Available online: http://www.mohurd.gov.cn/wjfb/201304/t20130412_213405.html (accessed on 18 September 2020).
65.   MOHURD. 13th Five-Year Plan for Building Energy Saving and Green Buildings Development. Available online: http://www.mohurd.gov.cn/wjfb/201703/t20170314_230978.html (accessed on 18 September 2020).
66.   Yang, X.; Zhang, S.; Xu, W. Impact of zero energy buildings on medium-to-long term building energy consumption in China. *Energy Policy* **2019**, *129*, 574–586. [CrossRef]
67.   Wu, Z.; Yu, A.T.W.; Shen, L. Investigating the determinants of contractor's construction and demolition waste management behavior in Mainland China. *Waste Manag.* **2017**, *60*, 290–300. [CrossRef]
68.   Geng, Y.; Ji, W.; Wang, Z.; Lin, B.; Zhu, Y. A review of operating performance in green buildings: Energy use, indoor environmental quality and occupant satisfaction. *Energy Build.* **2019**, *183*, 500–514. [CrossRef]
69.   Li, J.; Shui, B. A comprehensive analysis of building energy efficiency policies in China: Status quo and development perspective. *J. Clean. Prod.* **2015**, *90*, 326–344. [CrossRef]

International Journal of
Environmental Research and Public Health

MDPI

Article

Environmental, Economic and Social Impact Assessment: Study of Bridges in China's Five Major Economic Regions

ZhiWu Zhou *, Julián Alcalá and Víctor Yepes

Institute of Concrete Science and Technology (ICITECH), Universitat Politècnica de València, 46022 València, Spain; jualgon@cst.upv.es (J.A.); vyepesp@cst.upv.es (V.Y.)
* Correspondence: zhizh2@doctor.upv.es; Tel.: +34-96-387-9563

Abstract: The construction industry of all countries in the world is facing the issue of sustainable development. How to make effective and accurate decision-making on the three pillars (Environment; Economy; Social influence) is the key factor. This manuscript is based on an accurate evaluation framework and theoretical modelling. Through a comprehensive evaluation of six cable-stayed highway bridges in the entire life cycle of five provinces in China (from cradle to grave), the research shows that life cycle impact assessment (LCIA), life cycle cost assessment (LCCA), and social impact life assessment (SILA) are under the influence of multi-factor change decisions. The manuscript focused on the analysis of the natural environment over 100 years, material replacement, waste recycling, traffic density, casualty costs, community benefits and other key factors. Based on the analysis data, the close connection between high pollution levels and high cost in the maintenance stage was deeply promoted, an innovative comprehensive evaluation discrete mathematical decision-making model was established, and a reasonable interval between gross domestic product (GDP) and sustainable development was determined.

Keywords: sustainable development; LCIA; LCCA; SILA; cable-stayed bridge; GDP

Citation: Zhou, Z.; Alcalá, J.; Yepes, V. Environmental, Economic and Social Impact Assessment: Study of Bridges in China's Five Major Economic Regions. *Int. J. Environ. Res. Public Health* **2021**, *18*, 122. https://doi.org/10.3390/ijerph18010122

Received: 26 November 2020
Accepted: 22 December 2020
Published: 26 December 2020

Publisher's Note: MDPI stays neutral with regard to jurisdictional claims in published maps and institutional affiliations.

1. Introduction

The most common greenhouse gases in the Earth's atmosphere include water vapour (H_2O), carbon dioxide (CO_2), methane (CH_4), nitrous oxide (N_2O), ozone (O_3) and chlorofluorocarbons (CFC). The concentration of carbon dioxide in the atmosphere has a dominant influence on global warming [1,2]. According to predictions by the United Nations, the world's population will reach 9.8 billion in 2050 [3]. Population shifts will result in a massive consumption of resources and a rapid growth of energy requirements [4]. This makes the sustainable development of the construction industry, which accounts for 44% of all energy consumption, become more urgent [5,6]. What is the key to sustainable development? It is to reduce environmental, economic and social impacts [7]. Thus, the scope of research is expanded to the economic and social aspects, and the close correlation between producers and consumers is increased [8].

To avoid the serious consequences brought about by climate change, efforts should be made to substantially reduce the emission of greenhouse gases. Hansen et al. revealed that the concentration of carbon dioxide in the atmosphere must be less than 350 parts per million (ppm); otherwise, climate change will get worse [9]. The analysis of the latest global atmospheric observations by the World Meteorological Organisation shows that the global mean surface mole fractions of CO_2, CH_4, and N_2O reached new highs in 2015, i.e., 400.0 ± 0.1 ppm, 1845 ± 2 parts per billion (ppb), and 328.0 ± 0.1 ppb, respectively. These values constitute 144%, 256% and 121% of the pre-industrial levels (before 1750), respectively [10].

Low-carbon energy consumption and the reduction in greenhouse gas emissions from the construction industry are particularly critical [11]. Lin and Liu. cited the CO_2 emissions from commercial and residential buildings in China, surveyed by the Index Decomposition

Analysis (IDA), which concluded that emissions from the construction industry account for 30–50% of the total emissions [12]. Science researchers all over the world have proposed measures to reduce environmental pollution caused by the construction industry. For the accuracy and systematisms of the research, LCIA was introduced to solve problems facing the construction industry [13,14]. Standardised provisions for multiple systemic analysis methods were given in ISO 14040 and ISO 14044 [15].

Table 1 shows a comparative analysis of the latest research results of LCIA, LCCA and SILA.

Table 1. Recent statistics and analysis of some closely related achievements.

Methods	Description	Characteristic	Limitation	References
LCIA	Preventive design using 15 different methods of LCA concrete bridge deck.	How to reduce environmental pollution in the maintenance stage: Design and evaluation of 15 preventive measures.	The research content is relatively concentrated, single, and focuses on material replacement.	[16]
	Use LCIA to evaluate the rationality of the bridge design.	Use wooden bridges and alternative concrete to analyse the LCA impact of a cable-stayed bridge.	Ideal research design for the future. There are currently no large-span wooden bridges in operation. There are assumptions and uncertainties in the maintenance assessment of wooden bridges.	[17]
	Apply life cycle sustainability assessment to the superstructure of small span bridges.	The study was conducted using 27 bridges, and it was determined that a bridge composed of steel and concrete was the best indicator.	The LCA study of ordinary highway bridges, the conclusion is whether it is suitable for long-span special bridges.	[18]
	LCA was used to assess the environmental impact of the entire 60-year life span of the provincial highway.	The research structure has a complete range of tunnels, bridges, roadbeds, culverts, etc.	The road selection is in a remote area, and the research data are not representative.	[19]
	Several cases (schools, hospitals, commercial and residential buildings) were quantitatively studied using LCA.	There are many types of structures studied, and an evaluation model is established to quantitatively analyse emissions.	The research conclusions are poorly comparable, and the LCA data are highly uncertain.	[20]
LCCA	The article introduces a general framework for evaluating bridge life cycle performance and cost.	The focus is on analysis, prediction, optimization and decision-making under bridge uncertainty.	All the articles in this article are cost theory analysis, and there is no specific bridge case analysis.	[21]
	Research and develop an LCCA model to evaluate highway infrastructure investment.	Contributed to the systematic and informatised evaluation method of highway infrastructure investment.	Lack of case studies and model application research.	[22]
	The energy consumption cost of highway pavement is analysed based on LCCA and LCA.	Combining LCA and LCCA to determine the best pavement frame, road expansion projects are more practical.	Case application analysis of pavement concrete sustainability, no structural concrete evaluation.	[23]

Table 1. *Cont.*

Methods	Description	Characteristic	Limitation	References
	Quantify the life cycle environmental impact of the structure through environmental costs.	Calculate the environmental costs of materials, energy, transportation and construction equipment for the bridge structure.	The main research is the LCCA influence of the bridge girder structure.	[24]
	The LCC and LCA analysis of concrete bridges were discussed, and the optimization scheme was proposed.	Economic and environmental impact analysis of reinforced concrete and prestressed concrete bridges.	The bridge structure is simply a simply supported beam bridge across the river.	[25]
	Use SLCA to clarify the assessment (IA) methods and information covered in the different impact guidelines.	Use representational models to analyse the difference and connection between social influence and social performance.	All are written descriptions, without modeling and data analysis.	[26–28]
	Use SIA to study and practice all issues related to social issues in the entire project life cycle (before conception to after closure).	Analysed the overall social issues in the process of community and project management. Put forward that the biggest social problem management in the project is corruption.	Lack of case application analysis and discussion.	[29,30]
	SIA is undergoing a revolutionary force and revolutionary force for change.	SIA's unfamiliarity with social sciences and the concerns of practitioners' lack of competence.	Lack of case application analysis and discussion.	[31,32]
	EIA and SIA have technical flaws in analysis and evaluation.	Consider four conceptual elements in a sociological context of complexity and vitality.	Talked about the project SIA's attention to sensitive factors and the improvement of social responsibility. How to realize the scientific methodology needs to be developed.	[33,34]
	Evaluate the sustainability performance of different concrete and stone walls used in the building.	Multi-criteria decision analysis methods are used to evaluate and prioritise the alternative walls generated by LCA, LCC and S-LCA.	The research is sustainable and comprehensive, the evaluation structure is single, and recycling is not considered.	[35]
LCIA\LCCA\SILA	The study analysed the impact of different mixed timber building structures on three different categories of environment, economy and society.	The comparison of wood and concrete in the building structure has been analysed to improve sustainability.	There are few studies on the three pillars of sustainability. This article has the same research route and different structures.	[36]
	Three box-type concrete bridges were optimised and sustainable.	Researchers focus on the environmental pillar, while the social pillar has been slow to develop.	It mainly studies the process of sustainability assessment and briefly analyses three precast concrete bridges.	[37]
	Discussed the framework for assessing the sustainability of bridges, including related technical, economic, environmental and social issues.	The sustainability of four versions of the same bridge was studied, and the local details of the bridge were analysed.	There is a lack of sustainable research on regional and actual operating bridges.	[38]

First, this study aims to evaluate the impact of LCIA~LCCA~SILA (2L~1S) on six bridges in five different regions of China. This study will fill the gap in the research for bridges of similar structure and purpose across regions, provinces, and economic belts in this field. Secondly, the process of 2L~1S is digitised and visualised to display the research results more intuitively. Thirdly, this study also considers the mutual influence between 2L~1S and the regional economic belts, to obtain the optimal interval and scope of influence.

The main purpose of this research is to analyse and study the comprehensive impact of bridges of the same structure in different regional economic zones on the environment, economy and society (three pillars) throughout their life cycle through software. In addition, discussed the correlation between regional economic development and the three pillars through modelling.

The innovations of the research are as follows: (1) break through the usual sustainability research and only focus on textual descriptions, without accurate modelling data descriptions; (2) the selected research case represents the influence status between the main economic belts in China and has important guiding significance for the future planning of the government and related departments.

The rest of this work will be divided into the following sections: Section 2: Methods; Section 3: Results and Discussion; Section 4: Conclusions.

2. Methods

LCIA has become an international standardisation tool for environmental assessment [39,40]. Preliminary conditions need to be defined for every study: the functional unit and system boundary of the assessment were the six bridges and the SILAs of the corresponding communities. The assessment was conducted based on the LCIA, covering the whole of the life cycle. LCIA was analysed by using OpenLCA (Life cycle assessment) 1.10.1, LCCA by the budgetary estimate process, and SILA by OpenLCA1.10.3 (OpenLCA development team, Berlin, Germany) [14]. The three tools are relevant and systematic. The databases used in this study included Ecoinvent [41], Bedec [42], and Product Social Impact Life Cycle Assessment (PSILCA) [43]. See Sections 2.1 and 2.2 for detailed modelling.

2.1. Modeling Analysis

The construction industry is the most active sector in both developed and developing countries, forming a high global consistency [44]. LCIA was included as a sustainable survey method, because it can systematically assess the environment in all directions and complete the selection of friendly products [45]. ISO has issued a series of 14,040 standards and International Life Cycle Data (ILCD) manuals to promote sustainable development [15,46].

2.1.1. LCIA

The studied cases were six representative cable-stayed bridges, including South Tai Hu Lake Bridge (STHB), Shenzhen Bay Bridge (SZBB), New Bridge of Xishuangbanna Tropical Botanical Garden (BGNB), Cable-stayed Bridge of Changjiang West Road, Deyang City (CJWB), Hanjiang Highway Bridge, Xiantao City (XTHB), and Baishan Bridge, Baishan City (BSCB). Five of them adopted a reinforced concrete structure and one adopted a steel structure (the main beam of SZBB is constructed by welding and bolting steel components). All of them have a single tower. The length of the main bridge ranges from 136 to 410 m and all six bridges are Class I municipal highway bridges. Table 2 shows the detailed data.

Table 2. Cable-stayed bridge maintenance data statistics table.

Check Method	Inspection Cycle	Check Parts	Maintenance Cycle
Daily check Frequency check	Working day One time/every month	Pier foundation, cone slope, side wall of bridge abutment, pavement of bridge deck, drainage system, sidewalk, railing, guardrail, anti-collision wall of bridge deck, lighting system on bridge, expansion device, bridge head laying plate, sign, marking and traffic safety facilities, bridge installation sensors, wiring, cables, anchorage protection inspection, bridge damping device normal operation, support cleaning, rust and corrosion prevention.	Maintenance/year, Overhaul/5 years.
Regular check	One time/one to three years	Coating layer of exposed concrete.	Maintenance/year, Replacement/5 years.
		Bridge deck paving, waterproof layer.	Maintenance/year, Overhaul/2 years, Replacement/10 years.
		Anti-collision guardrail, expansion joint.	Maintenance/year, Overhaul/2-5 years, Replacement/15 years.
		Cable-stayed bridge cables, slings, tie rods, external damping devices.	Maintenance/year, Overhaul/5 years, Replacement/20 years.
		Main beams, steel supports, bridge floor drainage pipes, bridge floor lighting facilities.	Maintenance/year, Overhaul/5 years, Replacement/50 years.
		Basin type rubber bearing.	Maintenance/year, Overhaul/5 years, Replacement/25 years.
		Damping device between towers and beams.	Maintenance/year, Overhaul/5 years, Replacement/30 years.
		Main beams, steel supports, bridge floor drainage pipes, bridge floor lighting facilities.	Maintenance/year, Overhaul/5 years, Replacement/50 years.

According to ISO standards, and the requirement for the scope of strict assessment and examination of the life cycle of the bridge [47–49], the full life cycle of these six bridges was analysed in five stages: survey and design, material manufacturing, construction and installation, maintenance and operation, and disassembly and recycling. Since the cross section of the main girder of the bridge is variable, the calculation unit was based on 1 cubic meter. In order to achieve the rationality of the data comparison study and analysis, the study length of the six cable-stayed bridges was selected as a uniform 390 m to input relevant data (390 m including the main bridge and some auxiliary bridges).

Seven key impact categories, including energy use, ecotoxicity, acidification, eutrophication, climate change, particulate matter formation and ozone depletion, were determined through the comparative analysis of the oxidation separation of fossil materials and the European Union Product Environmental Footprint (EUPEF) [50–52]. Five of these seven categories were selected as the important goals of bridges' LCIA: global warming potential (GWP), acidification potential (AP), free-water eutrophication potential (FEP), particulate matter formation potential (PMFP), including fumes and dust, and waste potential (WP).

The assessment and modelling method of LCIA has a midpoint and endpoint. Huijbregts et al. made a clear distinction and explanation in their reports ReCiPe 2008 and 2016 LCIA [53,54]. By comparing the advantages and disadvantages of the two modelling approaches [55], it was found that the midpoint modelling is more appropriate for stages, while the end-point modelling is more appropriate for intervals.

Major modelling formulas of LCIS:

Environmental impact contribution of transport vehicle:

$$E_m = \sum_i^j \{K_{im} \times [\sum_i^j (K_i + K_2 + \cdots\cdots + K_j)] \times M \times (1 + \alpha) \times V_m \times \lambda_\mu + \cdots\cdots$$
$$+ K_{jm} \times [\sum_i^j (K_i + K_2 + \cdots\cdots + K_j)] \times M \times (1 + \beta) \times V_m \times \lambda_\mu\} \tag{1}$$

where E_m = Environmental impact contribution of transport vehicle (kg); K_{im}, K_{jm} = Fuel consumption of vehicles i, j (L/100 km); V_m = Quantity of surveying vehicles i, j; α, β = Engine fuel loss of different types of vehicles (%); and λ_μ = Physical and chemical environmental emission coefficient of fuel μ (kg/kg) [56].

Environmental impact contribution of mechanical equipment:

$$M_m = \sum_i^j \{[G_{im} \times (1 + \alpha) \times T_{im} \times (\lambda_\mu \oplus \lambda_v)] + \cdots\cdots + [G_{jm} \times (1 + \alpha) \times T_{jm} \times (\lambda_\mu \oplus \lambda_v)]\} \tag{2}$$

where M_m = Environmental impact contribution of mechanical equipment (kg); G_{im}, G_{jm} = Fuel consumption and power consumption of equipment i j (kg/h, kWh); T_{im} = Normal working hour of mechanical equipment (h); $\oplus$ = Logic "Or"; and λ_v = Physical and chemical environmental emission coefficient of electric energy v (kg/kg).

Environmental impact contribution of personnel:

$$P_m = W_m \times \lambda_p \times T_p \tag{3}$$

where P_m = Environmental impact contribution of personnel (kg); W_m = Total number of personnel (persons); λ_p = Environmental impact coefficient of personnel (kg/working day/person); T_p = Total working hours of personnel (working day).

Environmental impact contribution of office facilities:

$$W_m = \sum_i^j \{[F_{im} \times T_i \times (1 + L_i) \times \lambda_i] + \cdots\cdots [F_{jm} \times T_j \times (1 + L_j) \times \lambda_j]\} \tag{4}$$

where W_m = Environmental impact contribution of office facilities (kg); F_{im}, F_{jm} = Power consumption of office facilities i, j (Kwh); T_i, T_j = Working hours of office facilities i, j (h); and L_i, L_j = Electricity loss coefficient of facilities i, j (%).

SimaPro has been the world's leading life cycle assessment (LCA) software package for 30 years; it is trusted by industry and academics in more than 80 countries [57]. OpenLCA can access the social and economic impact of 15 different life cycles. The software has been widely used in various industries and research fields in Europe, the United States, Japan and the rest of the world; it is supported by databases such as Ecoinvent, Bedec, Soca, bridge design, construction drawings, and published research results.

2.1.2. LCCA

LCCA of bridges mainly includes initial cost, cost of maintenance, repair and replacement, casualties of personnel or loss of goods during operation, road use cost, and indirect loss of socio-economic benefits [58,59]. In order to accurately estimate these costs, it is necessary to clarify the degradation rate of bridge components and build a correct model for the designated fatigue life index [60,61]. Table 2 shows the maintenance cycle. The core elements of LCCA are financial factors, inter-generational responsibility, environmental aspects and sustainability, realising the optimal balance between safety, economic efficiency, and sustainability [62].

LCCA was conducted in accordance with the process of highway engineering in China, as shown in Figure 1. It was of equal importance to determine the life cycle cost, cost benefit, or cost risk by considering a variety of ways of calculating cost benefit [58].

$$E\left[C_T\left(\overline{x},\, T_{Ready}\right)\right] = C_i(\overline{x}) + \sum_{t=1}^{T_{Ready}} \left(\frac{\sum_{j=1}^{J} E\left[C_{Advisoryj}(\overline{x},t)\right] + \sum_{k=1}^{K} E[C_{Assess}(\overline{x},t)] + \sum_{l=1}^{L} E(C_{Mixed}(\overline{x},t))}{(1+r)^t} \right) \tag{5}$$

where $E\left[C_T\left(\overline{x},T_{Ready}\right)\right]$ = LCCA cost in the preparatory stage (Chinese Yuan: CNY); $C_i(\overline{x})$ = Direct cost in the preparatory stage (CNY); $\sum_{j=1}^{J} E\left[C_{Advisoryj}(\overline{x},t)\right]$ = Consulting fee of the development organisation (CNY); $\sum_{k=1}^{K} E[C_{Assess}(\overline{x},t)]$ = Impact assessment fee of the development organisation (CNY); $\sum_{l=1}^{L} E(C_{Mixed}(\overline{x},t))$ = Other costs incurred in the preparatory stage of the project, including expert review fee, transportation fee, approval procedure fee, office fee, labour fee for related personnel (CNY); and r = Discount rate (%).

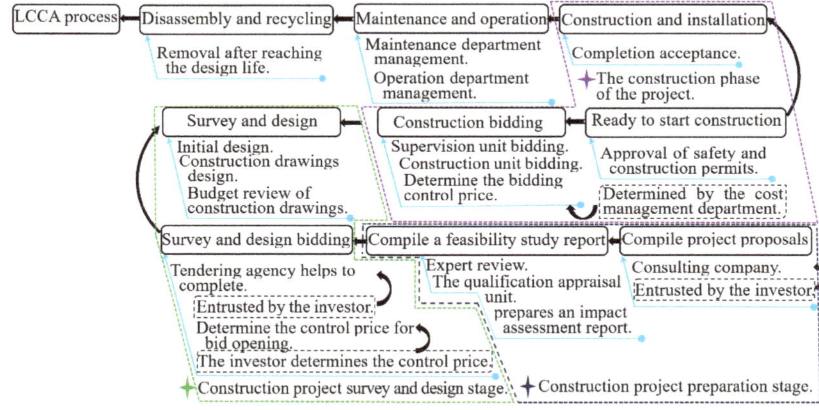

Figure 1. Basic procedure flow chart of highway engineering construction.

The service rate for the project-bidding agency issued by National Development and Reform Commission is given by [63]:

$$C_{Bidding\ Service} = \begin{cases} 500\ \text{million CNY} \leq C_{Build}(\overline{x},T_{End}) \leq 1000\ \text{million CNY} & C_{BS} = 0.035\% * C_{Build} \\ 1000\ \text{million CNY} < C_{Build}(\overline{x},T_{End}) \leq 5000\ \text{million CNY} & C_{BS} = 0.008\% * C_{Build} \\ 5000\ \text{million CNY} < C_{Build}(\overline{x},T_{End}) \leq 10,000\ \text{million CNY} & C_{BS} = 0.006\% * C_{Build} \\ 10,000\ \text{million CNY} < C_{Build}(\overline{x},T_{End}) & C_{BS} = 0.004\% * C_{Build} \\ C_{Bidding\ service} = \text{Maximum amount 3.5 million CNY} & 3.0\ \text{million CNY}, 4.5\ \text{million CNY} \end{cases} \tag{6}$$

Costs of survey and design:

$$C_{Design}(\overline{x},T_{End}) = \sum_{t=Start}^{T_{End}} \left\{ \frac{\left[C_{Survey}(\overline{x}) + \sum_{t_1}^{t_{End}}(\overline{x},t_{Survey})\right](1\pm\lambda_h) + \sum_{t_1}^{t_{End}}(\overline{x},t_{Survey}) + C_{Design}(\overline{x}) + \sum_{t_1}^{t_{End}}(\overline{x},t_{Design})}{(1+r)^t} \right\} [1 \pm (C_{Float})][1 \pm (R_t)] \tag{7}$$

where $C_{Design}(\overline{x},T_{End})$ = LCCA cost in the stage of survey and design (CNY); $C_{Driect}(\overline{x})$, $C_{Design}(\overline{x})$ = Direct cost in the stage of survey and design (CNY); $\sum_{t_1}^{t_{End}}(\overline{x},t_{Survey})$, $\sum_{t_1}^{t_{End}}(\overline{x},t_{Design})$ = Indirect cost in the stage of survey and design (CNY); R_t = National tax rate (%); C_{Float} = Adjustment range (%); and λ_h = Adjustment coefficient [64].

$$\lambda_h = \begin{cases} T_{Local} \geq 35\,^\circ C & \lambda_h = 1.2 \\ T_{Local} \leq -10\,^\circ C & \lambda_h = 1.2 \\ 2000\ \text{meters} \leq H_{Altitude} \leq 3000\ \text{meters} & \lambda_h = 1.1 \\ 3001\ \text{meters} \leq H_{Altitude} \leq 3500\ \text{meters} & \lambda_h = 1.2 \\ 3501\ \text{meters} \leq H_{Altitude} \leq 4000\ \text{meters} & \lambda_h = 1.3 \\ 4001\ \text{meters} \leq H_{Altitude} & \lambda_h \gg 1.3 (\text{Negotiated price}) \end{cases} \tag{8}$$

where T_{Local} = Ambient temperature of the place where the project locates (°C), and $H_{Altitude}$ = Altitude of the place where the project locates (m).

Concerning the rate for the design and examination of construction drawings [63], it is charged by the budgetary investment ratio, thus the rate should not be higher than 2‰ of the budget amount of the project.

Construction costs:

$$C_{Build}(\bar{x}, T_{End}) = \sum_{t=Start}^{T_{Warranty\ termination}} \left\{ \frac{C_{Direct\ cost} + C_{Extra\ charge} + \left[(C_{Direct\ cost} + C_{Extra\ charge}) * C_{Profit} \right]}{(1+r)^t} \right\} [1 \pm (R_t)] \tag{9}$$

where $C_{Build}(\bar{x}, T_{End})$ = LCCA cost in the stage of construction (CNY); $C_{Direct\ cost}$ = Direct cost of the project (CNY); $C_{Extra\ charge}$ = Indirect cost of the project (CNY); and C_{Profit} = Construction profits of the project (CNY).

Costs of maintenance and operation:

Global warming and extreme weather events have resulted in observable effects on people, the environment, and civil infrastructures [6]. Stewart et al. proposed four main factors for infrastructure corrosion and structural performance deterioration, including temperature [65]. Barbara Rossi et al. concluded that the total project cost decreases with the increase in the discount rate, and the period of investment return ranges between 18.5 and 24.2 years [66].

The six bridges are located in five economic belts. Climate, traffic density, traffic accidents, load effect of heavy-duty vehicles, and natural disasters (such as flooding, ice damage, freezing damage and mudslides) have different degrees of impact on the maintenance costs of bridges. The analysis was carried out according to the Chinese Code for Maintenance of Highway Bridges and Culverts (JTG H11-2004), as shown in Table 1 [67,68].

Costs of maintenance and repair: $C_{Maintenance}(\bar{x}, T_{100years}) =$

$$\begin{cases} \sum_{t=1year}^{t=100years} \left(\dfrac{C_{Direct\ cost} + C_{Extra\ charge}}{(1+r)^t} \right) [1 \pm (R_t)] \dfrac{T_{Total\ maintanence\ times}}{T_{Times\ of\ cycle}} & \text{Maintenance costs} \\[2ex] \sum_{t=1year}^{t=100years} \left(\dfrac{C_{Direct\ cost} + C_{Extra\ charge}}{(1+r)^t} \right) [1 \pm (R_t)] \dfrac{T_{Strengthening\ structure\ times}}{T_{Times\ of\ cycle}} & \text{Strengthening structure costs} \\[2ex] \sum_{t=1year}^{t=100years} \left(\dfrac{C_{Direct\ cost} + C_{Extra\ charge}}{(1+r)^t} \right) [1 \pm (R_t)] \dfrac{T_{Emergency\ repair\ times}}{T_{Times\ of\ cycle}} & \text{Emergency repair costs of road} \\[2ex] \sum_{t=1year}^{t=100years} \left(\dfrac{C_{Direct\ cost} + C_{Extra\ charge}}{(1+r)^t} \right) [1 \pm (R_t)] \dfrac{T_{Routine\ maintanence\ times}}{T_{Times\ of\ cycle}} & \text{Routine maintenance costs} \\[2ex] \sum_{t=1year}^{t=100years} \left(\dfrac{C_{Direct\ cost} + C_{Extra\ charge}}{(1+r)^t} \right) [1 \pm (R_t)] \dfrac{T_{Intermediate\ maintanence\ times}}{T_{Times\ of\ cycle}} & \text{Intermediate maintenance costs} \\[2ex] \sum_{t=1year}^{t=100years} \left(\dfrac{C_{Direct\ cost} + C_{Extra\ charge}}{(1+r)^t} \right) [1 \pm (R_t)] \dfrac{T_{Heavy\ maintanence\ times}}{T_{Times\ of\ cycle}} & \text{Heavy maintenance costs} \\[2ex] \sum_{t=1year}^{t=100years} \left(\dfrac{C_{Direct\ cost} + C_{Extra\ charge}}{(1+r)^t} \right) [1 \pm (R_t)] \dfrac{T_{Mad\ maintanence\ times}}{T_{Times\ of\ cycle}} & \text{Mad improvement costs} \end{cases} \tag{10}$$

where $C_{Maintenance}(\bar{x}, T_{100\ years})$ = Costs of maintenance and operation (CNY); $T_{Number\ of\ cycle}$ represents the days of each maintenance cycle (days); and $T_{Total\ maintanence\ times}$,

$T_{\text{Strengthening structure times}}$, $T_{\text{Emergency repair times}}$, $T_{\text{Routine maintanence times}}$, $T_{\text{Med maintanence times}}$, $T_{\text{Intermediate maintanence times}}$, and $T_{\text{Heavy maintanence times}}$ represent the total time for maintenance (days), the total time for strengthening (days), the total time of emergency repair (days), the total time for routine maintenance (days), the total time for intermediate maintenance (days), the total time for heavy maintenance (days), and the total time for overhaul maintenance (days), respectively.

Costs of traffic accidents:

Civilian car ownership in China reached 232,312,300 units in 2018, increasing by 42.7% since 2015 [69]. Wang et al. analysed the severity of traffic accidents in China from a macro perspective, finding that the total fatality rate and man-made injury rate of highway traffic accidents from 2000 to 2016 increased by 19.0% and 63.7% [70]. Vlegel et al. found that the average per capita health care cost was EU 8200 and the productivity cost was EU 5900 [71]. Rukaibi et al. estimated that the average cost of a traffic accident in Kuwait was 9122 KD/crash (equivalent to EU 25,333.02) [72]. According to the data in the China Statistical Yearbook-2019, there were 244,937 traffic accidents in 2018, resulting in 63,194 deaths, 258,532 injuries, and direct property losses of CNY 1385 million [69].

$$C_{\text{Traffic accident}}\left(\overline{x}, T_{100\text{years}}\right) =$$

$$\begin{cases} \sum_{t=1\text{year}}^{t=100\text{years}} \left(\dfrac{C_{\text{Human costs}} + C_{\text{Property damage}} + C_{\text{Other related losses}}}{(1+r)^t} \right) \left[(1+e)^t \right] \\[2ex] \sum_{t=1\text{year}}^{t=100\text{years}} C_{\text{Human costs}} = \sum_{t=1\text{year}}^{t=100\text{years}} \left(C_{\text{Loss of productivity}} + C_{\text{Quality of live costs}} + C_{\text{Medical costs}} \right) \\[2ex] \sum_{t=1\text{year}}^{t=100\text{years}} C_{C_{\text{Property damage}}} = \sum_{t=1\text{year}}^{t=100\text{years}} \left(C_{\text{Vehicle damage costs}} + C_{\text{Non vehicle damage costs}} \right) \\[2ex] \sum_{t=1\text{year}}^{t=100\text{years}} C_{\text{Other related losses}} = \sum_{t=1\text{year}}^{t=100\text{years}} \left(C_{\text{Adminstration costs}} + C_{\text{Enviromental costs}} + C_{\text{Travel delay costs}} \right) \end{cases} \tag{11}$$

where $C_{\text{Traffic accident}}\left(\overline{x}, T_{100\text{years}}\right)$ = Cost of traffic accidents (CNY); $C_{\text{Human cos ts}}$; $C_{\text{Property damage}}$; $C_{\text{Other related losses}}$ = Human costs (CNY); property damage (CNY); other related losses (CNY); and e = Economic growth rate (%).

The six bridges studied are municipal highway bridges and no traffic tolls were charged during the operation.

The total costs required in the stage of maintenance and operation are the sum of Equations (10) and (11).

Disassembly costs:

The cable-stayed bridges will be disassembled at the expiration of their designed service life. The modelling of incurred costs was subject to Eq. (4). The materials to be demolished include broken concrete, scrap steel and waste. Construction wastes dumped and stacked in the natural environment without authorisation are one of the sources of environmental pollution [73]. In recent years, countries all over the world have been using recycled materials for sustainable development and steel is re-smelted for recycling [74,75].

Recycling cost of waste and scraps:

$$C_{\text{Recycling}}\left(\overline{x}, T_{\text{Recycling}}\right) =$$

$$\sum_{t=\text{Secondary processing}}^{t=\text{New product}} \left[\frac{C_{\text{Waste concrete}} * u_{\text{Concrete}} * C_{\text{Post-processing costs}} + C_{\text{Waste steel}} * u_{\text{Steel}} * C_{\text{Steelmaking costs}}}{(1+r)^t} \right] \tag{12}$$

where $C_{\text{Recycling}}\left(\overline{x}, T_{\text{Recycling}}\right)$ = Recycling costs of waste and scraps (CNY); $C_{\text{Waste concrete}}$ = Quantity of demolished concrete waste (kg); $u_{\text{Concrete}}, u_{\text{steel}}$ = Recovery rate of concrete and steel waste (%); and $C_{\text{Steelmaking costs}}$, $C_{\text{Post-processing cost}}$ "51" = Cost of recycling and disposal (CNY/kg).

2.1.3. SILA

SILA witnessed its heyday from 1970 to 1980 and has been widely practiced in many fields around the world [76]. Social impact assessment comprises analysing, monitoring, and managing the social impacts of a project to bring about a more sustainable and equitable biophysical and human environment [77]. However, the assessment criteria and the quality of collected data are affected due to the limited resources of social assessment and the limited ability of regulatory agencies to control the management system [78,79].

PSILCA and USDA data and the Social Hotspots Database (SHDB) were used in this study to assess the research on sustainable social pillars [80,81]. The PSILCA database features the latest data sources, the original data sources and the quality assessment of risk data. Furthermore, the social contact messages from the PSILCA database can be associated with each other in the manner of SOCA (SOCA is an add-on for the Ecoinvent database, containing social inventory data based on PSILCA.) via Green Delta. The processes that are identical to those in environmental assessment can be used for social assessment, thus realising the coherence of the entire assessment (show in Figure 2). SILA uses input data from the LCIA for environmental and social assessment and determines 54 quantitative and qualitative indexes for 18 categories [82]. Five of the analysis indexes are closely related to the community stakeholders, according to the location where the six bridges are located and can be used as the factors for the social impact analysis. The five indexes are fatal accidents (FA), international migrant workers (IMW), youth illiteracy (YI), corruption (C), and sanitation coverage (SC).

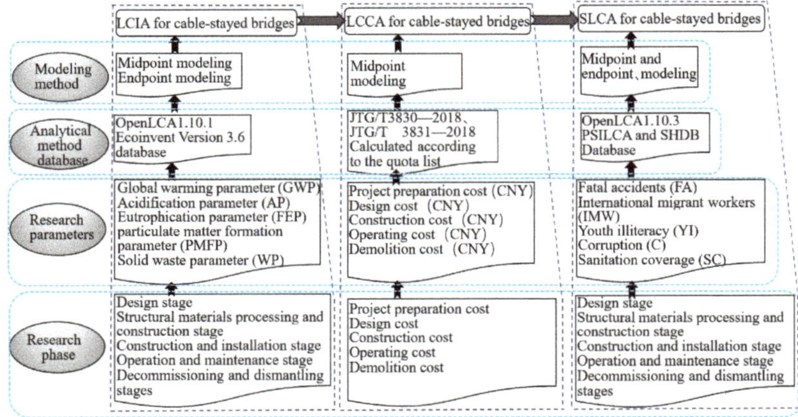

Figure 2. Schematic diagram of the LCIA, LCCA, and SILA analysis process.

According to the location of the six bridges in the region, the five indexes selected are closely linked to community stakeholders and can be used as factors for social impact analysis.

2.2. Research Process

The six cable-stayed bridges across five geographical zones of China (Northeast China, East China, Central China, South China, and Southwest China) and six provinces (Zhejiang, Guangdong, Sichuan, Hubei, Yunnan and Jilin) were selected as the objects of study [83]. They are important in terms of geographical location, economic value, environmental impact, and social assessment, becoming the strong backing for this study, as shown in Figure 3.

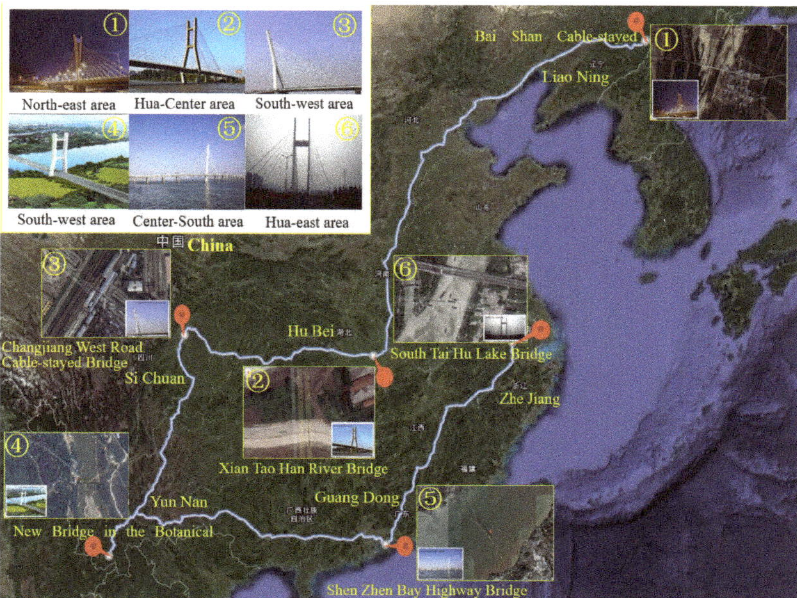

Figure 3. Schematic diagram of cable-stayed bridge regional distribution [83].

2.2.1. LCIA

General information about the six bridges is shown in Table 2. All of these bridges have been completed and put into operation. They are the main highway bridges of the cities where they are located.

The Chinese government classifies cities by criteria including the agglomeration degree of commercial resources, urban pivotability, resident activeness, lifestyle diversity and future plasticity [84]. Among these six cable-stayed bridges, STHB is located in a third-tier city, SZBB in a first-tier city, BGNB in a fourth-tier city, CJWB in a fourth-tier city, XTHB in a fifth-tier city, and BSCB in a fifth-tier city.

They were designed by six design institutes in different regions, which are between 84 and 2380 km away from the project sites. The surveying equipment used was self-owned, calibrated equipment with high precision, which needed to be transported by truck to the project site. The expressway is the preferred mode of transport, but rail travel should be adopted if the transport distance is larger than 500 km. The development organisation was not allowed to use self-produced concrete for cable-stayed bridges, because the bridges are municipal works. All concrete used for the cable-stayed bridges had to be purchased as commercial concrete. Concrete is classified into C55, C50, C40, C30, C25 and C20. SZBB is a steel bridge, using 374 m^3 of precast blocks of commercial concrete for the bridge deck.

During the construction, the materials were mainly transported and hoisted by a tower crane, a 25 T/50 T truck crane, and a floating crane (for the sections across the river), because the main tower of the cable-stayed bridge was too high. The main beam of SZBB is made of Q345-C low alloy steel and the accessory structures are made of Q235-B steel. The components and parts were connected by high-strength bolts and welding. The bridge was divided into 31 beam sections, which were manufactured in the factory and then transported to the bridge position by barges. The floating crane and land cranes worked together to lift and install these sections in the right place. The other five cable-stayed bridges adopted reinforced concrete structures. The main towers were subject to cast-in-place construction with creeping formwork by sections. The main beams were subject to cast-in-place construction with a sliding formwork using the full framing method. The details are shown in Table 3.

BSCB is located in Baishan City, Jilin Province. The construction environment is affected by the local climate. The local temperature in winter can be as low as −42 °C, with an annual average temperature of 4.6 °C [85]. Construction has to be stopped in October every year and can restart again by the end of April of the next year. The affected construction duration reaches 210 days a year.

The operation stage is the key period for the environmental impact contribution of bridges. A large number of vehicles will emit exhaust gases within the 100 years of service life, causing severe environmental pollution. Exhaust gas pollution is the key to research on LCIA. Dargay et al. concluded that the automobile saturation in China is 807 vehicles for every 1000 persons [86], which is set as the upper limit of the number of vehicles in each region. According to the study by Wu et al., car ownership will grow up to 4.8% in 2030, with the growth rate in 2050 being 2.9%, reaching 455 vehicles for every 1000 persons [87]. The traffic volume in 100 years is determined according to the comprehensive data analysis of the China Statistical Yearbook [88], as shown in Figure 4.

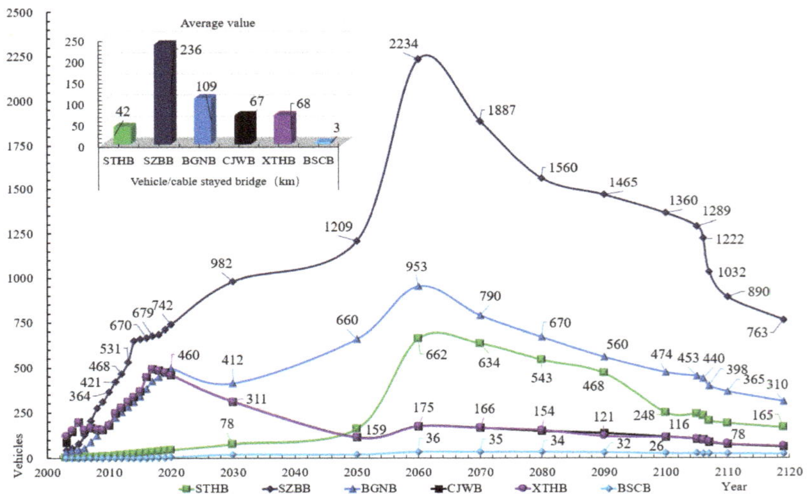

Figure 4. Schematic diagram of the number of vehicles driving on six cable-stayed bridges.

Table 3. Cable-stayed bridge engineering data statistics table.

Bridge Name	Regional Location	Basic Situation	Bridge Layout Drawing
South Tai Hu Lake Bridge (338 m)	East China, Huzhou in Zhejiang	The main bridge is a double-cable, plane H-shaped, single-tower, concrete, cable-stayed bridge with a span layout of 160 + 190 + 38 m, an urban expressway level, and a design speed of 60 Km/h. The standard section width of the bridge is 40.5 m. The main beam adopts the cross-section form of double main ribs, the building height is 3.055 m, the full width is 40.5 m, and the standard main rib is 2.7 m high and 1.7 m wide. The transverse partition is 0.28 m wide; the bridge deck is 28 cm thick, and each cable plane has 24 pairs of cables.	
Shenzhen Bay Bridge (345 m)	Central and South China, Shenzhen Bay	The North Channel Bridge adopts the "180 + 90 + 75 m" span layout, the main beam adopts bolt-welded streamlined steel box girder, the beam height is 4.12 m, the standard section length is 12 m, and the overall width is 38.6 m. The total height of the pylon is 139.053 m. The main beam adopts a single-box, four-chamber, thin-walled structure composed of steel box beams with cantilever arms. The top plate thickness of the bridge deck is 18 mm; the bottom plate is 12–20 mm. The bridge has a total of 12 pairs of stay cables with a cable distance of 3 m and a standard cable distance of 12 m.	

Table 3. *Cont.*

Bridge Name	Regional Location	Basic Situation	Bridge Layout Drawing
New Bridge of Xishuangbanna Tropical Botanical Garden (225 m)	Southwest China, Xishuangbanna Prefecture	The main bridge is an elliptical steel box with a concrete tower column, double cable plane, cable-stayed bridge with a span of 75 + 90 m and a total length of 165 m. The side span is 75 m and the main span is 90 m. The full width of the bridge deck is 14.2 m, the side main beam is 1.8 m high, the bottom width is 1.2 m, the outer top and bottom width is 1.55 m, and the bridge deck is 22 cm thick. The tower column of the cable-stayed bridge adopts a steel box concrete structure with a cross section of 2.5 * 4.0 m and a steel plate thickness of 20 mm.	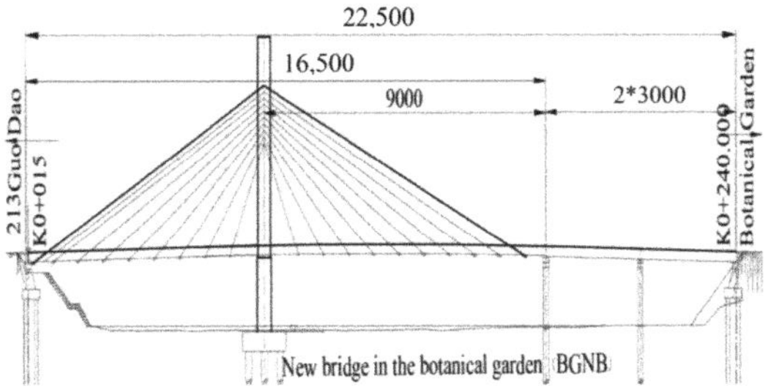
Cable-stayed Bridge of Changjiang West Road, Deyang City (136 m)	Southwest China, Deyang City	Single tower, single cable, plane cable-stayed bridge without back cable, main span 108 m, side span 27.7 m, harp-shaped cable surface, tower and beam consolidation. The standard cable distance on the beam is 8 m, the standard section is 8 m long and weighs about 300 Tons. The main beam adopts a pre-stressed concrete, single-chamber, three-box, flat, thin-walled box beam. The top plate of the box is 24 m wide; the bottom plate is 8.4 m wide, the beam height is 2.5 m, the top plate thickness is 24 cm, the bottom plate thickness is 30 cm, the inclined web plate thickness is 22 cm, and the vertical web plate thickness is 30 cm. A horizontal partition is set every 4 m with a thickness of 28 cm. The approach bridge adopts multi-span continuous beams, all of which are 20 m in span, and the main beam is a 1.4 m high box girder.	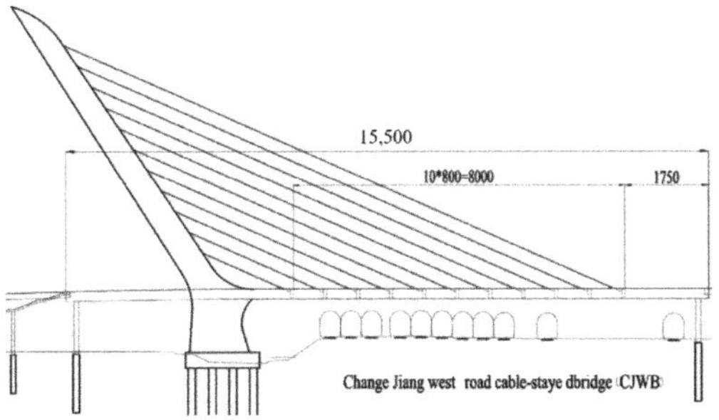

Table 3. *Cont.*

Bridge Name	Regional Location	Basic Situation	Bridge Layout Drawing
Hanjiang Highway Bridge in Xiantao City (312 m)	Central China, Xiantao City	The main bridge is a 50 + 82 + 180 m, three-span, single-tower, double-cable plane cable-stayed bridge, the main girder has a full cross-section width of 25.6 m, a basic section length of 8 m, a basic width of side ribs of 1.8 m, and a basic spacing of 8 m between the diaphragms. The roof thickness of the main beam is 0.30 m, and the beam height is 1.9 m.	
Baishan Bridge in Baishan City (410 m)	Northeast China, Baishan City	The main bridge is a two-span, single-cable, plane cable-stayed bridge with a span of 85 + 85 m. The main beam adopts a single box three-chamber section, the beam height is 2.0 m, the thickness of the top plate is 20 cm, and the thickness of the bottom plate is 40 cm. The section of the main tower adopts an "H" shaped cross-section concrete tower column. Oblique cable harp layout, single-cable deck bridge type, double-width layout with a net width of 15.5 m and a total width of 23.3 m.	

Establish a traffic flow analysis model:

$$N_V\left(x, T_{SO}^{SD}\right) =$$

$$\begin{cases} T_C^D & T_C = \text{Completion report query}, T_D = \text{Design time}(100\ Y) \\ N_{P_{T_C}}^{P_{T_D}}(\lambda 1_{GR}) & P_C = [88]\ I, P_D = [96]\ I\left(P_{2050\ Y}^{1.42\ B}, P_{2060\ Y}^{1.365\ B}, P_{2110\ Y}^{1.0\ B}\right) \\ N_H\left[T_{\text{Quantity of the Y}}^{\text{Urban H in 100 Y}}(\lambda 2_{GR})\right] & T_{\text{Quantity of the Y}} = [96]I, T_{\text{Urban H in 100 Y}} = \text{CA based on }\lambda \\ N_{N\text{ of }V}\left[v_C^{100\ Y}(\lambda 3_{GR})\right] & V_N = [86]\ I\left(GA = \frac{455\ V}{1000\ \text{persons}}[87]\right), \lambda 3_{GR}\left(\lambda 3_{2030\ Y}^{4.8\%}, \lambda 3_{2050\ Y'}^{2.9\%}\right) \\ N_{1000\ \text{persons}}^{N\text{ of }V} & N = \text{The above }CA \\ N_{\text{Every Y}}^{\text{The N of V passing on the bridge}} = \times L_{\text{Bridge}} & N = \text{The above }CA \end{cases} \quad (13)$$

where B = Billion; CA = Calculated; C = Completed; D = Disassembly; GA = Greatest amount; GR = Growth rate; H = Highways; I = Inquiry; N = Number; P = Population; SD = Start disassembly; SO = Start operation; V = Vehicles; and Y = Years. (Note: this abbreviation is only used in Equation (13)).

Figure 4 and Equation (13) show that the traffic volume of SZBB and BGNB is 2 to 5 times that of the traffic volume of the other four bridges, which will affect the subsequent environmental pollution data of the bridges. After 2000, infrastructure expenditure in China accounted for approximately 6.5% of gross domestic product (GDP), much higher than the average level of 4% in other developing countries. After 2009, coastal provinces and cities increased investment in infrastructure (including energy, transportation, telecommunications, water and sewage treatment), reaching 15–20% of GDP [89].

After the expiration of the operation stage, the cable-stayed bridges enter the disassembly stage. These bridges will be demolished by mechanical disruption because blasting demolition has many safety-impacting factors and these bridges are all located in urban areas. The scrapped steel materials will be transported to steel works for recycling. Concrete blocks will be transported to the production plants of reclaimed materials for crushing and classification. All of the remaining waste will be transported to the waste treatment plant for recycling.

2.2.2. LCCA

All of these cable-stayed bridges are municipal works, so the construction costs are analysed based on Engineering Standards for China's Transportation Industry, JTG 3830-2018 Measures for Preliminary Estimate/Budgeting of Highway Projects, and JTG/T 3831-2018 Norms for Preliminary Estimate of Highway Projects [90].

The construction cost is first calculated by Equation (9), in accordance with design drawings, bill of quantities, and norms for preliminary estimates of highway projects. As shown in Table 4, the construction costs of the cable-stayed bridges were: CNY 72,055,116.25 for STHB, CNY 103,996,538.70 for SZBB, CNY 18,803,871.58 for BGNB, CNY 24,721,480.22 for CJWB, CNY 47,164,942.89 for XTHB, and CNY 37,812,245.23 for BSCB, respectively.

Table 4. Statistical table of construction cost of six cable-stayed bridge projects ([91]). Unit: CNY.

Number	Cost Incurred	Ratio	Calculation Method	STHB	SZBB	BGNB	CJWB	XTHB	BSCB
1	Direct project cost			63,392,933.82	92,208,319.2	15,353,271.88	20,691,737.1	40,938,707.24	32,501,337.6
2	Insurance fee			1,901,788.015	2,766,249.576	460,598.1564	620,752.114	1,228,161.217	975,040.129
2-1	Project insurance stipulated in the contract	2.50%	1*2(2-1)	1,584,823.346	2,305,207.98	383,831.797	517,293.428	1,023,467.681	812,533.441
2-2	Third-party liability insurance stipulated in the contract	0.50%	1*2(2-2)	316,964.6691	461,041.596	76,766.3594	103,458.686	204,693.5362	162,506.688
3	Completion Files.	500,000	Constant cost	500,000	500,000	500,000	500,000	500,000	500,000
4	Construction environmental protection fees	1,000,000	Constant cost	1,000,000	1,000,000	1,000,000	1,000,000	1,000,000	1,000,000
5	Safety production fees	1.50%	1*5	950,894.0074	1,383,124.788	230,299.0782	310,376.057	614,080.6085	487,520.064
6	Engineering management software (temporary estimate)	100,000	Constant cost	100,000	100,000	100,000	100,000	100,000	100,000
7	Application fee for building information model technology	100,000	Constant cost	100,000	100,000	100,000	100,000	100,000	100,000
8	Temporary road construction, maintenance and dismantling fees			101,428.6941	147,533.3107	24,565.23501	33,106.7794	65,501.93158	52,002.1402
8-1	Fees for the construction, maintenance and dismantling of the original roads	0.08%	1*8(8-1)	50,714.34706	73,766.65536	12,282.6175	16,553.3897	32,750.96579	26,001.0701
8-2	Construction, maintenance and dismantling fees of temporary steel trestle and wharf	0.08%	1*8(8-2)	50,714.34706	73,766.65536	12,282.6175	16,553.3897	32,750.96579	26,001.0701
9	Temporarily occupying land and occupying the river	0.25%	1*9	158,482.3346	230,520.798	38,383.1797	51,729.3428	102,346.7681	81,253.3441

Table 4. *Cont.*

Number	Cost Incurred	Ratio	Calculation Method	STHB	SZBB	BGNB	CJWB	XTHB	BSCB
10	Erection, maintenance and dismantling of temporary power supply facilities	0.08%	1*10	50,714.34706	73,766.65536	12,282.6175	16,553.3897	32,750.96579	26,001.0701
11	Provision, maintenance and dismantling of telecommunications facilities	0.08%	1*11	50,714.34706	73,766.65536	12,282.6175	16,553.3897	32,750.96579	26,001.0701
12	Water supply and sewage facilities	0.08%	1*12	50,714.34706	73,766.65536	12,282.6175	16,553.3897	32,750.96579	26,001.0701
13	The construction fee of the contractor's project department	0.42%	1*13	266,250.3221	387,274.9406	64,483.74189	86,905.2959	171,942.5704	136,505.618
14	Provisional expenses.	5.00%	(1 + 2 + 3 + 4 + 5 + 6 + 7 + 8 + 9 + 10 + 11 + 12 + 13)*14	3,431,196.012	4,952,216.129	895,422.4561	1,177,213.34	2,245,949.661	1,800,583.11
	The total fees of the project		1 + … + 14	72,055,116.25	103,996,538.7	18,803,871.58	24,721,480.2	47,164,942.89	37,812,245.2

In the operation stage, aging parts and components need to be repaired and replaced in the bridges. Table 1 presents the maintenance and repair cycles of the main components. The costs generated by multiple replacements will be included in the costs for the construction stage, and the economic growth coefficient can then be considered.

The costs of traffic accidents are mainly used to analyse losses caused by traffic accidents and related expenses. According to the Chinese transportation statistics [32], the incidence of traffic accidents from 2001 to 2018 dropped by 25.7%, resulting in the reduction in property losses by 29.3%. After 2014, the annual reduction rate of traffic accidents stayed between 0.4% and −0.7%, and the property losses remained at CNY 5600 per accident.

As shown in Table 5, LCCA was conducted in three stages. The first stage covered the years from 2003 to 2018. The costs of traffic accidents were analysed based on the existing data. The coefficient for the growth or reduction rate of traffic accidents in 15 years, and the annual average number of traffic accidents were also determined. The second stage covered the years from 2019 to 2030. In 2030, the population of China will reach its peak and so will the level of car ownership (Figure 4). The population and car ownership will begin to decline after 2031 and the accident rate will tend to be stable.

Table 5. Statistical table of loss from traffic accidents of six bridges during operation ([32]).

Bridge Name	Time Period (Years)	Accident Loss (CNY/Time)	Times of Accidents	Comprehensive Loss Fee (CNY)
STHB	2006~2018, 2019~2030, 2031~2105	3866	693\659\460	7,005,192
SZBB	2007~2018, 2019~2030, 2031~2106	3259	268\255\179	2,287,818
BGNB	2006~2018, 2019~2030, 2031~2105	4831	301\286\201	3,806,828
CJWB	2005~2018, 2019~2030, 2031~2104	8706	1070\1019\718	24,437,742
XTHB	2003~2018, 2019~2030, 2031~2102	6885	262\250\175	4,730,682
BSCB	2019~2030, 2031~2118	7213	456\434\306	8,626,748

2.2.3. SILA

As shown in Figure 2, SILA was also conducted in five stages. The impact of the bridges on communities was analysed for all aspects, from the design stage to the final disassembly stage. The International Finance Corporation's Performance Standards on Social and Environmental Sustainability (IFC 2012a) was taken as the reference. These Standards has become globally recognised good practice for handling environmental and social risk management and has been adopted by more than 80 leading banks as the "gold standard" for guiding project development [92,93]. The Standards formulate eight performance standards, including social and environmental assessment and management systems, labour and working conditions, pollution prevention and abatement, community health, safety and security, land acquisition and involuntary resettlement, biodiversity conservation and sustainable natural resource management, indigenous peoples, and cultural heritage. Based on the characteristics of Chinese communities (aboriginals will not be considered, because there are no aboriginals in the communities where cable-stayed bridges are located, and cultural heritage will also not be considered, because there is no newly-built cultural heritage in the construction areas), and the latest assessment factors in the PSILCA database, five assessment standards were selected as the research parameters, in accordance with the conclusions of comprehensive analysis (see Figure 2).

3. Results and Discussion

3.1. LCIA

According to our findings (shown in Table 6), the GWPs of six bridges are the main sources of environmental pollution, accounting for over 92% of the total pollution of each bridge. This is why the authors chose these five parameters in the long-term research. Effective control of GWP is the top priority for alleviating global pollution.

Table 6. Life cycle assessment (LCA) statistical tables for six cable-stayed bridges. Unit: kg.

Bridge Name	GWP	AP	FEP	PMFP	WP
STHB	285,792,121.03	758,359.05	778,387.38	2,755,862.99	4,202,670.97
SZBB	75,192,817.81	538,510.86	445,853.55	1,469,182.83	3,451,343.80
BGNB	69,261,736.42	214,170.43	251,077.34	756,768.56	1,397,595.57
CJWB	80,429,187.06	236,629.18	264,255.94	845,577.45	1,414,549.54
XTHB	167,606,486.66	424,005.32	502,313.61	1,559,831.83	2,530,246.34
BSCB	151,598,681.32	322,031.97	424,120.38	1,219,842.08	1,917,809.39

Figure 5 shows the environmental impact contributions of the six cable-stayed bridges, in the maintenance and operation stage, as follows: STHB = 209,488.94 tonnes > XTHB = 133,511.65 tonnes > BSCB = 126,010.36 tonnes > CJWB = 648,518 tonnes > BGNB = 49,735.66 tonnes > SZBB = 1230.24 tonnes.

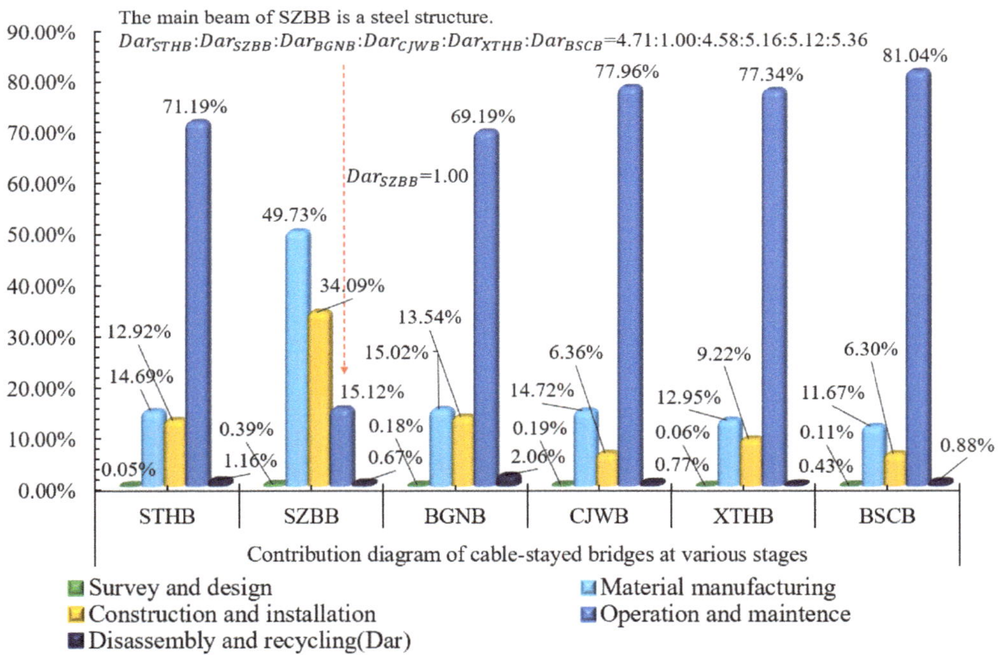

Figure 5. Environmental impact contribution diagrams of six cable-stayed bridges at various stages.

An interesting research finding is that the main beam of SZBB is a steel structure, *Environmental impact contribution*$_{material\ manufacturing\ stage}$ > *Maintenance and operation stage*$_{material\ manufacturing\ stage}$, which is 40,327.22 tonnes and accounts for 49.73% of the total contribution of SZBB. This finding also proves that the environmental impact contribution of the steel bridge mainly comes from the material manufacturing stage and the construction and installation stage, accounting for 83.82% of the total contribution. Although there is a huge difference between the environmental impact contribution of a steel bridge and that of a concrete bridge, the total environmental impact contribution of the two kinds of bridges are approximate to each other.

3.2. Comparison

The differences in the durability of building materials and standards between Europe and China result in a difference in the life span of bridges, and the difference is mainly manifested in the service life of concrete; the warranty period of concrete for stay cables in Europe is 100 years, while in China, it is 20 or 50 years [67,94].

Thus, a large amount of maintenance and replacement work is required, resulting in great changes in environmental pollution values during the maintenance period.

Table 7 shows the environmental impact contribution values of five impact factors in the maintenance stage. Subject to the European and Chinese design standards, the maximum value falls on $GWP_{European\ standard}$ = 5343.68 tonnes for SZBB and $GWP_{Chinese\ standard}$ = 19,736.99 tonnes for STHB. Interestingly, the value of SZBB's steel structure under the European standard is 10,824.72 tonnes greater than that under the Chinese standard. The difference in the design life of the materials leads to 33- to 73-fold differences, in terms of the environmental pollution value in the maintenance stage, and this is just a comparison analysis for one stage.

Table 7. Environmental pollution data in Europe and China during the maintenance phase. Unit: kg.

Bridge Name	Quantity Analysed According to Chinese Standards	Quantity Analysed According to European Standards
STHB	202,577,714.70	4,060,953.15
SZBB	8,469,275.96	5,413,303.55
BGNB	46,427,579.22	1,264,900.09
CJWB	61,909,222.65	1,857,067.35
XTHB	127,556,952.20	3,689,371.79
BSCB	120,405,196.80	1,648,154.08

Figure 6 shows the difference in the environmental pollution value for the six bridges under five environmental impact factors and subject to two standards. The replacement times of the exposed stable cables and concrete of the cable-stayed bridges in the 100 years of the service life increases with time, resulting in an increase in GWP by 3249~15761 tonnes, particularly the GWP of the steel bridge at SZBB, which reduces by 4568 tonnes. The pollution contributions of the six cable-stayed bridges increase by 549,412.2 tonnes in total, which is an amazing figure.

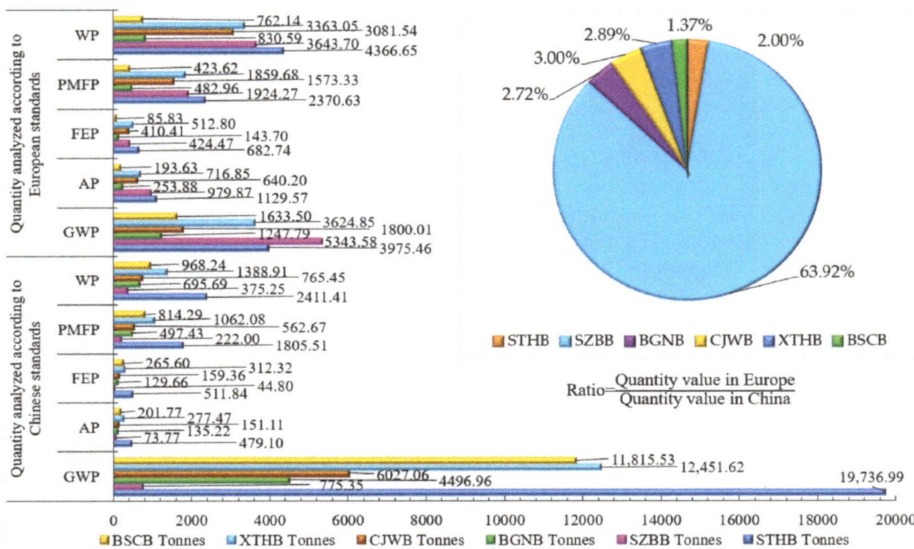

Figure 6. Environmental impact contribution diagrams of six cable-stayed bridges at various stages.

3.3. LCCA

The conclusions of LCCA are shown in Table 8. The bridges selected in the case analyses are located in China, so the norms for Chinese highways were used in each analysis. For the cable-stayed bridges with reinforced concrete structures, the cost ratio of the maintenance and operation stage remains between 49% and 64%. However, the cost of steel bridges in the construction stage accounts for 63.2% of the total expenses because of the high investment cost. The maintenance cost of the steel bridge is 30% lower than that of the concrete bridge. The main reason is that the steel structure is superior to the concrete structure in terms of durability.

Table 8. Statistical table of the cost ratio of 6 cable-stayed bridges.

Cost Name	STHB	SZBB	BGNB	CJWB	XTHB	BSCB
Cost of project preparation	0.01%	0.02%	0.01%	0.01%	0.01%	0.01%
Survey and design costs	0.07%	0.13%	0.06%	0.05%	0.06%	0.06%
Project construction costs	33.63%	63.20%	28.29%	24.39%	30.40%	29.99%
Maintenance and operating costs	60.57%	33.56%	63.69%	49.94%	63.78%	60.00%
Accident costs	3.27%	1.39%	5.73%	24.11%	3.05%	6.84%
Demolition stage costs	2.45%	1.69%	2.23%	1.50%	2.69%	3.10%

As shown in Figure 7, the maintenance cost of STHB is CNY 120 million, which is 1.8 times the construction cost. The maintenance costs of BGNB, CJWB, XTHB and BSCB are 2.0 to 2.3 times their construction costs. For the cable-stayed bridges with the reinforced concrete structure, the stay cables and concrete need to be replaced two to five times, because their service life and durability ranges between 20 and 50 years. Costs for multiple replacement events are the primary reason for the excessive maintenance costs, so the key to reducing costs is to improve the service life of materials.

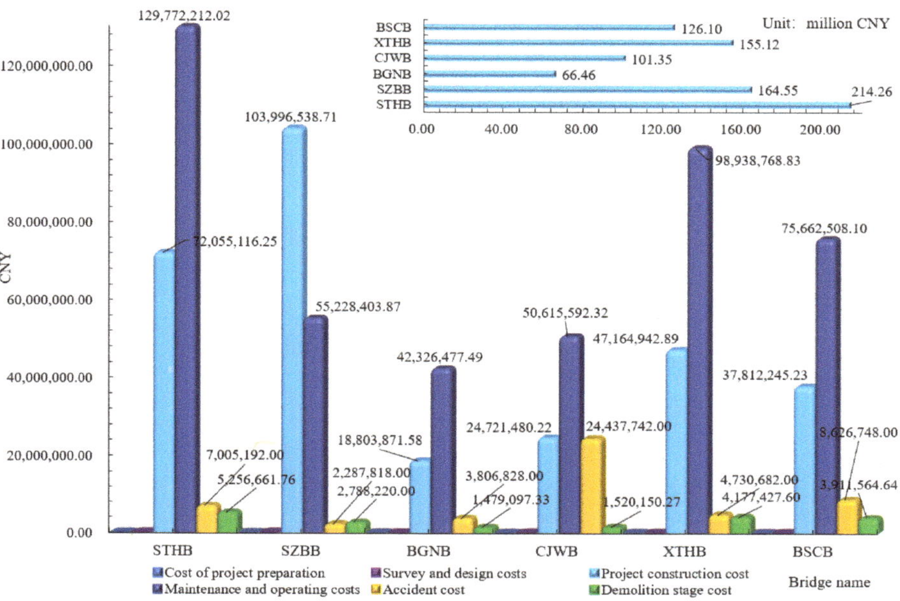

Figure 7. The cost diagram of six cable-stayed bridges at different stages.

3.4. SILA

SILA was conducted for the six cable-stayed bridges from four categories, including the population impact, community system, social resources and economic development. Five impact factors were selected according to the classification.

Table 9 shows some of the SILA values for the six bridges. For each cable-stayed bridge, corruption > sanitation coverage > fatal accidents > international migrant workers > youth illiteracy.

Table 9. Statistical table of five social environmental impact data for 6 cable-stayed bridges. Unit: med risk hours.

Bridge Name	Fatal Accidents	International Migrant Workers	Youth Illiteracy	Corruption	Sanitation Coverage
STHB	55,792,892.84	31,765,165.76	28,624,476.33	118,864,998.3	88,496,114.86
SZBB	47,282,293.11	26,919,734.79	24,258,123.41	1,007,33434	74,996,993.87
BGNB	6,502,779.89	3,702,297.38	3,336,243.37	13,853,967.44	10,314,409.72
CJWB	9,202,951.4	5,239,614.97	4,721,563.11	19,606,597.66	14,597,297.3
XTHB	28,358,724.3	16,145,776.5	14,549,409.27	60,417,367.61	44,981,301.3
BSCB	14,063,615.15	8,006,988.77	7,215,320.78	29,962,088.48	22,307,058.1

As shown in Figure 8, the values of five impact factors in each stage of the six cable-stayed bridges are ranked as follows:

$Numbers_{Construction\ and\ installation\ stage} > Numbers_{Decommissioning\ and\ dismantling\ stage} >$
$Numbers_{Structural\ materials\ processing\ and\ construction\ stage} >$
$Numbers_{Design\ stage} > Numbers_{Maintenance\ and\ operation\ stage}.$

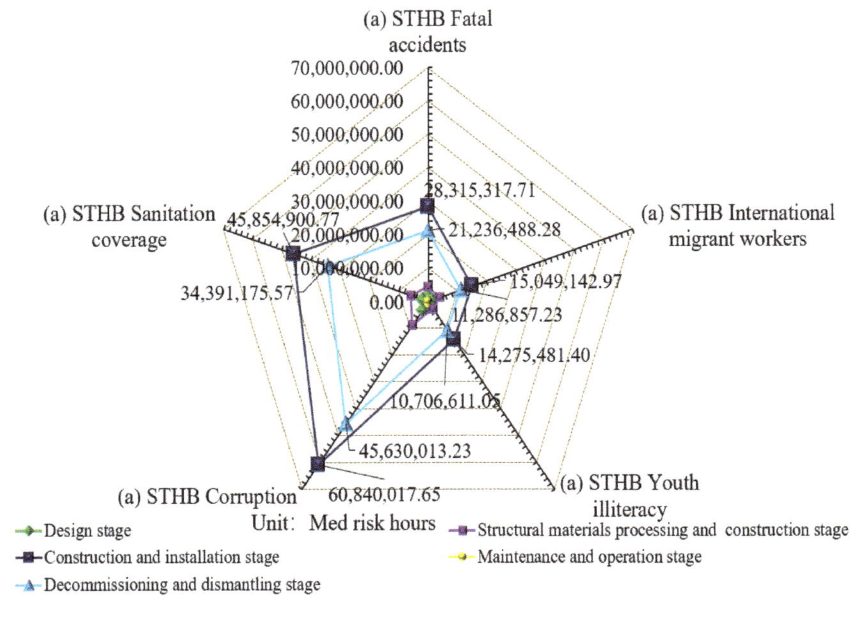

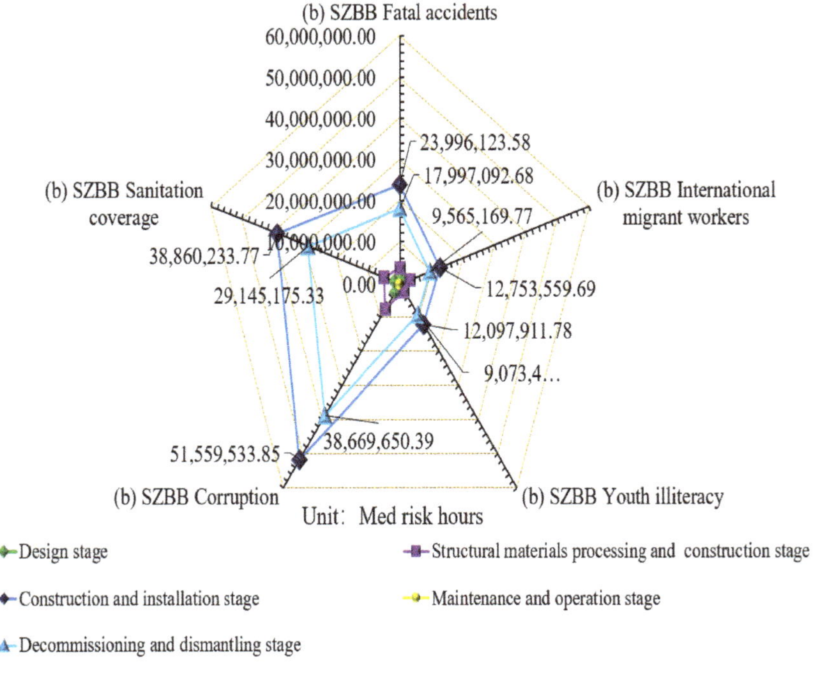

Figure 8. *Cont.*

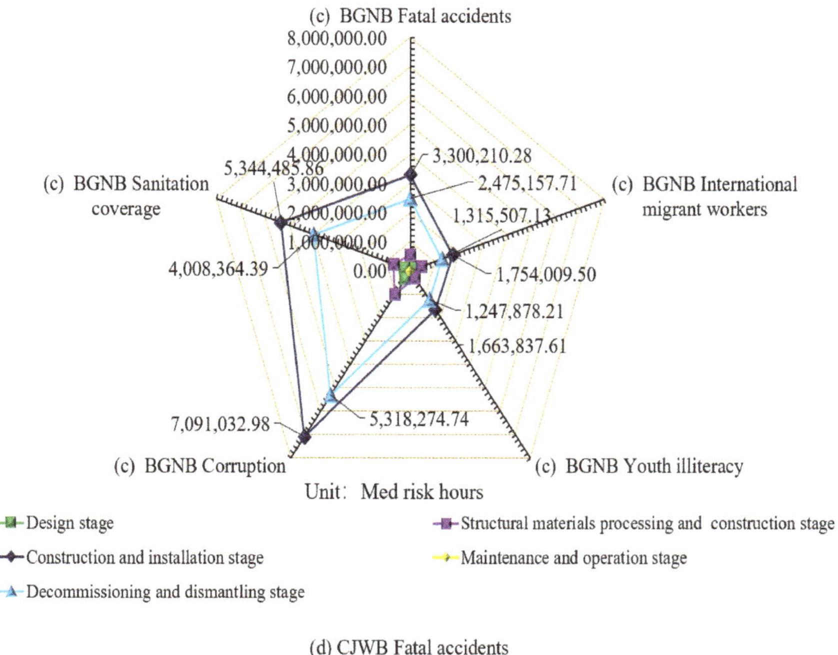

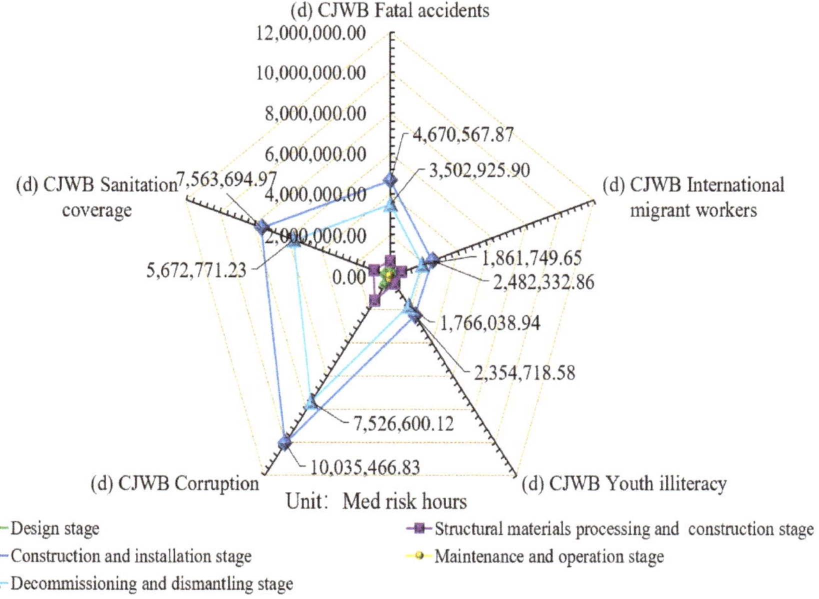

Figure 8. *Cont.*

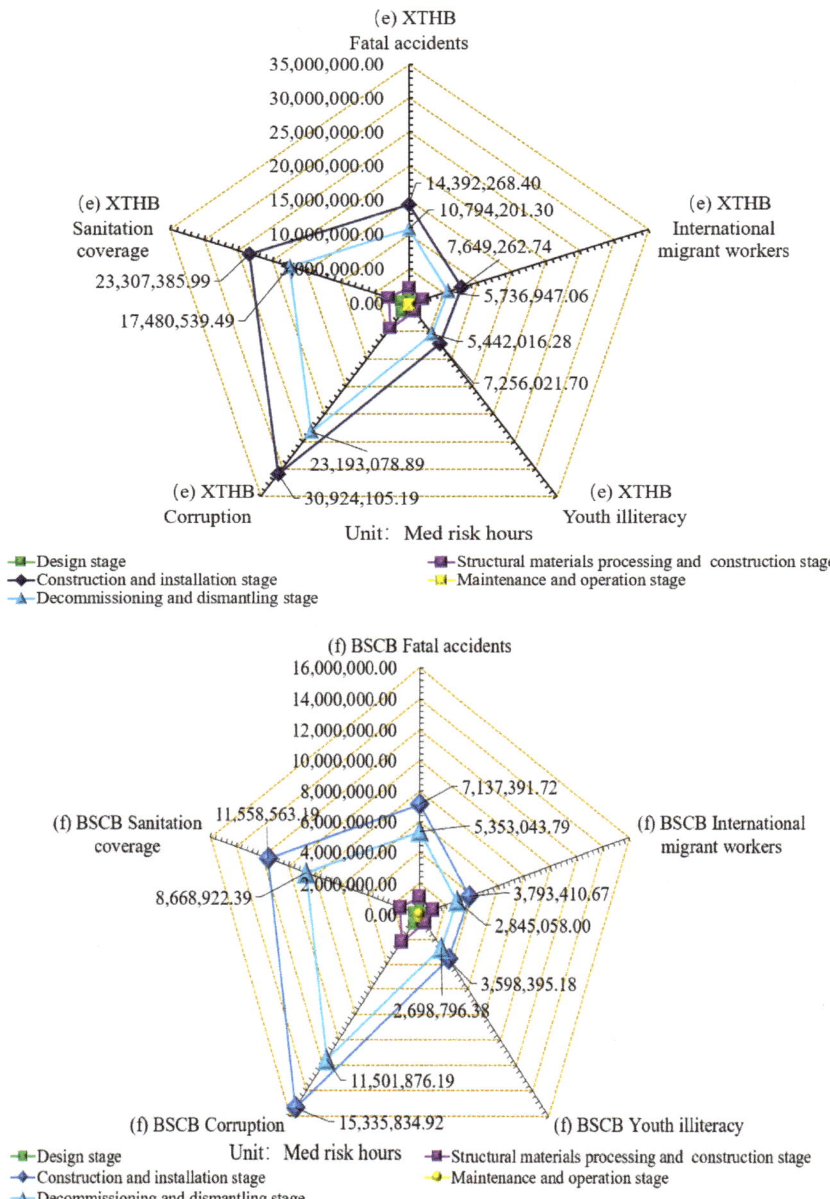

Figure 8. (**a**) The content in the first panel is the description of the five SILA factors of STHB; (**b**) The content in the second panel is the description of the five SILA factors of SZBB; (**c**) The content in the third panel is the description of the five SILA factors of BGNB; (**d**) The content in the fourth panel is the description of the five SILA factors of CJWB; (**e**) The content in the fifth panel is the description of the five SILA factors of XTHB; (**f**) The content in the sixth panel is the description of the five SILA factors of BSCB.

3.5. Deepen the Analysis

3.5.1. Economic Evaluation

As shown in Figure 9, the bridges with the peak value of GDP in 10 years are SZBB and STHB (Government, n.d.); the bridges with the peak value of LCIA are STHB and XTHB; the bridges with the peak values of LCCA and SLCA are STHB and SZBB. The analysis concludes that the environmental pollution, production cost and social impact generated by infrastructure in developed regions increase accordingly. In particular, there is a complementary relationship between GDP and the emissions of environmentally polluting gases. The constant emission load of environmental pollution gases under GDP growth signifies that the current energy technologies must be replaced with renewable energy resources, and/or more energy-efficient production technologies must be adopted [95].

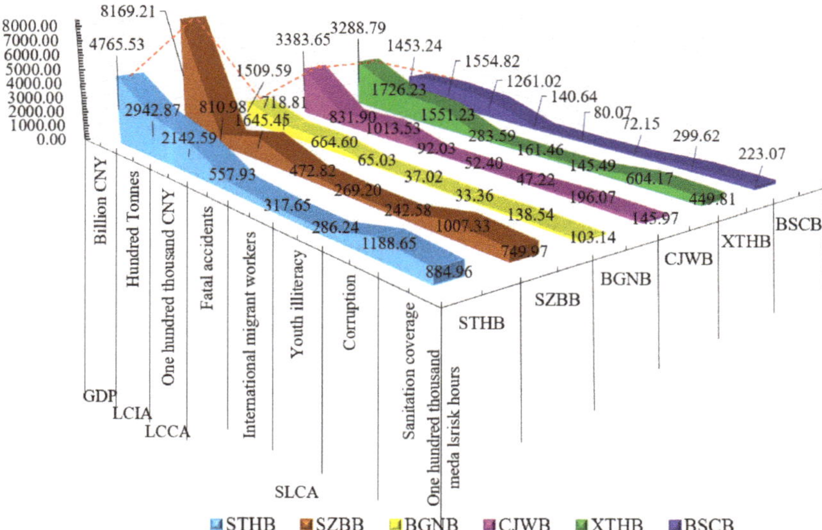

Figure 9. Schematic diagram of gross domestic product (GDP) [96], LCIA, LCCA, and SILA data in the area where the six cable-stayed bridges are located.

3.5.2. Modelling and Discussion

Definition of Markov chain: assuming that $X_1, X_2, \cdots\cdots X_n$ is the discrete sequence of random influence variables, abbreviated as $\{X_n\}$, the state space of the entire $\{X_n\}$ is denoted as $E = \{x_1, x_2, \cdots\cdots x_n\}$; if any impact factor is subject to $x_{i1} x_{i2}, \cdots\cdots x_{in} E$, then $P(X_{n+1}) = (x_{i_{n+1}} \mid X_1 = x_{i_1}, \cdots\cdots X_n = x_{i_n})$.

The impact matrix is established based on the definition,

$$K_h = \begin{cases} x_{11}(h_1) & x_{12}(h_1) \cdots\cdots & x_{1m}(h_1) & h_1 = GDP_{variables} \\ x_{21}(h_2) & x_{22}(h_2) \cdots\cdots & x_{2m}(h_2) & h_2 = GWP_{variables} \\ \vdots & \vdots & \vdots & \vdots \\ \vdots & \vdots & \vdots & \vdots \\ x_{n1}(h_k) & x_{n2}(h_k) \cdots\cdots & x_{nm}(h_k) & h_k = H_{variables} \end{cases} \qquad (14)$$

where K_h = conclusion of the infrastructure's comprehensive impact assessment.

According to Equation (13),

$$K_{\text{Six bridges}} = \begin{bmatrix} \text{GDP} & \text{LCIA} & \text{LCCA} & \text{SLCA}-1 & \text{SLCA}-2 & \text{SLCA}-3 & \text{SLCA}-4 & \text{SLCA}-5 \\ 4766 & 2943 & 2143 & 559 & 318 & 284 & 1189 & 885 \\ 8169 & 811 & 1645 & 473 & 269 & 243 & 1007 & 750 \\ 1510 & 718 & 665 & 65 & 37 & 33 & 139 & 103 \\ 3384 & 832 & 1014 & 92 & 52 & 47 & 196 & 146 \\ 3289 & 1726 & 1551 & 283 & 162 & 146 & 604 & 450 \\ 1453 & 1555 & 1261 & 141 & 80 & 72 & 300 & 223 \end{bmatrix}$$

$$K_{\text{Six bridges1}} = \begin{bmatrix} \text{GDP} & \text{LCIA} & \text{LCCA} & \text{SLCA}-1 & \text{SLCA}-2 & \text{SLCA}-3 \\ 4766 & 2943 & 2143 & 559 & 318 & 284 \\ 8169 & 811 & 1645 & 473 & 269 & 243 \\ 1510 & 718 & 665 & 65 & 37 & 33 \\ 3384 & 832 & 1014 & 92 & 52 & 47 \\ 3289 & 1726 & 1551 & 283 & 162 & 146 \\ 1453 & 1555 & 1261 & 141 & 80 & 72 \end{bmatrix}$$

$$K_{\text{Six bridges1}} = \begin{bmatrix} 4766 & 2943 & 2143 & 559 & 318 & 284 \\ 8169 & 811 & 1645 & 473 & 269 & 243 \\ 1510 & 718 & 665 & 65 & 37 & 33 \\ 3384 & 832 & 1014 & 92 & 52 & 47 \\ 3289 & 1726 & 1551 & 283 & 162 & 146 \\ 1453 & 1555 & 1261 & 141 & 80 & 72 \end{bmatrix}$$

$$\text{Assuming } \left| K_{\text{Six bridges1}} - \lambda_1 E \right| = \begin{vmatrix} 4766-\lambda_1 & 2943 & 2143 & 559 & 318 & 284 \\ 8169 & 811-\lambda_1 & 1645 & 473 & 269 & 243 \\ 1510 & 718 & 665-\lambda_1 & 65 & 37 & 33 \\ 3384 & 832 & 1014 & 92-\lambda_1 & 52 & 47 \\ 3289 & 1726 & 1551 & 283 & 162-\lambda_1 & 146 \\ 1453 & 1555 & 1261 & 141 & 80 & 72-\lambda_1 \end{vmatrix} = \Sigma_1^6 K_{\text{bridges1}}$$

If the diagonal method is used, then (14) = $(4766 - \lambda_1) \times (811 - \lambda_1) \times (655 - \lambda_1) \times (92 - \lambda_1) \times (162 - \lambda_1) \times (72 - \lambda_1) - 4332878707784\lambda_1 - 5454599392867510 = 0$, $\lambda_1 = \Sigma_1^7 (12588 + 4766 + 811 + 665 + 92 + 162 + 72)/7 = 2736.7 \approx 2737$.

$$K_{\text{Six bridges2}} = \begin{bmatrix} 4766 & 2943 & 2143 & 559 & 1189 & 885 \\ 8169 & 811 & 1645 & 473 & 1007 & 750 \\ 1510 & 718 & 665 & 65 & 139 & 103 \\ 3384 & 832 & 1014 & 92 & 196 & 146 \\ 3289 & 1726 & 1551 & 283 & 604 & 450 \\ 1453 & 1555 & 1261 & 141 & 300 & 223 \end{bmatrix}$$

$$\text{Assuming } \left| K_{\text{Six bridges2}} - \lambda_2 E \right| = \begin{bmatrix} 4766-\lambda_2 & 2943 & 2143 & 559 & 1189 & 885 \\ 8169 & 811-\lambda_2 & 1645 & 473 & 1007 & 750 \\ 1510 & 718 & 665-\lambda_2 & 65 & 139 & 103 \\ 3384 & 832 & 1014 & 92-\lambda_2 & 196 & 146 \\ 3289 & 1726 & 1551 & 283 & 604-\lambda_2 & 450 \\ 1453 & 1555 & 1261 & 141 & 300 & 223\lambda_2 \end{bmatrix} = \sum_1^6 K_{\text{bridges2}} \quad (15)$$

If the diagonal method is used, then (15) $(4766 - \lambda_2) \times (811 - \lambda_2) \times (665 - \lambda_2) \times (92 - \lambda_2) \times (604 - \lambda_2) \times 223\lambda_2 - 1478251935682100000\lambda_2 - 1046549405522410 = 0$, $\lambda_2 = \Sigma_1^7 (82565 + 4766 + 811 + 665 + 92 + 604 + 223)/7 = 12818$.

Based on Equations (14) and (15), we can conclude that the most reasonable impact range is $2737 < K_{\text{Six bridges}} < 12818$.

According to Figure 10, five-point positions are located in the reasonable comprehensive evaluation range. The five points are Point ② and ⑤ of STHB, Point ① of SZBB, Point ③ of CJWB, and Point ④ of XTHB.

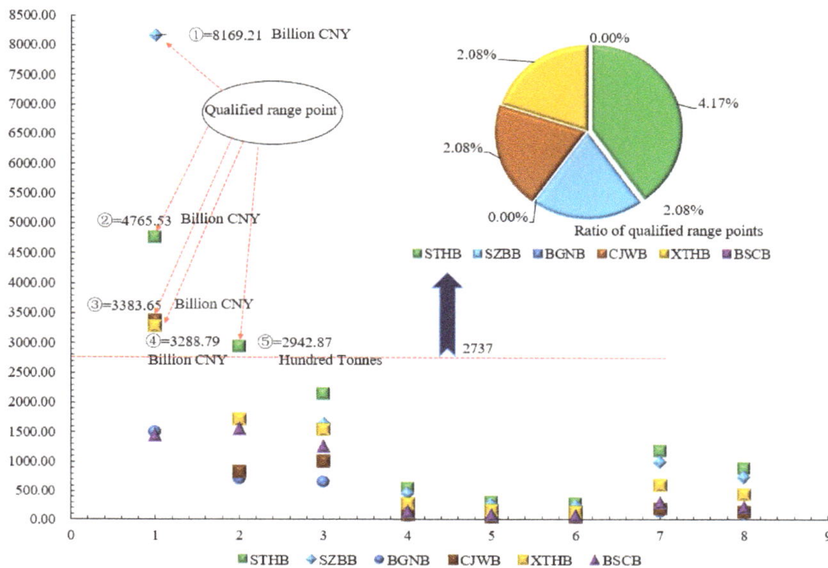

Figure 10. Schematic diagram of discrete points for comprehensive evaluation.

4. Conclusions

The manuscript proposes a comprehensive and effective sustainability assessment method and establishes an assessment framework and modelling theory for complex structural bridges (cable-stayed bridges) in terms of environment, economy, and social impact. Through the comprehensive evaluation of six highway cable-stayed bridges in five provinces of China in the whole life cycle (from cradle to grave), the conclusion is drawn. GWP is the main source of environmental pollution in LCIA, accounting for more than 92% of the emissions of each bridge, which are concentrated in the maintenance stage. In LCCA, the proportion of maintenance stage cost is 49–64%. In SILA, the corruption value has the greatest influence, accounting for 36% of the total amount. The SZBB steel structure bridge is special: GWP accounts for 50% in the LCIA material stage and 63.2% in the LCCA construction stage.

In view of the high pollution and high cost in the maintenance stage, the conclusion shows that it is closely related to the design standard and service life of the materials. It is found that the difference in LCIA between Europe and China is 33~73-fold, which is due to the difference in the replacement period between the main girder and stay cable of 80 years and 50 years/cycle. More interestingly, the LCIA value of SZBB in Europe is higher than that in China by 10,824.7 tonnes, because the maintenance period of steel structure differs by 15 years/cycle. The differences in the above conclusions are closely related to regional population density, vehicle ownership and traffic frequency, which is one of the research directions in the future.

Finally, to obtain the relationship between GDP and sustainable impact, the comprehensive evaluation coefficient of the influence matrix is established by using discrete mathematics for multi factor decision-making, and the reasonable range of 2737~12,818 (The theoretical judgment standard of innovation) between China's five major economic regional bridges and regional GDP is analysed.

This study aims to propose a complete method for assessing the sustainability of bridges. This article provides important knowledge for preliminary decisions in the construction of bridges as well as how to mitigate the loads of the three pillars. The limitation of the study is that there is no questionnaire survey in the social impact assessment, and it is

Int. J. Environ. Res. Public Health **2021**, 18, 122

impossible to compare and analyse whether there is a big difference between the conclusion and the actual impact. Future research directions need to strengthen the resilience analysis of evaluating the impact of the construction industry on society, and the mutual promotion and optimization of the GDP influencing factors and sustainable development.

Author Contributions: Investigation, Z.Z. and J.A.; methodology, Z.Z., J.A., and V.Y.; supervision, J.A. and V.Y.; validation, Z.Z. and V.Y.; writing—original draft, Z.Z.; writing—review and editing, J.A. and V.Y. All authors have read and agreed to the published version of the manuscript.

Funding: This research was funded by the Spanish Ministry of Economy and Competitiveness, along with FEDER (Fondo Europeo de Desarrollo Regional), project grant number: BIA2017-85098-R.

Conflicts of Interest: The authors declare no potential conflicts of interest with respect to the research, authorship, and/or publication of this article.

References

1. ISO 14044:2006/AMD 1:2017. Environmental Management-Life Cycle Assessment-Requirements and Guidelines. ISO. 2006. Available online: https://www.iso.org/standard/72357.html (accessed on 20 November 2020).
2. Wuni, Y.; Shen, G.Q.P.; Osei-Kyei, R. Scientometric review of global research trends on green buildings in construction journals from 1992 to 2018. *Energy Build.* **2019**, *190*, 69–85. [CrossRef]
3. United Nations. World Population in 2050. Available online: https://www.un.org/development/desa/en/news/population/world-population-prospects-2017.html (accessed on 20 November 2020).
4. Huisingh, D.; Zhang, Z.; Moore, J.C.; Qiao, Q.; Li, Q. Recent advances in carbon emissions reduction: Policies, technologies, monitoring, assessment and modeling. *J. Clean. Prod.* **2015**, *103*, 1–12. [CrossRef]
5. Zhang, X. Toward a regenerative sustainability paradigm for the built environment: From vision to reality. *J. Clean. Prod.* **2014**, *65*, 3–6. [CrossRef]
6. IPCC. Summary for Policymakers, Climate Change 2014: Mitigation of Climate Change. 2014. Available online: https://www.buildup.eu/en/practices/publications/ipcc-2014-climate-change-2014-mitigation-climate-change-contribution-working (accessed on 20 November 2020).
7. Dong, Y.H.; Ng, S.T. A social life cycle assessment model for building construction in Hong Kong. *Int. J. Life Cycle Assess.* **2015**, *20*, 1166–1180. [CrossRef]
8. Hellweg, S.; Canals, L.M.I. Emerging approaches, challenges and opportunities in life cycle assessment. *Science* **2014**, *344*, 1109–1113. [CrossRef]
9. Hansen, J.; Sato, M.; Kharecha, P.; Beerling, D.; Berner, R.; Masson-Delmotte, V.; Pagani, M.; Raymo, M.; Royer, D.L.; Zachos, J.C. Target Atmospheric CO: Where should Humanity Aim? *Open Atmos. Sci. J.* **2008**, *2*, 217–231. [CrossRef]
10. World Meteorological Organization. WMO Statement on the State of the Global Climate in 2016. 2017. Available online: https://library.wmo.int/doc_num.php?explnum_id=3414 (accessed on 20 November 2020).
11. Muntean, M.; Guizzardi, D.; Schaaf, E.; Crippa, M.; Solazzo, E.; Olivier, J.G.J.; Vignati, E. *Fossil CO_2 Emissions of All World Countries: 2018 Report*; EU Science Hub, European Commission, Joint Research Centre: Ispra (VA), Italy, 2018; ISBN 9789279972409. [CrossRef]
12. Lin, B.; Liu, H. CO_2 emissions of China's commercial and residential buildings: Evidence and reduction policy. *Build. Environ.* **2015**, *92*, 418–431. [CrossRef]
13. Kim, T.H.; Tae, S.H. Proposal of environmental impact assessment method for concrete in South Korea: An application in LCA (life cycle assessment). *Int. J. Environ. Res. Public Health* **2016**, *13*, 1074. [CrossRef]
14. Hildenbrand, J.; Michael Srocka, A.C. OpenLCA 1.10. 2020. Available online: http://www.openlca.org/openlca/ (accessed on 23 November 2020).
15. ISO,14044:2006/AMD 2:2020, Environmental Management-Life Cycle Assessment-Requirements and Guidelines. ISO. 2006. Available online: https://www.iso.org/standard/76122.html (accessed on 23 November 2020).
16. Navarro, I.J.; Yepes, V.; Martí, J.V.; González-Vidosa, F. Life cycle impact assessment of corrosion preventive designs applied to prestressed concrete bridge decks. *J. Clean. Prod.* **2018**, *196*, 698–713. [CrossRef]
17. O'Born, R. Life cycle assessment of large scale timber bridges: A case study from the world's longest timber bridge design in Norway. *Transp. Res. Part. D Transport. Environ.* **2018**, *59*, 301–312. [CrossRef]
18. Milani, C.J.; Kripka, M. Evaluation of short span bridge projects with a focus on sustainability. *Struct. Infrastruct. Eng.* **2020**, *16*, 367–380. [CrossRef]
19. Trunzo, G.; Moretti, L.; D'Andrea, A. Life cycle analysis of road construction and use. *Sustainability* **2019**, *11*, 377. [CrossRef]
20. Li, H.; Deng, Q.; Zhang, J.; Xia, B.; Skitmore, M. Assessing the life cycle CO_2 emissions of reinforced concrete structures: Four cases from China. *J. Clean. Prod.* **2019**, *210*, 1496–1506. [CrossRef]
21. Frangopol, D.M.; Dong, Y.; Sabatino, S. Bridge life-cycle performance and cost: Analysis, prediction, optimisation and decision-making. *Struct. Infrastruct. Eng.* **2017**, *13*, 1239–1257. [CrossRef]

22. Goh, K.C.; Goh, H.H.; Chong, H.-Y. Integration Model of Fuzzy AHP and Life-Cycle Cost Analysis for Evaluating Highway Infrastructure Investments. *J. Infrastruct. Syst.* **2019**, *25*, 04018045. [CrossRef]

23. Heidari, M.R.; Heravi, G.; Esmaeeli, A.N. Integrating life-cycle assessment and life-cycle cost analysis to select sustainable pavement: A probabilistic model using managerial flexibilities. *J. Clean. Prod.* **2020**, *254*. [CrossRef]

24. Wang, Z.; Yang, D.Y.; Frangopol, D.M.; Jin, W. Inclusion of environmental impacts in life-cycle cost analysis of bridge structures. Sustain. *Resilient Infrastruct.* **2020**, *5*, 252–267. [CrossRef]

25. Cadenazzi, T.; Dotelli, G.; Rossini, M.; Nolan, S.; Nanni, A. Life-Cycle Cost and Life-Cycle Assessment Analysis at the Design Stage of a Fiber-Reinforced Polymer-Reinforced Concrete Bridge in Florida. *Adv. Civ. Eng. Mater.* **2019**, *8*, 20180113. [CrossRef]

26. ESMS. Social Impact Assessment (SIA). 2016. Available online: https://www.iucn.org/sites/dev/files/iucn_esms_sia_guidance_note.pdf (accessed on 23 November 2020).

27. Zhang, A.; Zhong, R.Y.; Farooque, M.; Kang, K.; Venkatesh, V.G. Blockchain-based life cycle assessment: An implementation framework and system architecture. *Resour. Conserv. Recycl.* **2020**, *152*, 104512. [CrossRef]

28. Parent, J.; Cucuzzella, C.; Revéret, J.P. Impact assessment in SLCA: Sorting the sLCIA methods according to their outcomes. *Int. J. Life Cycle Assess.* **2010**, *15*, 164–171. [CrossRef]

29. Vanclay, F. Reflections on Social Impact Assessment in the 21st century. *Impact Assess. Proj. Apprais.* **2020**, *38*, 126–131. [CrossRef]

30. Zamarrón-Mieza, I.; Yepes, V.; Moreno-Jiménez, J.M. A systematic review of application of multi-criteria decision analysis for aging-dam management. *J. Clean. Prod.* **2017**, *147*, 217–230. [CrossRef]

31. Parsons, R. Forces for change in social impact assessment. *Impact Assess. Proj. Apprais.* **2020**, *38*, 278–286. [CrossRef]

32. Vanclay, F. Statistical Analysis on the Number of Traffic Accidents in China. *Impact Assess. Proj. Apprais.* **2003**, *21*, 3–4. [CrossRef]

33. Domínguez-Gómez, J.A. Four conceptual issues to consider in integrating social and environmental factors in risk and impact assessments. *Environ. Impact Assess. Rev.* **2016**, *56*, 113–119. [CrossRef]

34. Fischer, T.B.; Jha-Thakur, U.; Fawcett, P.; Clement, S.; Hayes, S.; Nowacki, J. Consideration of urban green space in impact assessments for health. *Impact Assess. Proj. Apprais.* **2018**, *36*, 32–44. [CrossRef]

35. Balasbaneh, A.T.; Marsono, A.K. Bin Applying multi-criteria decision-making on alternatives for earth-retaining walls: LCA, LCC, and S-LCA. *Int. J. Life Cycle Assess.* **2020**, *25*, 2140–2153. [CrossRef]

36. Balasbaneh, A.T.; Marsono, A.K.B.; Khaleghi, S.J. Sustainability choice of different hybrid timber structure for low medium cost single-story residential building: Environmental, economic and social assessment. *J. Build. Eng.* **2018**, *20*, 235–247. [CrossRef]

37. Penadés-Plà, V.; Martínez-Muñoz, D.; García-Segura, T.; Navarro, I.J.; Yepes, V. Environmental and Social Impact Assessment of Optimized Post-Tensioned Concrete Road Bridges. *Sustainability* **2020**, *12*, 4265. [CrossRef]

38. Ali, M.S.; Aslam, M.S.; Mirza, M.S. A sustainability assessment framework for bridges—A case study: Victoria and Champlain Bridges, Montreal. *Struct. Infrastruct. Eng.* **2016**, *12*, 1381–1394. [CrossRef]

39. Kloepffer, W. Life cycle sustainability assessment of products (with Comments by Helias A. Udo de Haes, p. 95). *Int. J. Life Cycle Assess.* **2008**, *13*, 89–95. [CrossRef]

40. Hu, M. Building impact assessment—A combined life cycle assessment and multi-criteria decision analysis framework. *Resour. Conserv. Recycl.* **2019**, *150*, 104410. [CrossRef]

41. Association, E.; Ecoinvent Datebase. Ecoinvent. 2019. Available online: https://www.ecoinvent.org/database/database.html (accessed on 23 November 2020).

42. Bedec Datebase. The Regional Catalan Government. 2010. Available online: https://en.itec.cat/database/ (accessed on 23 November 2020).

43. Psilca Greendatebase. Available online: https://psilca.net/ (accessed on 23 November 2020).

44. Ortiz, O.; Castells, F.; Sonnemann, G. Sustainability in the construction industry: A review of recent developments based on LCA. *Constr. Build. Mater.* **2009**, *23*, 28–39. [CrossRef]

45. Asdrubali, F.; Baldassarri, C.; Fthenakis, V. Life cycle analysis in the construction sector: Guiding the optimization of conventional Italian buildings. *Energy Build.* **2013**, *64*, 73–89. [CrossRef]

46. European Union. *International Reference Life Cycle Data System (ILCD) Handbook*; Springer Science + Business Media B.V.: Berlin, Germany, 2011; p. 2011933605. Available online: https://link.springer.com/chapter/10.1007/978-94-007-1899-9_11 (accessed on 23 November 2020).

47. Ramesh, T.; Prakash, R.; Shukla, K.K. Life cycle energy analysis of buildings: An overview. *Energy Build.* **2010**, *42*, 1592–1600. [CrossRef]

48. Cabeza, L.F.; Rincón, L.; Vilariño, V.; Pérez, G.; Castell, A. Life cycle assessment (LCA) and life cycle energy analysis (LCEA) of buildings and the building sector: A review. *Renew. Sustain. Energy Rev.* **2014**, *29*, 394–416. [CrossRef]

49. Chau, C.K.; Leung, T.M.; Ng, W.Y. A review on life cycle assessment, life cycle energy assessment and life cycle carbon emissions assessment on buildings. *Appl. Energy* **2015**, *143*, 395–413. [CrossRef]

50. Baker, L. Of embodied emissions and inequality: Rethinking energy consumption. *Energy Res. Soc. Sci.* **2018**, *36*, 52–60. [CrossRef]

51. Chen, L.; Pelton, R.E.O.; Smith, T.M. Comparative life cycle assessment of fossil and bio-based polyethylene terephthalate (PET) bottles. *J. Clean. Prod.* **2016**, *137*, 667–676. [CrossRef]

52. Walker, S.; Rothman, R. Life cycle assessment of bio-based and fossil-based plastic: A review. *J. Clean. Prod.* **2020**, *261*, 121158. [CrossRef]

53. Recipe. 2008. Available online: https://www.researchgate.net/publication/230770853_Recipe_2008 (accessed on 30 November 2020).

54. New Version Recipe. New Version ReCiPe 2016 to Determine Environmental Impact | RIVM. 2016. Available online: https://www.rivm.nl/en/news/new-version-recipe-2016-to-determine-environmental-impact (accessed on 30 November 2020).

55. Penadés-Plà, V.; Martí, J.V.; García-Segura, T.; Yepes, V. Life-cycle assessment: A comparison between two optimal post-tensioned concrete box-girder road bridges. *Sustainability* **2017**, *1864*, 1864. [CrossRef]

56. Zhou, Z.; Alcalá, J.; Yepes, V. Bridge Carbon Emissions and Driving Factors Based on a Life-Cycle Assessment Case Study: Cable-Stayed Bridge over Hun He River in Liaoning, China. *Int. J. Environ. Res. Public Health* **2020**, *17*, 5953. [CrossRef]

57. SimaPro. 2020. Available online: https://simapro.com/about/ (accessed on 23 November 2020).

58. Lee, K.M.; Cho, H.N.; Cha, C.J. Life-cycle cost-effective optimum design of steel bridges considering environmental stressors. *Eng. Struct.* **2006**, *28*, 1252–1265. [CrossRef]

59. Navarro, I.J.; Penadés-Plà, V.; Martínez-Muñoz, D.; Rempling, R.; Yepes, V. Life cycle sustainability assessment for multi-criteria decision making in bridge design: A review. *J. Civil Eng. Manag.* **2020**, *26*, 690–704. [CrossRef]

60. García-Segura, T.; Penadés-Plà, V.; Yepes, V. Sustainable bridge design by metamodel-assisted multi-objective optimization and decision-making under uncertainty. *J. Clean. Prod.* **2018**, *202*, 904–915. [CrossRef]

61. Jang, B.; Mohammadi, J. Impact of fatigue damage from overloads on bridge life-cycle cost analysis. *Bridge Struct.* **2019**, *15*, 181–186. [CrossRef]

62. Matos, J.; Solgaard, A.; Santos, C.; Silva, M.S.; Linneberg, P.; Strauss, A.; Casas, J.; Caprani, C.; Akiyama, M. Life cycle cost, as a tool for decision making on concrete infrastructures. In *Proceedings of the High-Tech Concrete: Where Technology and Engineering Meet, Maastricht, The Netherlands, 12–14 June 2017*; Springer International Publishing: Berlin, Germany, 2017; pp. 1832–1839. [CrossRef]

63. Edited by the Policy Research Office, National Development and Reform Commission, 2011. National Development and Re-Form Commission Notified to Reduce Fees for Some Construction Projects. Available online: https://www.ndrc.gov.cn/xwdt/xwfb/201103/t20110323_956782.html (accessed on 23 November 2020).

64. Edited by the Ministry of Construction, National Development and Reform Commission, 2002. Engineering Survey and Design Charging Standards. Available online: https://wenku.baidu.com/view/3fa74a62effdc8d376eeaeaad1f34693daef1088.html (accessed on 23 November 2020).

65. Stewart, M.G. *Analysis of Climate Change Impacts on the Deterioration of Concrete Infrastructure—Part. 1: Mechanisms, Practices... Analysis of Climate Change Impacts on the Deterioration of Concrete Infrastructure Part. 1: Mechanisms, Practices, Modelling and Sim*; CSIRO: Canberra, Australia, 2016; ISBN 9780643103658; pp. 1–88. Available online: https://apo.org.au/sites/default/files/resource-files/2011-02/apo-nid25469_5.pdf (accessed on 23 November 2020).

66. Rossi, B.; Marquart, S.; Rossi, G. Comparative life cycle cost assessment of painted and hot-dip galvanized bridges. *J. Environ. Manag.* **2017**, *197*, 41–49. [CrossRef]

67. Ministry of Communications of China, 2004; Code for Maintenance of Highway Bridge and Culvers. Available online: http://www.chhca.org.cn/web_html/001/HYDT03.asp (accessed on 23 November 2020).

68. Wang, H.; Schandl, H.; Wang, X.; Ma, F.; Yue, Q.; Wang, G.; Wang, Y.; Wei, Y.; Zhang, Z.; Zheng, R. Measuring progress of China's circular economy. *Resour. Conserv. Recycl.* **2020**, *163*, 105070. [CrossRef]

69. National Bureau of Statistics-2019. Available online: http://www.stats.gov.cn/tjsj/ndsj/ (accessed on 23 November 2020). (In Chinese)

70. Wang, D.; Liu, Q.; Ma, L.; Zhang, Y.; Cong, H. Road traffic accident severity analysis: A census-based study in China. *J. Saf. Res.* **2019**, *70*, 135–147. [CrossRef]

71. Van der Vlegel, M.; Haagsma, J.A.; de Munter, L.; de Jongh, M.A.C.; Polinder, S. Health Care and Productivity Costs of Non-Fatal Traffic Injuries: A Comparison of Road User Types. *Int. J. Environ. Res. Public Health* **2020**, *17*, 2217. [CrossRef]

72. Al-Rukaibi, F.; AlKheder, S.; AlOtaibi, N.; Almutairi, M. Traffic crashes cost estimation in Kuwait. *Int. J. Crashworthiness* **2020**, *25*, 203–212. [CrossRef]

73. Jiménez, J.R.; Ayuso, J.; Agrela, F.; López, M.; Galvín, A.P. Utilisation of unbound recycled aggregates from selected CDW in unpaved rural roads. *Resour. Conserv. Recycl.* **2012**, *58*, 88–97. [CrossRef]

74. Tavira, J.; Jiménez, J.R.; Ayuso, J.; Sierra, M.J.; Ledesma, E.F. Functional and structural parameters of a paved road section constructed with mixed recycled aggregates from non-selected construction and demolition waste with excavation soil. *Constr. Build. Mater.* **2018**, *164*, 57–69. [CrossRef]

75. Sangiorgi, C.; Lantieri, C.; Dondi, G. Construction and demolition waste recycling: An application for road construction. *Int. J. Pavement Eng.* **2015**, *16*, 530–537. [CrossRef]

76. Prenzel, P.V.; Vanclay, F. How social impact assessment can contribute to conflict management. *Environ. Impact Assess. Rev.* **2014**, *45*, 30–37. [CrossRef]

77. Vanclay, F. International principles for social impact assessment. *Impact Assess. Proj. Apprais.* **2003**, *21*, 5–12. [CrossRef]

78. Esteves, A.M.; Franks, D.; Vanclay, F. Social impact assessment: The state of the art. *Impact Assess. Proj. Apprais.* **2012**, *30*, 34–42. [CrossRef]

79. Sierra, L.A.; Pellicer, E.; Yepes, V. Method for estimating the social sustainability of infrastructure projects. *Environ. Impact Assess. Rev.* **2017**, *65*, 41–53. [CrossRef]

80. Navarro, I.J.; Yepes, V.; Martí, J.V. Social life cycle assessment of concrete bridge decks exposed to aggressive environments. *Environ. Impact Assess. Rev.* **2018**, *72*, 50–63. [CrossRef]
81. Shab-Home. 2020. Available online: http://www.socialhotspot.org/ (accessed on 23 November 2020).
82. Ciroth, A.; Eisfeldt, F. PSILCA-A Product Social Impact Life Cycle Assessment Database Database Version 1.0. 2016. Available online: https://www.openlca.org/wp-content/uploads/2016/08/PSILCA_documentation_v1.1.pdf (accessed on 23 November 2020).
83. Wiki. Geographical Division of China-Wikiwand. Wiki. 2020. Available online: https://www.wikiwand.com/en/Geography_of_China (accessed on 23 November 2020).
84. List of New Cities in China-Wikiwand. 2020. Available online: https://m.sohu.com/n/486287408/ (accessed on 23 November 2020).
85. Baishan City People's Government. 2020. Available online: http://www.cbs.gov.cn/sq/sthj/201805/t20180519_323816.html. (accessed on 23 November 2020).
86. Dargay, J.; Gately, D.; Sommer, M. The saturation is 807 cars per 1000 people. *Energy J.* **2007**, *28*, 143–170. [CrossRef]
87. Wu, T.; Zhang, M.; Ou, X. Analysis of Future Vehicle Energy Demand in China Based on a Gompertz Function Method and Computable General Equilibrium Model. *Energies* **2014**, *7*, 7454–7482. [CrossRef]
88. Huzhou Statistical Yearbook-2019. Available online: http://tjj.huzhou.gov.cn/hustats/html/tjnj2019/05.html (accessed on 20 November 2020).
89. Shi, Y.; Guo, S.; Sun, P. The role of infrastructure in China's regional economic growth. *J. Asian Econ.* **2017**, *49*, 26–41. [CrossRef]
90. Announced by the Ministry of Transport of the People's Republic of China, M. of T. of the P.R. of Highway Engineering Construction Project Estimate Budget Preparation Method (JTG 3830-2018). Ministry of Transport of the People's Republic of China. 2019. Available online: http://xxgk.mot.gov.cn/2020/jigou/glj/202006/t20200623_3313130.html (accessed on 24 November 2020). (In Chinese)
91. Department of Science and Technology, Ministry of Transport. Announced by the Ministry of Transport of the People's Republic of China, M. of T. of the P.R. of. Highway Engineering Industry Standard. China Communications Press Co.,Ltd. 2018. Available online: http://www.mot.gov.cn/ (accessed on 24 November 2020).
92. Vanclay, F.; Franks, D.M. International Association for Impact Assessment Purpose and Intended Readership. 2015, pp. 1–74. Available online: https://www.socialimpactassessment.com/documents/IAIA%202015%20Social%20Impact%20Assessment%20guidance%20document.pdf (accessed on 24 November 2020).
93. Appiah-Opoku, S. Land access and resettlement: A guide to best practice, by Gerry Reddy, Eddie Smyth, and Michael Steyn. *Impact Assess. Proj. Apprais.* **2015**, *33*, 290. [CrossRef]
94. Fomento, T. Virtual de Publicaciones del M. de. 2010. Code on Structural Concrete (EHE-08) Articles and Annexes. Available online: http://asidac.es/asidac-en/wp-content/uploads/2016/07/EHE-ENG.pdf (accessed on 26 November 2020).
95. Suzuki, S.; Nijkamp, P. An evaluation of energy-environment-economic efficiency for EU, APEC and ASEAN countries: Design of a Target-Oriented DFM model with fixed factors in Data Envelopment Analysis. *Energy Policy* **2016**, *88*, 100–112. [CrossRef]
96. Government, C. Data of GDP. Available online: http://www.gov.cn/shuju/index.htm (accessed on 24 November 2020).

International Journal of
Environmental Research and Public Health

Article

An Indicator System for Evaluating Operation and Maintenance Management of Mega Infrastructure Projects in China

Dan Chen [1], Pengcheng Xiang [1,2,3,*], Fuyuan Jia [1], Jian Zhang [1] and Zhaowen Liu [4]

[1] School of Management Science & Real Estate, Chongqing University, Chongqing 400045, China; 20180301001@cqu.edu.cn (D.C.); 20180301006@cqu.edu.cn (F.J.); jason2016.cqu@foxmail.com (J.Z.)
[2] International Research Center for Sustainable Built Environment, Chongqing University, Chongqing 400045, China
[3] Construction Economics and Management Research Center, Chongqing University, Chongqing 400045, China
[4] Faculty of Civil Engineering and Geosciences, Delft University of Technology, 2628CN Delft, The Netherlands; z.liu-8@tudelft.nl
* Correspondence: pcxiang@cqu.edu.cn; Tel.: +86-2365-120-848

Received: 18 November 2020; Accepted: 16 December 2020; Published: 21 December 2020

Abstract: Mega infrastructure projects provide a basic guarantee for social development, economic construction, and livelihood improvement. Their operation and maintenance (O&M) management are of great significance for the smooth operation and the realization of the value created by the projects. In order to provide an approach for effectively evaluating O&M management, this study develops a holistic indicator system using a mixed-review method from the national macro perspective in China. In this study, literature analysis, policy texts, expert interviews, and grounded theory were used to collect relevant data at home and abroad, and establish an initial evaluation indicator system with 23 indicators covering two dimensions and five aspects. Then the questionnaire survey and factor analysis were used to score and categorize the indicators, and finally an evaluation indicator system for O&M management of mega infrastructure projects was formed. The results show that social relations, environmental benefits, macro policy, and operational capacities play an important role in the evaluation of the O&M of mega infrastructure projects. This study helps the management team to avoid negative impacts in the O&M management of mega infrastructure projects and lays a theoretical foundation for future research. The indicator system in this study is based on the Chinese context, and it remains to be verified whether the indicator system is applicable to other countries due to the differences in political and cultural backgrounds in different regions.

Keywords: mega infrastructure projects; operation and maintenance management; assessment level; indicator system

---

1. Introduction

At present, megaprojects all over the world have entered the "trillions era" [1]. During the past few decades, China has carried out the largest infrastructure activities in history, which has become a powerful driving force to promote China's urbanization construction, economic development, and improvement of people's living standards [2]. By the end of 2019, the country's infrastructure investment exceeded CNY 18.5 trillion, the total operating mileage of high-speed railways exceeded 35,000 km, accounting for more than two thirds of the total mileage of high-speed rail in the world, and the mileage of expressways exceeded 140,000 km, ranking first all over the world [3]. Compared with general projects, mega infrastructure projects, as large-scale social activities, involve a wide range of scope and construction difficulty, consume a lot of capital, and have significant complexity in technology, environment, organization, and management [4]. This kind of project has higher requirements for decision-making, design, construction, and operation and maintenance (O&M). Once the project has problems, not only the project itself, but also the economy, environment, and society closely related to the project will cause great harm [5]. According to the data from the Gartner Group, 80% of the lifecycle cost of projects occurs in the O&M management phase, and 40% of the projects never resume operation again [6]. At the same time, the impact of the O&M of mega infrastructure projects on society and the environment has also been widely a cause for concern for academics. For example, the Three Gorges Dam, the world's largest hydroelectric project and a symbol of China's confidence in high-risk technological solutions, has been heavily criticized for the threats it poses to the environment, animal species, and migrants [7–9]. The report by McKinsey shows that by 2030, the investment in mega infrastructure projects will reach $ 57 trillion, which means that the annual investment will reach $ 6-9 trillion, accounting for 8% of the global GDP [10]. In the context of global sustainable development, the sustainable development of mega infrastructure projects has become a hot topic in academia.

Issues related to the O&M management of mega infrastructure projects have been a concern from different perspectives. Researchers in the organization and management field have emphasized the importance of O&M management throughout the lifecycle management (LM). These studies cover a variety of topics, such as input–output of funds [11], resource consumption [12], and satisfaction of all participants [13]. In the field of project management, the complexity, ambiguity [14], politics [15], risk, and ambition [16] of mega infrastructure projects have been explored. It also involves the management of stakeholders from different levels [17]. The long-term impact of O&M management on society, the economy, and the natural environment is also considered along with the measurement of the distribution of benefits [18]. In addition, economists and policy experts described the macro impact of the projects. These issues are related to economic growth [19], labor market [20], regional health [21], social stability [22], market integration [23], and national productivity [24]. In practice, the O&M management of mega infrastructure projects mainly relies on the experiences and intuition of managers, lacks a theoretical basis, and the conclusions are relatively subjective. The indicator system in published studies is usually based on the project and is used to measure performance [14] and technical [25] and teamwork capabilities [20], and these indicators can only meet the general project requirements. Therefore, there is an urgent need for a complete indicator system to quantitatively assess the O&M management of mega infrastructure projects.

Nowadays, the evaluation of the O&M of infrastructure projects mostly focuses on transportation infrastructure. The indicators are only for one or some types of infrastructure [19,23], and some only measure one aspect of the infrastructure, such as environmental [26,27] or economic factors [28]. There are few indicators considering the macro impact of O&M on the government and society, which cannot systematically and comprehensively evaluate the O&M of mega infrastructure projects. Based on this, this study uses a combination of quantitative and qualitative analysis to construct the indicators for the O&M of mega infrastructure projects so as to lay a theoretical foundation for the quantitative analysis of O&M management.

The indicator system constructed in this paper is tailored to the O&M of mega infrastructure projects in the Chinese context. These indicators can effectively evaluate the O&M of mega infrastructure projects and pave the way for quantifying them. The structure of this study is as follows: The second section provides a literature review on the O&M of mega infrastructure projects. The third section introduces the research methodology. The fourth part describes the process of the indicator system. Last, the fifth section summarizes the findings and future direction of the research, and the contributions and limitations of this study.

2. Literature Review

Since the 1990s, as an important part of LM, O&M management has been a hot topic in academia. O&M management generally involves space management, public safety management, asset management, and energy consumption management. Because of its long duration and large cost,

O&M management plays an important role in LM. Previous academic studies are mostly found in the field of power and telecommunication engineering to evaluate and discuss the cost control, resource scheduling, and efficiency evaluation of projects [25]. In the field of engineering, O&M management research mainly focuses on the application, innovation, and deficiencies of BIM (Building Information Modeling) [26], the Internet of Things, 3D printing [27] and other technologies in the O&M management of projects.

However, the O&M management of mega infrastructure projects is different from that of general projects. The U.S. Presidential Commission for Critical Infrastructure Protection (PCCIP) [28] made an authoritative definition of critical infrastructure projects in its report: Critical infrastructure refers to the facilities that can weaken a country's national defense strength and have a significant impact on economic security after failure or damage. Eight key infrastructures are defined in the report: communication system, power system, oil and gas, banking and finance, transportation system, water supply system, government service, and emergency service system. The object of this study was mega infrastructure projects, which have the following characteristics: (1) They are key projects in national or local government economic development planning; (2) they concern strategic and public welfare; (3) they have far-reaching impacts on the national or regional economy, society, and ecological environment; (4) they are of great concern to the public. They are not a combination of general projects, and the purpose of them is often to change the world [14]. The O&M management of this kind of project has a great impact on society, the economy, and the ecological environment, and may cause social chaos and inequality in certain areas [25]. In the context of global economic integration, mega infrastructure projects have become a research hotspot because of changing production and lifestyle, and have brought a huge and permanent impact to the Earth [1]. In terms of the management of mega infrastructure projects, scholars have conducted exploratory studies from different perspectives. From a corporate perspective, scholars have focused on aspects such as green construction [29], building safety [30], environmental management [31], and public pressure [32]. From a project point view, the O&M management of mega infrastructure projects that span multiple geographic regions is also a key part of project management. In the process of O&M, it is proposed to strengthen the links between the central government and local governments, local governments at all levels, and social organizations and citizens [33] so as to achieve a win-win situation for all parties through the joint participation of multiple subjects [34]. From a social point of view, the impact of mega infrastructure projects goes far beyond the projects themselves—for example, the impact of the Qinghai–Tibet Railway on the ecological aspects of the Tibetan region and the impact on the living standards and lifestyles of the residents of the Tibetan Plateau. There is evidence that the construction and O&M of mega infrastructure projects can enhance national productivity, promote regional economic development, and create employment opportunities for the local people [35]. Establishing indicators for the O&M management of mega infrastructure projects is not suitable due to the complexity and impact of the operation process. First, existing indicators are based on the project perspective and examine O&M in terms of quality, cost, schedule, and safety [36], which cannot reflect the social benefits and project sustainability. Second, some indicators originate from the enterprise perspective, which only focuses on the input–output of the enterprise and specific aspects of the enterprise's resources [37], benefits, and services [38]. Many scholars, experts, and institutions have researched and constructed various types of indicator systems, ranging from social benefits to economic benefits to a comprehensive system of indicators for assessing the usefulness of projects [39]. The existing indicators and criteria are only applicable to enterprises in the construction industry, and in such an indicator system, the social and ecological impact of the operation is excluded from consideration. Third is a system of indicators from the micro level. This kind of study usually covers a category of infrastructure projects, such as highway [19] and railway [40], which are limited in their indicators and cannot reflect the overall O&M management of mega infrastructure projects. Moreover, these fragmented indicators are difficult to aggregate due to different perspectives. Although the current sustainability initiatives, strategies, frameworks, and processes for the O&M management of mega infrastructure projects emphasize broader national aspirations and strategic goals [41], there are still serious challenges in integrating these national strategic objectives (such as the economy, society, and the environment) into the micro level. Given the current state of the research, there is an urgent need for methods and techniques to evaluate the O&M management of mega infrastructure projects, and to assess making decisions about unexpected problems in the process.

In recent years, scholars have used various methods to construct evaluation indicators for different research objects. Commonly used methods include literature review [42], questionnaire survey [43], case analysis [44], semi-structured interview [45], the Delphi method [46], multi-criteria decision-making [47], and grounded theory [48]. Based on the complexity of the projects, the methods used in this study include literature analysis, policy texts, expert interviews, grounded theory, questionnaire survey, and factor analysis to comprehensively and systematically sort out, summarize, and filter the O&M management indicators of mega infrastructure projects so as to build a complete, effective, and comprehensive indicator system.

3. Methodology

This study aims to filter the indicator system and clarify the evaluation indicators in the O&M management of mega infrastructure projects. To end this, this article adopts the "mixed-review method" [49,50]. Generally speaking, this method includes qualitative analysis (i.e., literature analysis, policy texts, etc.) and quantitative analysis (i.e., questionnaire survey and factor analysis) so it can eliminate biased conclusions and subjective interpretations while providing an in-depth understanding of domain knowledge.

The mixed-review method refers to the combination of quantitative analysis and qualitative analysis. This method can not only reduce the subjective judgment influence brought by qualitative analysis, but also deepen the understanding by quantitative analysis so as to make up for the shortcomings of a single method. As a practical application of mixed-method research, the mixed-review method combines the quantitative evaluation method and qualitative review method in the process [51]. It can reduce the impact of subjective judgment of manual qualitative review methods, and improve the depth and understanding of the results of quantitative study methods [52]. This study takes literature analysis, policy texts, and grounded theory as the qualitative analysis, and questionnaire survey and factor analysis as the quantitative analysis. The mixed-review method is divided into three stages (see Figure 1). The first stage is to collect the indicators for evaluation of the O&M of mega infrastructure projects, and then sort out the relevant indicators through literature analysis, policy texts, and expert interviews in this stage. During this stage, we can comprehensively understand the research in this field, and collect, identify, and organize previous studies from all aspects. The second stage is the identification of indicators, in which the information collected in the first stage is sorted out and the preliminary list of indicators for measuring the O&M management of mega infrastructure projects is identified by combining it with the grounded theory. The purpose of this stage is to summarize, analyze, and evaluate the research subject. The third stage is the construction of the indicator system, which will determine the key indicators of the O&M management of mega infrastructure projects through questionnaire survey and factor analysis, and systematically construct the final evaluation indicator system of the O&M management of mega infrastructure projects.

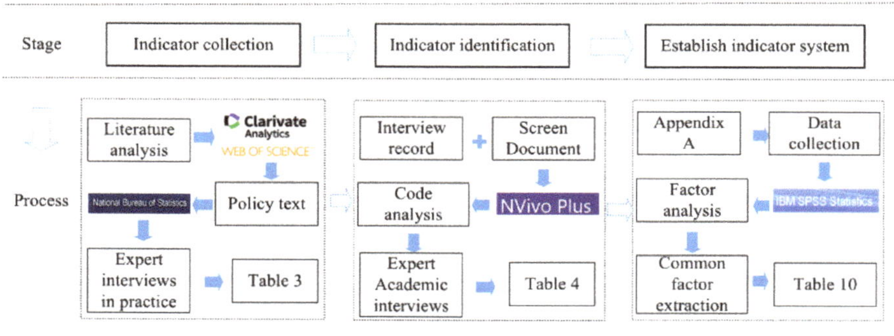

Figure 1. Research framework and process.

3.1. Step 1. Indicator Collection

Literature review is a method of collecting, identifying, and organizing previous studies to form a scientific understanding of the target domain [42]. It is the most basic method that provides a valuable theoretical basis for in-depth analysis in the field of the research. In addition, due to the uniqueness of mega infrastructure projects, the government (including the central government and local government) has the absolute right in LM [53], so it is necessary to organize and summarize the policies, systems, laws, and other documents issued by the government to analyze the influence of the government on the O&M management of mega infrastructure projects from the macro level. Indicator collection is crucial in this study, so the academic literature and policy texts must be carefully selected. As shown in Figure 1, the data acquisition channel (i.e., the selection of academic literature and policy texts) is first selected for the analysis below. (1) Academic literature: With the help of keywords such as "large-scale project," "large-scale infrastructure," "complex project," "megaproject," "operation and maintenance management," "indicator," and "evaluation" in the core collection of Google Scholar, Web of Science, CNKI (China National Knowledge Infrastructure), and school digital library, study was carried out in the databases of the science core collection. In order to cover the literature in the research field as comprehensively as possible, a subject search selected to define the publication time range was defined as 2010–2020. (2) Policy text: Central and local government networks, official websites such as the NDRC (National Development and Reform Commission) and the Ministry of Housing and Urban–Rural Development, documentary reports of mega infrastructure projects, official websites, special websites, news reports, and related sustainability reports and initiatives.

In addition, as the world has entered a new phase of globalization, an increasing number of megaprojects have attracted extensive attention from scholars. Universities and institutions are enhancing their research on megaprojects by establishing independent research centers. Therefore, in addition to academic literature and policy texts, this study complements the constructed indicators for evaluating the O&M management of mega infrastructure projects through expert interviews. The expert interviews can directly obtain valuable interview data through face-to-face interaction with the interviewees, which is helpful for clarifying ideas and identifying problems [54]. In particular, the expert interview process divided into two parts. (1) Interviews with experts in practice. Interviews were conducted with eight practice experts (sample size determined according to the principle of theoretical saturation) who had participated in mega infrastructure projects [45], and their views on project O&M management were solicited from the interviewees as a supplement to the initial indicators, taking into account the actual situation of specific megaprojects. The interviewed experts had a range of 5–30 years of experience in project management of megaprojects, including middle and senior management of the three owners, two constructors, and three project management consultants (two of whom are also part-time staff at research institutions) (the background information of the interviewed experts is in Appendix B). (2) Interviews with academic experts. The compiled initial indicators were discussed with academic experts to solicit their opinions on the reasonableness and scientific validity of the setting and presentation of the indicators in the initial scale and to determine the initial indicators for the evaluation of the O&M management of mega infrastructure projects. All three academic experts were senior professors and doctoral supervisors in the field of engineering management with rich experience in both research and practice.

3.2. Step 2. Indicator Identification

The indicator system needed to include systematicality, validity, independence, scientificity, and operability. In order to ensure the applicability of the indicators and the comprehensive evaluation of mega infrastructure project O&M management, we used the grounded theory to analyze the elements of the indicator system. Grounded theory is a qualitative analysis method that leads the heory by combing the development of a phenomenon on the basis of collecting a large amount of data [55]. Based on the literature and the situation of this study, the main steps were theoretical sampling, data coding (including open coding, principal axis coding, and selective coding), theoretical saturation testing, and results [48]. The theoretical sampling was based on the literature analysis, policy texts, and the records of eight expert interviews. The data coding was based on the results of the literature and texts. The theoretical saturation test was based on the expert interview records (Figure 2). So as to establish relationships between conceptual classes to construct an initial indicator system for the O&M management of mega infrastructure projects, laying the groundwork for the final indicator system was the next stage.

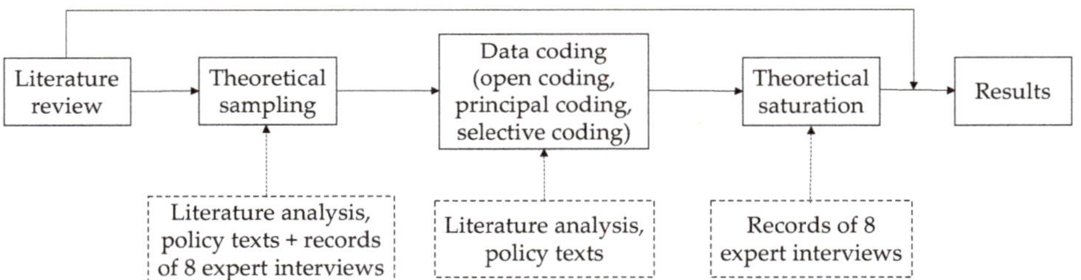

Figure 2. Process of Grounded theory.

3.3. Step 3. Establish Indicator System

Based on the results of first two steps, the questionnaire was administered under the theme of "evaluation of the O&M management of mega infrastructure projects." In order to achieve the research objectives, the questionnaire was divided into two parts: The first part was the basic information survey, including the respondents' background information (i.e., personnel statistics) and organization information (basic information of the project). The second part focused on the evaluation of the O&M management of mega infrastructure projects. The questionnaire was designed with the Likert scale, and assigned 1–5 points according to the degrees of "totally disagree," "disagree," "neutral," "agree," and "totally agree," respectively [45]. In order to ensure the reliability of the questionnaire, on the one hand it was necessary to carefully consider the way to express the questions in the questionnaire, which needed to be clear, objective, and easy to understand. On the other hand, the questionnaire did not clearly reflect the content and logic of the study, so as to prevent the interviewees from getting the possible implication of a causal relationship and affecting the questionnaire responses. The questionnaire was sent to engineers and managers with mega infrastructure project experience, including contractors, supervisors, and operators (Appendix A for details of the questionnaire). They were asked to rate each indicator in relation to their own work experience. In order to improve the authenticity and validity of the questionnaire, we collected the questionnaires through both online and face-to-face distribution. Finally, a total of 197 questionnaires were collected and 184 were valid, with an effective rate of 93.4%. T recovery rate exceeded 80%, which was statistically significant. [45].

Considering that there could be strong correlation between some indicators of the O&M management of mega infrastructure projects, the 23 initial indicators identified above were classified by the factor analysis method. The categories of indicators that have a significant impact on the O&M management of mega infrastructure projects were summarized and the indicators were classified according to the metric weight so as to establish the indicator system and provides a basis for the evaluation of the O&M management of mega infrastructure projects, which follows up on research on the O&M management of mega infrastructure projects.

4. Results and Discussion

4.1. Collection of Evaluation Indicators

4.1.1. Indicator Collection Based on Literature Analysis

In order to develop a suitable and comprehensive indicator system for evaluating the O&M management of mega infrastructure projects, the first stage was to identify the aspects. A review of the literature revealed that there were few empirical studies on the O&M of projects, and even fewer from the perspective of project management, specifically focusing on the O&M management of mega infrastructure projects. Therefore, the O&M indicator system studied in this paper is essentially a complete system set up to achieve the ultimate goal of the operating entity of mega infrastructure projects—that is, the level of O&M is identified by judging the extent to which mega infrastructure projects achieve their operational goals. Studies have shown that the O&M of mega infrastructure projects can have an obvious and significant impact on the economy, society, and environment. The O&M of mega infrastructure projects can be judged from both internal and external aspects, and the indicators were extracted and summarized to obtain the indicator system shown in Table 1.

Table 1. Indicators from studies.

Aspect	Indicator	Related Literature
Economy	Resource allocation	Ahola et al. (2014) [50]
	Resource exploitation and utilization	Li et al. (2016) [51]
	Economic exchange	Tauterat (2015) [38]
	Industrial distribution	Pollack et al. (2017) [15]
	Industrial structure	Mok et al. (2017) [17]
	Consumption structure	Li et al. (2018) [52]
	Labor structure	Qi et al. (2016) [32]
	Incomes	Chen et al. (2014) [53]
	Government revenue	Bernardo (2014) [49]
	Impact on GDP	Biesenthal et al. (2014) [54]
Social/Political	Policy guarantee	Li (2010) [55]
	Legal protection	Ma et al. (2006) [16]
	Rules and regulations	Xue et al. (2015) [27]
	Maintaining social stability	Matten (2008) [56]
	Meet social needs	McWilliams (2001) [57]
	Social benefits	Peloza (2009) [58]
	Social resource	Zeng et al. (2015) [24]
Technology	Quality performance	Ogata (2015) [37]
	Technology innovation	Lin et al. (2014) [29]
	Technology maturity	Tidd (2017) [41]
	Technical difficulties	Davies (2009) [59]
	Technical requirement	Bstieler (2015) [60]

Table 1. *Cont.*

Aspect	Indicator	Related Literature
Environment	Ecological protection level	Levitt (2007) [14]
	Ecological impact	Lin et al. (2015) [31]
	Resource assurance level	Du (2015) [39]
	Resource utilization	Peng (2007) [40]
Participant	Payback period	Mok et al. (2017) [13]
	Investment yield period	Moodley (2008) [18]
	Happy operation	Daily et al. (2003) [61]
	Reasonable profit margin	Zhai (2009) [62]
	Meeting the scheduled tax	Aguinis (2012) [63]
	Meet the special needs	Aguinis (2012) [63]
	Satisfaction with the project	Li et al. (2016) [64]
	Acceptance of the project	Li et al. (2015) [65]

4.1.2. Indicator Collection Based on Policy Texts

Most of the mega infrastructure projects are invested and constructed by the government, which plays a leading role in the LM of the projects. This ensures the successful realization of their functional, economic, social, and environmental benefits [66]. Therefore, the construction and O&M management of mega infrastructure projects are also the reflection of government policies and systems. Based on this, it is necessary to evaluate the O&M management of mega infrastructure projects from the political dimension to reflect the uniqueness. Since there are few studies that deal with the political aspects and it is impossible to obtain the influencing factors objectively through literature analysis, this study researched policy texts by collecting publicly available textual materials of mega infrastructure projects, such as from the State Council's policy document library, "China South-to-North Water Diversion Network," the official website of Gezhouba Company, the main operator of the Three Gorges Project, the official website of the Hong Kong–Zhuhai–Macau Bridge Authority, the news feature "The Full Completion of the Qinghai–Tibet Railway," and in news features, newspapers, and other channels. By sorting out news reports, policy documents, case studies, and relevant reviews on the O&M management of mega infrastructure projects, nearly 50,000 words of textual materials were compiled. Through keyword induction and extraction, the political indicators of the O&M management of mega infrastructure projects were finally obtained. The detailed results are shown in Table 2.

Table 2. Indicators from policy texts.

Aspect	Indicator
Politics	Consistency with policy, legal, and strategic approaches
	Meeting the requirements of national defense construction
	Strong sustainability
	Adjustment of policy differences related to the project
	Coordination of social contradictions among regions related to the project
	Narrowing of regional economic differentials in relation to the project

4.1.3. Indicator Collection Based on Expert Interviews

This study used a combination of interviews with practicing experts and academic experts to explore the factors influencing the O&M management of mega infrastructure projects [57]. Most of the expert interviews were conducted with no more than 20 people [46], so eight practice experts who participated in mega infrastructure projects were interviewed and asked for their opinions on the factors as a supplement to the initial indicators as follows: (1) economic aspects. Experts agreed that the impact of mega infrastructure projects on the economy should be analyzed from the perspective of changing the industrial structure of the economy and improving the income level of residents, and experts from government departments also mentioned that the O&M management of mega infrastructure projects contributed to an increase in government financial revenue. The significance of economic structure and economic function was unclear in the evaluation, and was suggested to be deleted. The economic indicators changed significantly, eliminating the secondary indicator and adding three indicators: changing the economic and industrial structure, improving residents' income level, and increasing the government's fiscal revenue. (2) Social aspects. Experts believed that the O&M management of mega infrastructure projects in the context of China could be evaluated separately from political and social aspects, and political factors should be mainly analyzed by policy texts, so the original indicator system was split. (3) Technology aspects. Experts believed that the technical level and technical risk were influential factors that enhanced or constrained the O&M management of projects, rather than evaluation indicators used to judge the O&M of mega infrastructure projects, so it was suggested that the whole system be deleted. The technical aspect was therefore deleted. (4) Participant aspects. Experts believed that the O&M management of projects was not only evaluated from the overall benefits to the external environment, but was also an important indicator to evaluate the O&M management of mega infrastructure projects themselves. The satisfaction degree of investors and the public overlapped with the indicators of social and economic aspects, which was suggested to be deleted.

The second round of expert interviews: The indicator system constructed in the first two stages was submitted to three experts in academia for their opinion in order to test the reasonableness of the indicator system and the accuracy of the question formulation on the basic scale. Experts discussed and revised the structural design and terminology of the indicators, and finally reached a consensus as shown in Table 3.

Table 3. Indicators from expert interviews.

Dimension	Aspect	Indicator
Operational Benefits	Politics	Consistency with policy, legal, and strategic approaches Meeting the requirements of national defense construction Strong sustainability Adjustment of policy differences related to the project Coordination of social contradictions among regions related to the project Narrowing of regional economic differentials in relation to the project
	Economy	Changing the economic industrial structure Improving residents' income level Government revenue increasing
	Social	Maintaining social stability Significant social benefits Making full use of social resources Meeting the needs of social development
	Environment	Meeting the needs of environmental protection Protecting the regional resources Making full use of local resources Having a positive impact on the local ecological balance
Operational Capability	Operation Team	Education level of staff Experiences of staff Number of practitioners participating in projects

4.2. Initial Indicator System

With the assistance of the mentor and team members, the abovementioned channels (3.1) were used to collect relevant information on mega infrastructure projects from June 2020 to August 2020, and 70,000 words of literature were compiled. Through the extensive reading of the literature, a deeper understanding of the concept, scope, and performance of O&M management was obtained, which formed the theoretical knowledge of O&M management as a criterion for text search and target selection. The coding analysis of the materials using NVivo yielded the following results: (1) open coding. After comparing the literature and policy texts, the indicators with the same or similar meanings were combined. A total of 52 tags and 163 initial sentences were sorted out from the documents. The evaluation indicators were extracted from the literature and policy texts, and the existing indicators in the documents were deleted. After repeated reorganization, integration, and refinement, 23 subcategories were obtained. (2) Spindle coding. The 52 labels and 23 subcategories obtained from the open code were carefully compared, and these labels and categories were placed in the context of mega infrastructure projects. Combined with the actual situation in the process of project O&M management, five main categories of evaluation indicators were obtained. (3) Selective coding. This study selected "O&M management of mega infrastructure projects" as the core coding, and adopted the idea of a storyline to guide coding [48]. First of all, combined with the comparative analysis of raw materials and codes at all levels, the story context was defined as the O&M management of mega infrastructure projects evaluated by political, economic, social, environmental, and operational teams. Then, these indicator aspects were taken as the concrete embodiment of the O&M management of mega infrastructure projects by connecting and comparing the main category and the subcategory. Finally, it combed the relationship between the core category and other category levels. After systematic analysis, the core category, the main category, and the subcategory constituted a whole.

In order to improve the reliability and validity of the findings, a theoretical saturation test was conducted. In-depth interviews were conducted with eight practice experts on the consistency and completeness of the descriptions of the indicators selected in 4.1. Nearly 50,000 words of interview records were compiled, and 13 labels were obtained by coding and analyzing the interview transcripts. Further coding and analysis did not form a new core category and relationship, and no new theory was found in the main category. Therefore, the model integration process was reliable and theoretical saturation was realized. The mega infrastructure projects were constructed through grounded theory. The initial indicator system of O&M management evaluation is shown in Table 4.

4.3. Screen Key Indicators

4.3.1. Descriptive Statistics

Descriptive statistical analysis is a statistical method that uses mathematical language to describe sample characteristics or explain the relationship between variables [43]. A total of 184 valid samples were statistically analyzed, and the results are shown in Table 5. The number of male and female respondents in the survey was 148 and 36, respectively, with a ratio close to 4:1, which is consistent with the industry characteristics in that there are more men than women in the projects. From the perspective of an education background, half of the respondents had bachelor's degrees, and the proportion of master's degrees and doctor's degrees accounted for 26.09% and 19.57%, respectively, which indicates that the educational background of those working in the construction industry is constantly increasing. A total of 50% of the respondents had been involved in the O&M management of mega infrastructure projects for 6–10 years. Only 6.52% of the respondents had more than 10 years of work experience, indicating that the O&M management of mega infrastructure projects is still in its infancy. A total of 100 of the respondents had been involved in the O&M management of highways, 36 in the O&M of bridges, 24 in the O&M of railways, 16 had participated in the O&M of water conservancy projects, and only 8 respondents had been involved in the O&M of power grid projects, which reflects the ratio of different types of projects in mega infrastructure to some extent. Due to the particularity of the project, the respondents were located in research institutes, government agencies, and operation management departments. The units issued by the questionnaire included most of the participants in O&M management, which was a good way to achieve multi-party participation and ensure the quality of the questionnaire. The fact that almost all of these parties belonged to the state indicates that the operation process of mega infrastructure projects is under the supervision of the government. The distribution of these data basically reflects the actual proportion of researchers in mega infrastructure projects in China. These proportions basically reflect the actual distribution of O&M management in China. The data also suggest that the respondents were experienced and knowledgeable about the issues under this study, which increased the confidence in the data quality.

4.3.2. Factor Analysis

Factor analysis is a process of identifying relatively few factor groups, which is used to represent the relationship between multiple groups of interrelated variables [67]. In order to confirm the applicability of the data collected from the questionnaire survey, two tests were needed: (1) the sample size and the strength of the relationship between the indicators [68], and (2) the number of samples collected being at least five times the number of indicator variables [69]. A total of 184 valid questionnaires were used to rate 23 indicators. The sample size was eight times the indicator variable to meet the demand. The valid questionnaire results were imported into SPSS 23.0 (IBM corporation, Stanford, CA, USA), and the data were standardized. The KMO (Kaiser-Meyer-Olkin) value was 0.816 (>0.7) [69], and the significance level of the Bartlett spherical test was <0.0001 [70], as shown in Table 6, which shows that the sample size was sufficient and the correlation coefficient matrix indicates strong correlation between the indicators, which could be used for factor analysis. The common degree of indicators was close to or greater than 0.8, indicating that the common factors had a strong explanatory power for variables and were suitable for factor analysis. The results of the KMO and Bartlett spherical test indicate that the initial indicators for evaluating the O&M management of mega infrastructure projects have good construction validity.

Table 4. Initial indicator system.

Dimension	Aspect	Indicator
Operational Benefits	Politics (A)	Consistency with policy, legal, and strategic approaches (A_1) Meeting the requirements of national defense construction (A_2) Strong sustainability (A_3) Adjustment of policy differences related to the project (A_4) Coordination of social contradictions among regions related to the project (A_5) Narrowing of regional economic differentials in relation to the project (A_6)
	Economy (B)	Changing the economic industrial structure (B_1) Improving residents' income level (B_2) Government revenue increasing (B_3) Promoting regional GDP growth (B_4) Meeting the tax target of the project (B_5)
	Social (C)	Maintaining social stability (C_1) Meeting the needs of social development (C_2) Significant social benefits (C_3) Improving employment (C_4) Making full use of social resources (C_5)
	Environment (D)	Meeting the needs of environmental protection (D_1) Protecting the regional resources (D_2) Making full use of local resources (D_3) Having a positive impact on the local ecological balance (D_4)
Operational Capability	Operation Team (E)	Education level of staff (E_1) Experiences of staff (E_2) Number of practitioners participating in projects (E_3)

Table 5. Summary of the statistics of the interviewees.

Indicators	Number	Proportion
Male	148	80.43%
Female	36	19.57%
Average age	35.31	
Education		
PhD	36	19.57%
Master's	48	26.09%
Bachelor's	92	50%
Other	8	4.35%
Work experience		
≤5 years	80	43.48%
6–10 years	92	50%
>10 years	12	6.52%
Job position		
Top manager	26	14.13%
Middle manager	48	26.09%
Other	110	59.78%
Type		
Highway	100	54.35%
Railway	24	13.04%
Water conservancy	16	8.7%
Power grid	8	4.35%
Bridge	36	19.57%
Institute		
Government	48	26.09%
Research institutes	64	34.8%
Operating unit	72	39.11%

Table 6. KMO and Bartlett spherical test.

KMO Sampling Appropriateness		**0.816**
Bartlett Spherical Test	The approximate chi-square	1080.214
	Degree of freedom	253
	Significant	0.000

According to the standard extraction factor, eigenvalues reater than 1 when the characteristic value is less than 1 [71], and the total variance interpretation is shown in Table 7. It can be seen that a total of four common factors were extracted from the 23 evaluation indicators, and the cumulative total variance explained by the eigenroots of the factor was 74.439%, which is greater than 60% [72], indicating that the extracted common factors can effectively reflect the O&M management of mega infrastructure projects.

Table 7. Explanation of total variance.

Component	Initial Eigenvalue			Sum of Squares of Rotating Loads		
	Characteristic Value	Variance%	Cumulative%	Characteristic Value	Variance%	Cumulative%
1	12.236	53.200	53.200	12.236	53.200	12.236
2	2.472	10.750	63.950	2.472	10.750	2.472
3	1.365	5.936	69.886	1.365	5.936	1.365
4	1.047	4.553	74.439	1.047	4.553	1.047
5	0.965	4.196	78.439			
6	0.730	3.175	81.810			

Since the typical representative variables of each common factor in the initial factor solution were not prominent and not convenient for the analysis of the actual problems, for this consideration, the rotation was carried out using the maximum variance method to obtain the well-defined common factors. After four iterative convergence cycles, the rotated factor load matrix was obtained, and the 23 indicators were reduced to four common factors, as shown in Tables 8 and 9.

Table 8. Rotating posterior factor load matrix.

Indicator	Common Factor 1	Common Factor 2	Common Factor 3	Common Factor 4
A_1	0.589	0.010	0.179	0.158
A_2	0.586	0.270	0.391	0.070
A_3	0.434	0.341	0.542	0.324
A_4	0.788	0.161	0.256	0.185
A_5	0.810	0.109	0.144	0.350
A_6	0.697	0.018	0.362	0.259
B_1	0.820	0.214	0.120	0.228
B_2	0.582	0.158	0.588	0.212
B_3	0.411	0.149	0.821	0.137
B_4	0.576	0.124	0.671	−0.002
B_5	0.335	0.252	0.821	0.139
C_1	0.662	0.554	0.130	0.043
C_2	0.734	0.324	0.188	−0.076
C_3	0.502	0.529	0.318	−0.017
C_4	0.726	0.322	0.249	−0.084
C_5	0.716	0.497	0.094	0.092
D_1	0.262	0.843	0.187	0.141
D_2	0.171	0.833	0.310	0.276
D_3	0.139	0.783	-0.026	0.287
D_4	0.210	0.816	0.228	0.290
E_1	0.015	0.424	0.501	0.514
E_2	0.127	0.321	0.243	0.728
E_3	0.324	0.370	0.013	0.784

Table 9. Public factor extraction after dimensionality reduction.

Common Factors	1	2	3	4
Indicator	A_1, A_2, A_4, A_5, A_6, B_1, C_1, C_2, C_4, C_5	C_3, D_1, D_2, D_3, D_4	A_3, B_2, B_3, B_4, B_5	E_1, E_2, E_3
Naming	Social relations	Environmental benefit	Macro policy	Operational capability

Common factor 1 included indicators A_1, A_2, A_4, A_5, A_6, B_1, C_1, C_2, C_4, and C_5, which were five political indicators, four indicators for social aspects, and one indicator for economic aspects. These indicators all reflected the impact of mega infrastructure projects on social relations, so they were named "social relations." Common factor 2 included indicators C_3, D_1, D_2, D_3, and D_4, namely, four environmental indicators and one social indicator. These indicators reflected environmental benefits, so they were named "environmental benefits." Common factor 3 included A_3, B_2, B_3, B_4, and B_5, namely, four economic indicators and one political indicator, which reflected the impact of project O&M management on macro policy, so they were called "macro policy." Common factor 4 included E_1, E_2, and E_3, which was completely consistent with the indicators of operation capacity, so the name "operation capacity" was retained.

4.3.3. Results Discussion

Through factor analysis, four factors of O&M management of mega infrastructure projects were concluded, which were social relationship, environmental benefit, macro policy, and operation capability. According to the indicator contribution rate ranking, it can be seen that social relations play a crucial role in the O&M management of mega infrastructure projects. The indicators of each factor are described in detail below.

Factor 1: Social Relations

The social relations factor included 10 indicators: consistent with policies, legal and strategic approaches, meeting the requirements of national defense construction, adjustment of policy differences related to the project, coordinating of social contradictions among regions related to the project, narrowing of regional economic differentials in relation to the project, changing the economic and industrial structure, maintaining social stability, meeting the needs of social development, improving employment, and making full use of social resources. The essence of any infrastructure project is to serve society [73], and China is no exception. In social relations, respondents generally believed that the O&M management of mega infrastructure projects had a high score on maintaining community relations, meeting the needs of social development, and improving regional employment, with an average score of more than 4, indicating that the O&M management of mega infrastructure projects played an important role in promoting these three aspects. The ratings were lower for coordinating policy differences and reducing economic disparities between regions related to the projects. It shows that the role of mega infrastructure projects in reducing regional political and economic disparities has not been fully exploited. Mega infrastructure projects generally span two or more provincial administrative regions [74]. There are also differences in economic level and guidelines in the policies for project O&M management between different regions. The O&M of mega infrastructure projects requires unified standards and collaborative governance across different provinces to maximize social benefits and better serve society. During the interview, a policy advisor from the Transport Bureau said that the core purpose of the construction, delivery, and operation of mega infrastructure projects was to ensure social stability. The central government and local governments at all levels are committed to increasing the proportion of government spending on infrastructure projects so as to ensure the quality and avoid the emergence of "jerry-built" projects, which have a significant negative impact on society. Academic experts added that the O&M management of mega infrastructure projects plays a crucial role in improving the resilience of society as a whole.

Factor 2: Environmental Benefits

The second factor environmental benefits consisted of five indicators: significant social benefits, meeting the needs of environmental protection, protecting the regional resources, making full use of local resources, and having a positive impact on the local ecological balance. Respondents generally gave low scores to environmental benefits, with the lowest score among the 23 indicators being for satisfying environmental needs, conserving local resources, and adequate and rational use of local resources. Respondents generally believed that the O&M management of mega infrastructure projects has little effect on the protection of resources and environment. More than half of the suggestions in the self-selected section of the questionnaire indicated that at this stage, all stakeholders are trying their best to ensure that the completion and the O&M management of the project does not damage the surrounding natural environment and ecological balance, but the positive impact on the surrounding environment is almost non-existent. Respondents also indicated that the environmental protection problems in the O&M management of mega infrastructure projects have not been given sufficient attention, and the green operation concept has not been well implemented. Interviewees from NGOs (non-governmental organizations) said that throughout the LM of mega infrastructure projects, there is a certain degree of impact on the ecological environment, and the vast majority of the impact is negative. University experts said that infrastructure should adhere to the concept of green and low-carbon from the design, construction, operation, and final demolition in the process of project delivery, energy conservation, water conservation, rational use of resources, and other contents. Government representatives said that in recent years, the concept of green, low-carbon, and sustainable development has indeed been paid attention to in the construction industry, and has also been applied throughout the whole lifecycle management of megaprojects. The negative impact of projects on the surrounding environment has gradually decreased. Moreover, governments at all levels have special funds to support the introduction of advanced green technologies from abroad.

Factor 3: Macro policy

The third type of factor, macro policy, included five indicators: strong sustainability, improving residents' income level, government revenue increasing, promoting regional GDP growth, and meeting the tax target of the project. In terms of macro policy, the respondents generally believed that the O&M management of mega infrastructure projects has a positive effect on the sustainable development of society and the improvement of economic level. Among the indicators, the highest score of strong sustainability was 4.02, which indicates that the sustainability of mega infrastructure project operation receives recognition, and almost all respondents believed that the O&M management of mega infrastructure project contributes to regional development, and vice versa. The scores of each indicator show that the economic benefits of the project are considerable and have been recognized by the public. As the respondents from the government said, local governments generally hold a positive attitude towards the construction of mega infrastructure projects. In addition to being in line with the national macro policy, local governments can also accelerate the development of regional industries such as tourism, increase employment opportunities, and enhance the regional economy as a whole. A university professor said that the construction of mega infrastructure project requires the local government to take a stand and weigh the realities of the region and deliver mega infrastructure projects without considering the actual needs, which may bring heavy economic burden to the local government and people.

Factor 4: Operation Capability

The fourth factor, operational capacity, consisted of three indicators: education level of staff, experiences of staff, and the number of practitioners participating in projects. It was not difficult to find that the average age of mega infrastructure project O&M management personnel is 35.31 years old, and the project operation time is generally between 1–5 years. The frontline and grassroots operation staff are mainly young employees. Although the entire project lifecycle is planned and designed as early as the planning period, the professional quality of the operation team is also needed to deal with the unexpected situations that arise during the actual operation. Mega infrastructure projects, especially transportation infrastructure, cause huge losses and social influence when accidents occur. Data shows that the role of staff in the O&M of transport infrastructure is much greater than 26% [75], so improving the skills of staff in monitoring and handling malfunctions is an effective way to ensure the safe and smooth operation of infrastructure. According to the research of safety expert Heinrich, there is an "88:10:2" rule in the operation safety of infrastructure [75]. That is, in 100 safety accidents, 88 are caused by human causes, 10 are caused by people and objects, and only 2 are caused by "natural disasters" that are difficult to prevent. Both human factors and equipment factors have to be resolved by the staff. Therefore, it is very important to continuously improve the ability of employees to perform their duties and the rational allocation of the skill structure of the staff team, and to improve the team's ability to deal with failures. An official from the transportation bureau said that they recruit employees openly every year and train them intensively on a regular basis. University representatives said that to improve the professional level of the operations team, not only should we consider the matching of people and positions, but also the rationalization of the team's professional skill structure. Optimized team structure and reasonable manpower allocation are conducive to improving the operational efficiency of infrastructure safety.

4.4. Construction of Indicator System

The results of the factor analysis show that the initial indicator system of the O&M management of the mega infrastructure projects has a reasonable category of indicators, and some of the categories can be further refined. The indicators were ranked according to the influence on O&M management, and the final evaluation indicator system for the O&M management of mega infrastructure projects is shown in Table 10.

Table 10. Final indicator system.

Dimension	Indicator
Social Relations	Consistency with policy, legal, and strategic approaches Meeting the requirements of national defense construction Adjustment of policy differences related to the project Coordination of social contradictions among regions related to the project Narrowing of regional economic differentials in relation to the project Changing the economic industrial structure Maintaining social stability Meeting the needs of social development Improving employment Making full use of social resources
Environmental Benefits	Significant social benefits Meeting the needs of environmental protection Protecting the regional resources Making full use of local resources Having a positive impact on the local ecological balance
Macro Policy	Strong sustainability Improving residents' income level Government revenue increasing Promoting regional GDP growth Meeting the tax target of the project
Operational Capability	Education level of staff Experiences of staff Number of practitioners participating in projects

5. Conclusions

5.1. Findings and Contribution

To fuel rapid economic growth, a number of impressive mega infrastructure projects have been built all over the world. These planned mega infrastructure projects are unprecedented, and they have exacted a great influence on economic construction, social development, and people's livelihood. However, it should also be noted that this construction takes a toll on local communities and the natural environment. The consequences have gone beyond the project itself and have evolved into a series of social problems because of lack of O&M management. In this context, it is important to evaluate the O&M management of mega infrastructure projects and identify the problems in the operation process. Based on the macro perspective, this study determined 23 initial indicators of the O&M management of mega infrastructure project from two dimensions and five aspects by using literature analysis, policy texts, expert interviews, and grounded theory, and evaluated the relative importance of indicators from the participants' perspective through questionnaire survey. Finally, 23 indicators were grouped into four factors by factor analysis, namely, social relations, environmental benefits, macro policy, and operation capacity, which resulted in an indicator system for evaluating the O&M management of mega infrastructure projects.

This study provides contributions in three ways. First, the indicator system of O&M management of mega infrastructure projects aims to provide a basis for the evaluation and decision-making of the O&M management of megaprojects. For these projects, the operation directly determines the performance of the project and the impact on society and the economy. Paying attention to the O&M management of megaprojects helps to improve the operation performance. Second, the indicator system not only includes the social, economic, and environmental aspects of the previous research, but also innovatively includes the political aspects, which is different from the general project operated for economic benefit. The O&M management of mega infrastructure projects is aimed at serving society [76]. By evaluating the role of project operation in adjusting the disparity of political, economic, and social development in different regions to judge the role of the project in national macro control, not only can we clearly judge the operation performance of the project, but also promote the sustainable development of the project. Finally, this study adopts a quantitative approach to measure the impact of project O&M management on different indicators by calculating the mean score of respondents for different indicators. This method can visually reflect the intuitive feedback of the respondents and overcome the obstacles of considering different levels of issues, which helps to objectively reflect the real sense of the respondents.

5.2. Limitations and Future Directions

This study has several limitations, but also provides several promising areas for future research. First of all, our indicator system was established based on the context of China. The O&M management models adopted by different countries and regions are significantly different due to differences in regional economic development and cultural backgrounds. Generalization must be validated in the future to find out whether our indicator system can be appropriate for general engineering projects and for other countries around the world. Second, it is important to acknowledge that there are contextual factors that may affect the presence and scoring of these O&M management indicators. For example, national circumstances, cultural characteristics, and different types of projects could greatly influence O&M management (highway, railway, water conservancy, power grid, and bridge). An in-depth discussion of these problems is beyond the scope of this study. Furthermore, because the context is constantly changing, the indicator system should be dynamically adjusted according to the time and the national environment in which it is applied.

This study provides new ideas for the field of project O&M management and lays a theoretical foundation for the quantitative analysis of the O&M management of mega infrastructure projects. In the next stage, we will focus on the evaluation and improvement of the O&M management of mega infrastructure projects and carry out empirical studies for one or some types of mega infrastructure projects.

Author Contributions: Conceptualization, D.C. and P.X.; methodology, D.C.; software, D.C.; validation, D.C., F.J., and J.Z.; formal analysis, P.X.; investigation, Z.L.; resources, Z.L.; data curation, D.C.; writing—original draft preparation, D.C.; writing—review and editing, D.C., P.X., F.J., and J.Z. All authors have read and agreed to the published version of the manuscript.

Funding: This research was funded by the General Project of Chongqing Natural Science Foundation, grant number cstc2020jcyj-msxmX0107 and the Fundamental Research Funds for the Central Universities, grant numbers 2020CDJSK03YJ06 and 2020CDJ-LHSS-007.

Conflicts of Interest: The authors declare no conflict of interest.

Appendix A. The Questionnaire on the Operation and Maintenance of Mega Infrastructure Projects

Dear Sir/Madam,

Thank you very much for taking the time to fill in our questionnaire. We are a research team from the School of Management Science and Real Estate, Chongqing University. This study is an important part of the "Operation and Maintenance Management of Mega Infrastructure Projects," which aims to construct an indicator system for the operation and maintenance level of mega infrastructure projects in China, and further explore the mechanism of their influence on the current situation of operation and maintenance.

Mega infrastructure projects are a kind of national strategic project with far-reaching impact on the development of political, economic, social, scientific and technological, environmental, and other fields, and are the indispensable material basis for national prosperity, as well as the basic platform for the high-speed development of modern society. In general, mega infrastructure projects are characterized by large investment scale, long construction period, and many participants, etc., which make project operation and maintenance management more difficult and are of high concern for the government and society. The mega infrastructure projects covered in this study include the Three Gorges Project, South-to-North Water Diversion Project, West-to-East Gas Transmission Project, and other infrastructure of energy and water conservancy, as well as transportation infrastructure projects such as the Qinghai–Tibet Railway, Zhengwan High Speed Rail, and Hong Kong–Zhuhai–Macau Bridge. There is no right or wrong answer to the questions in this research questionnaire. Please select a mega infrastructure project you have recently participated in or are participating in as a reference and elaborate on the actual situation of project operation and maintenance. All the information you provide in this questionnaire is only for the use of academic research, and the whole analysis process is confidential, please rest assured.

Contact:

E-mail:

Part 1: Basic Information and Project Characteristics

Conditions	Options
1. Gender	□Male □Female
2. Age	□≤30 □31-40 □41-50 □>50
3. Education	□Others □Bachelor's □Master's □ PhD
4. Work experience	□<3 □3–5 □6–10 □>10
5. Positions	□Top manager □Middle manager □Others
6. Types of projects	□Highway □Railway □Water conservancy □Power grid □Bridge □Others:_______
7. Institute	□Government □Research institutes □Operating unit □Others:_______

Note: Depending on your situation, check √ in the options you think are appropriate.

Part 2: Survey on Project Operations and Maintenance

Dimension	Indicator	Totally Disagree	Disagree	Neutral	Agree	Totally Agree
Politics	Consistency with policy, legal, and strategic approaches	1	2	3	4	5
	Meeting the requirements of national defense construction	1	2	3	4	5
	Strong sustainability	1	2	3	4	5
	Adjustment of policy differences related to the project	1	2	3	4	5
	Coordination of social contradictions among regions related to the project	1	2	3	4	5

Dimension	Indicator	Totally Disagree	Disagree	Neutral	Agree	Totally Agree
	Narrowing of regional economic differentials in relation to the project	1	2	3	4	5
Economy	Changing the economic industrial structure	1	2	3	4	5
	Improving residents' income level	1	2	3	4	5
	Government revenue increasing	1	2	3	4	5
	Promoting regional GDP growth	1	2	3	4	5
	Meeting the tax target of the project	1	2	3	4	5
Social	Maintaining social stability	1	2	3	4	5
	Meeting the needs of social development	1	2	3	4	5
	Significant social benefits	1	2	3	4	5
	Improving employment	1	2	3	4	5
	Making full use of social resources	1	2	3	4	5
Environment	Meeting the needs of environmental protection	1	2	3	4	5
	Protecting the regional resources	1	2	3	4	5
	Making full use of local resources	1	2	3	4	5
	Having a positive impact on the local ecological balance	1	2	3	4	5
Operational Capability	Education level of staff	1	2	3	4	5
	Experiences of staff	1	2	3	4	5
	Number of practitioners participating in projects	1	2	3	4	5

Note: Please answer the question about the operation and maintenance management of mega infrastructure projects according to your actual experience and personal feelings, and check √ in the options you think are appropriate.

Finally, we welcome your comments and suggestions on this survey, as well as your ideas and suggestions on the operation and maintenance of mega infrastructure projects in China.

> Your suggestions for this survey:

The end of the questionnaire, thank you for participating!

Appendix B. The Background Information of the Interviewed Experts

Numbers	Job Position	Age	Education	Field	Engaged in Engineering Construction Industry	Engaged in Mega Infrastructure Project
1	Project-level leadership	43	Bachelor's	Project supervision and consultation	11–20 years	6–10 years
2	Middle manager	45	Bachelor's	Party A's management	11–20 years	1–5 years
3	Middle manager	38	Bachelor's	Project supervision and consultation	11–20 years	11–20 years
4	Top manager	35	Master's	Party A's management	6–10 years	1–5 years
5	Top manager	48	Master's	Party A's management	21–30 years	11–20 years
6	Top manager	52	Bachelor's	Project supervision and consultation	21–30 years	11–20 years
7	Project-level leadership	43	Master's	Project supervision and consultation	6–10 years	6–10 years
8	Middle manager	47	Master's	Construction management	21–30 years	11–20 years

References

1. Flyvbjerg, B. What you Should Know about Megaprojects and Why: An Overview. *Proj. Manag. J.* **2014**, *45*, 6–19. [CrossRef]
2. Xiang, P.C.; Liu, Y.L. Research on the risk coupling mechanism of major engineering projects across regions. *Constr. Econ.* **2018**, *39*, 97–101.
3. Statistical Bulletin. Available online: http://www.stats.gov.cn/tjsj/sjjd/202002/t20200228. (accessed on 28 February 2020).
4. Flyvbjerg, B.; Turner, J.R. Do classics exist in megaproject management? *Int. J. Proj. Manag.* **2018**, *36*, 334–341. [CrossRef]
5. Zeng, S.X. Management of megaproject. *Sci. Obser.* **2018**, *13*, 45–47.
6. Eastman, C.; Teicholz, P.; Sacks, R. *BIM Handbook: A Guide to Building Information Modeling for Owners, Managers, Designers, Engineers and Contractors*; John Wiley & Sons, Inc.: Hoboken, NJ, USA, 2011; pp. 1–30.
7. Stone, R. Three Gorges Dam: Into the Unknown. *Science* **2008**, *321*, 628–632. [CrossRef] [PubMed]
8. Wu, J.; Huang, J.; Han, X.; Xie, Z.Q.; Gao, X.M. Three-Gorges Dam: Experiment in habitat fragmentation? *Science* **2003**, *300*, 1239–1240. [CrossRef] [PubMed]
9. Xie, P.; Wu, J.; Huang, J.; Han, X. Three-Gorges Dam: Risk to Ancient Fish. *Science* **2003**, *302*, 1149–1151. [CrossRef]
10. Dobbs, R.; Pohl, H.; Lin, D. *Infrastructure Productivity: How to Save $1 Trillion a Year*; McKinsey Global Institute: Chicago, IL, USA, 2013.
11. Peloza, J. The Challenge of Measuring Financial Impacts From Investments in Corporate Social Performance. *J. Manag.* **2009**, *35*, 1518–1541. [CrossRef]

12. Tam, C.M.; Tam, V.W.; Tsui, W.S. Green construction assessment for environmental management in the construction industry of Hong Kong. *Int. J. Proj. Manag.* **2004**, *22*, 563–571. [CrossRef]

13. Mok, K.Y.; Shen, G.Q.; Yang, J. Stakeholder management studies in mega construction projects: A review and future directions. *Int. J. Proj. Manag.* **2015**, *33*, 446–457. [CrossRef]

14. Levitt, R.E. CEM Research for the Next 50 Years: Maximizing Economic, Environmental, and Societal Value of the Built Environment. *J. Constr. Eng. Manag.* **2007**, *133*, 619–628. [CrossRef]

15. Pollack, J.; Biesenthal, C.; Sankaran, S.; Clegg, S. Classics in megaproject management: A structured analysis of three major works. *Int. J. Proj. Manag.* **2017**, *36*, 372–384. [CrossRef]

16. Ma, H.; Zeng, S.; Lin, H.; Chen, H.; Shi, J.J. The societal governance of megaproject social responsibility. *Int. J. Proj. Manag.* **2017**, *35*, 1365–1377. [CrossRef]

17. Mok, K.Y.; Shen, G.Q.; Yang, R.J.; Li, C.Z. Investigating key challenges in major public engineering projects by a network-theory based analysis of stakeholder concerns: A case study. *Int. J. Proj. Manag.* **2017**, *35*, 78–94. [CrossRef]

18. Moodley, K.; Smith, N.; Preece, C.N. Stakeholder matrix for ethical relationships in the construction industry. *Constr. Manag. Econ.* **2008**, *26*, 625–632. [CrossRef]

19. Ghani, E.; Goswami, A.G.; Kerr, W.R. Highway to success: The impact of the golden quadrilateral project for the location and performance of Indian manufacturing. *Econ. J.* **2015**, *126*, 317–357. [CrossRef]

20. Leigh, A.; Neill, C. Can national infrastructure spending reduce local unemployment? Evidence from an Australian roads program. *Econ. Lett.* **2011**, *113*, 150–153. [CrossRef]

21. Guikema, S.D. Infrastructure Design Issues in Disaster-Prone Regions. *Science* **2009**, *323*, 1302–1303. [CrossRef]

22. Fernald, J.G. Roads to Prosperity? Assessing the Link Between Public Capital and Productivity. *Am. Econ. Rev.* **1999**, *89*, 619–638. [CrossRef]

23. Faber, B. Trade integration, market size, and industrialization: Evidence from China's National Trunk Highway System. *Rev. Econ. Stud.* **2014**, *81*, 1046–1070. [CrossRef]

24. Zeng, S.; Ma, H.Y.; Lin, H.; Zeng, R.C.; Tam, V.W.Y. Social responsibility of major infrastructure projects in China. *Int. J. Proj. Manag.* **2015**, *33*, 537–548. [CrossRef]

25. Weiss, W.H. Management by Exception in Operations and Maintenance. *Chem. Eng.* **1977**, *84*, 151–166.

26. Wang, T.; Wang, S.; Zhang, L.; Huang, Z.; Li, Y. A major infrastructure risk-assessment framework: Application to a cross-sea route project in China. *Int. J. Proj. Manag.* **2016**, *34*, 1403–1415. [CrossRef]

27. Xue, X.; Zhang, R.; Zhang, X.; Yang, R.J.; Li, H. Environmental and social challenges for urban subway construction: An empirical study in China. *Int. J. Proj. Manag.* **2015**, *33*, 576–588. [CrossRef]

28. PCCIP. *Presidential Commission for Critical Infrastructure Protection*; President's Commission on Critical Infrastructure Protection: Washington, DC, USA, 1997.

29. Lin, H.; Zeng, S.X.; Ma, H.Y.; Qi, G.Y.; Tam, V.W.Y. Can political capital drive corporate green innovation? Lessons from China. *J. Clean. Prod.* **2014**, *64*, 63–72. [CrossRef]

30. Fang, D.P.; Huang, X.Y.; Hinze, J. Benchmarking Studies on Construction Safety Management in China. *J. Constr. Eng. Manag.* **2004**, *130*, 424–432. [CrossRef]

31. Lin, H.; Zeng, S.X.; Ma, H.Y.; Chen, H.Q. Does commitment to environmental self-regulation matter? An empirical examination from China. *Manag. Decis.* **2015**, *53*, 932–956. [CrossRef]

32. Qi, G.Y.; Jia, Y.; Zeng, S.; Shi, J.J. Public participation in China's project plans. *Science* **2016**, *352*, 1065. [CrossRef]

33. Aguinis, H. Organizational responsibility: Doing good and doing well. In *APA Handbook of Industrial and Organizational Psychology: Maintaining, Expanding, and Contracting the Organization*; Zedeck, S., Ed.; American Psychological Association: Washington, DC, USA, 2011; Volume 3, pp. 855–879. [CrossRef]

34. Abednego, M.P.; Ogunlana, S.O. Good project governance for proper risk allocation in public–private partnerships in Indonesia. *Int. J. Proj. Manag.* **2006**, *24*, 622–634. [CrossRef]

35. Badewi, A.; Shehab, E. The impact of organizational project benefits management governance on ERP project success: Neo-institutional theory perspective. *Int. J. Proj. Manag.* **2016**, *34*, 412–428. [CrossRef]

36. Goh, B.H.; Sun, Y. The development of life-cycle costing for buildings. *Build. Res. Inf.* **2015**, *44*, 319–333. [CrossRef]

37. Ogata, M. Management and Technology Innovation in Rail Industry as Social Infrastructure for Improved Quality of Life. *IEICE Trans. Commun.* **2015**, *99*, 778–785. [CrossRef]

38. Tauterat, T. Development of a Method for the Economic Evaluation of Predictive Maintenance. In *Lecture Notes in Business Information Processing*; Fernandes, J., Machado, R., Wnuk, K., Eds.; Springer: Cham, Switzerland, 2015; Volume 210, pp. 179–185. [CrossRef]

39. Du, X. Is Corporate Philanthropy Used as Environmental Misconduct Dressing? Evidence from Chinese Family-Owned Firms. *J. Bus. Ethics* **2015**, *129*, 341–361. [CrossRef]

40. Peng, C.; Ouyang, H.; Gao, Q.; Jiang, Y.; Zhang, F.; Li, J.; Yu, Q. Building a "Green" Railway in China. *Science* **2007**, *316*, 546–547. [CrossRef]

41. Tidd, J. Innovation management in context: Environment, organization and performance. *IEEE Eng. Manag. Rev.* **2017**, *45*, 43–55. [CrossRef]

42. Aarseth, W.; Ahola, T.; Aaltonen, K.; Økland, A.; Andersen, B. Project sustainability strategies: A systematic literature review. *Int. J. Proj. Manag.* **2017**, *35*, 1071–1083. [CrossRef]

43. Politis, I.; Papaioannou, P.; Basbas, S. Integrated Choice and Latent Variable Models for evaluating Flexible Transport Mode choice. *Res. Transp. Bus. Manag.* **2012**, *3*, 24–38. [CrossRef]

44. Eisenhardt, K.M. Building theories from case study research. *Acad. Manag. Rev.* **1989**, *14*, 532–550. [CrossRef]

45. Chambers, W.J.; Puig-Antich, J.; Hirsch, M.; Paez, P.; Ambrosini, P.J.; Tabrizi, M.A.; Davies, M. The Assessment of Affective Disorders in Children and Adolescents by Semistructured Interview. *Arch. Gen. Psychiatry* **1985**, *42*, 696–702. [CrossRef]

46. Rowe, G.; Wright, G. The Delphi technique as a forecasting tool: Issues and analysis. *Int. J. Forecast.* **1999**, *15*, 353–375. [CrossRef]

47. Yager, R.R. On Ordered Weighted Averaging Aggregation Operators in Multicriteria Decisionmaking. *IEEE Trans. Syst. Man Cybernetics* **1993**, *18*, 80–87. [CrossRef]

48. Anselm, S.; Juliet, C. Grounded theory methodology: An overview. *Handb. Qual. Res.* **1994**, *17*, 273–285.

49. Bernardo, M.D.R. Project Indicators for Enhancing Governance of Projects. *Procedia Technol.* **2014**, *16*, 1065–1071. [CrossRef]

50. Ahola, T.; Ruuska, I.; Artto, K.; Kujala, J. What is project governance and what are its origins? *Int. J. Proj. Manag.* **2014**, *32*, 1321–1332. [CrossRef]

51. Yin, X.; Liu, H.; Chen, Y.; Al-Hussein, M. Building information modelling for off-site construction: Review and future directions. *Autom. Constr.* **2019**, *101*, 72–91. [CrossRef]

52. Oraee, M.; Hosseini, M.R.; Papadonikolaki, E.; Palliyaguru, R.; Arashpour, M. Collaboration in BIM-based construction networks: A bibliometric-qualitative literature review. *Int. J. Proj. Manag.* **2017**, *35*, 1288–1301. [CrossRef]

53. Li, Y.; Lu, Y.; Ma, L.; Kwak, Y.H. Evolutionary Governance for Mega-Event Projects (Meps): A Case Study of the World Expo 2010 in China. *Proj. Manag. J.* **2018**, *49*, 57–78. [CrossRef]

54. Chen, L.; Manley, K. Validation of an Instrument to Measure Governance and Performance on Collaborative Infrastructure Projects. *J. Constr. Eng. Manag.* **2014**, *140*, 04014006. [CrossRef]

55. Biesenthal, C.; Wilden, R. Multi-level project governance: Trends and opportunities. *Int. J. Proj. Manag.* **2014**, *32*, 1291–1308. [CrossRef]

56. Li, W.; Zhang, R. Corporate Social Responsibility, Ownership Structure, and Political Interference: Evidence from China. *J. Bus. Ethics* **2010**, *96*, 631–645. [CrossRef]

57. Matten, D.; Moon, J. "Implicit" and "explicit" CSR: A conceptual framework for a comparative understanding of corporate social responsibility. *Acad. Manag. Rev.* **2008**, *33*, 404–424. [CrossRef]

58. McWilliams, A.; Siegel, D. Corporate social responsibility: A theory of the firm perspective. *Acad. Manag. Rev.* **2001**, *26*, 117–127. [CrossRef]

59. Flyvbjerg, B. Introduction: The iron law of megaproject management. In *Oxford Handbook of Megaproject Management*; Oxford University: Oxford, UK, 2017; pp. 1–18.

60. Davies, A.; Gann, D.M.; Douglas, T. Innovation in Megaprojects: Systems Integration at London Heathrow Terminal 5. *Calif. Manag. Rev.* **2009**, *51*, 101–125. [CrossRef]

61. Bstieler, L.; Hemmert, M. The effectiveness of relational and contractual governance in new product development collaborations: Evidence from Korea. *Technovation* **2015**, *46*, 29–39. [CrossRef]

62. Daily, C.M.; Dalton, D.R.; Cannella, A.A. Corporate governance: Decades of dialogue and data. *Acad. Manag. Rev.* **2003**, *28*, 371–382. [CrossRef]

63. Zhai, L.; Xin, Y.; Cheng, C. Understanding the value of project management from a stakeholder's perspective: Case study of mega-project management. *Proj. Manag. J.* **2009**, *40*, 99–109. [CrossRef]

64. Aguinis, H.; Glavas, A. What we know and don't know about corporate social responsibility: A review and research agenda. *J. Manag.* **2012**, *38*, 932–968. [CrossRef]

65. Li, T.H.; Ng, S.T.T.; Skitmore, M.; Li, N. Investigating stakeholder concerns during public participation. *Proc. Inst. Civ. Eng. Munic. Eng.* **2016**, *169*, 199–219. [CrossRef]
66. Li, H.; Ng, S.T.; Skitmore, M. Modeling Multi-Stakeholder Multi-Objective Decisions during Public Participation in Major Infrastructure and Construction Projects: A Decision Rule Approach. *J. Constr. Eng. Manag.* **2016**, *142*, 04015087. [CrossRef]
67. Bosch-Rekveldt, M.; Jongkind, Y.; Mooi, H.; Bakker, H.; Verbraeck, A. Grasping project complexity in large engineering projects: The TOE (Technical, Organizational and Environmental) framework. *Int. J. Proj. Manag.* **2011**, *29*, 728–739. [CrossRef]
68. Yang, J.; Shen, G.Q.; Ho, M.; Drew, D.S.; Chan, A.P.C. Exploring critical success factors for stakeholder management in construction projects. *J. Civ. Eng. Manag.* **2009**, *15*, 337–348. [CrossRef]
69. Linstone, H.A.; Turoff, M. *The Delphi Method: Techniques and Applications*; Addison-Wesley Reading: Boston, MA, USA, 1975.
70. Pallant, J. *SPSS Survival Manual*; McGraw-Hill Education: London, UK, 2013.
71. Bartlett, M.S. A Note on the Multiplying Factors for Various χ^2 Approximations. *J. R. Stat. Soc.* **1954**, *16*, 296–298. [CrossRef]
72. Chan, E.; Lee, G.K.L. Critical factors for improving social sustainability of urban renewal projects. *Soc. Indic. Res.* **2008**, *85*, 243–256. [CrossRef]
73. Dong, D.; Zhang, Y.C. Separation of public goods' double value orientations: Resolving dispute over 'people-benefit projects 'and 'vanity projects'. *J. Nanjing Normal Univ. Soc. Sci. Ed.* **2012**, *6*, 20–26.
74. Heinrich, H.W. *Industrial Accident Prevention: Scientific Approach*; McGraw Hill: New York, NY, USA, 1959.
75. Qiu, Y.M.; Chen, H.Q.; Sheng, Z.H.; Cheng, S.P. Governance of institutional complexity in megaproject organizations. *Int. J. Proj. Manag.* **2019**, *37*, 425–443. [CrossRef]
76. Sierra, L.A.; Yepes, V.; García-Segura, T.; Pellicer, E. Bayesian network method for decision-making about the social sustainability of infrastructure projects. *J. Clean. Prod.* **2018**, *176*, 521–534. [CrossRef]

Publisher's Note: MDPI stays neutral with regard to jurisdictional claims in published maps and institutional affiliations.

International Journal of
*Environmental Research
and Public Health*

Article

A New Methodology to Design Sustainable Archimedean Screw Turbines as Green Energy Generators

Mar Alonso-Martinez *, José Luis Suárez Sierra, Juan José del Coz Díaz and Juan Enrique Martinez-Martinez

Construction and Manufacturing Engineering Department, University of Oviedo, EDO-7 Campus de Viesques, 33204 Gijón (Asturias), Spain; suarezsjose@uniovi.es (J.L.S.S.); juanjo@constru.uniovi.es (J.J.d.C.D.); quique@constru.uniovi.es (J.E.M.-M.)
* Correspondence: mar@constru.uniovi.es; Tel.: +34-985182353

Received: 1 November 2020; Accepted: 8 December 2020; Published: 10 December 2020

Abstract: Current energy demand and climate target plans are leading to green energy facilities which are efficient and sustainable. Archimedean screw turbines (ASTs) are used to generate hydroelectricity in low heads. They have been manufactured and installed worldwide. However, there is a lack of knowledge about how to design them efficiently. In this study, the performance of ASTs is analyzed using an analogy between ASTs and bucket elevators. Based on this analogy, a theoretical hypothesis on how to produce efficient ASTs is proposed. The new methodology for the design of ASTs is based on two considerations: the filling level of the AST buckets must be 85% and the increase of leakage losses must be minimized. This hypothesis is numerically and experimentally studied. Two experimental prototypes were developed and installed in the north of Spain. The numerical and experimental results are provided. A discussion comparing the results of this work and other results from the literature is presented. Finally, conclusions are drawn from this work that contribute to the improvement of AST technology as a sustainable facility to generate green energy.

Keywords: Archimedean screw turbine; green energy; sustainable construction; resilient structures

---

1. Introduction

High energy consumption is one of the worries of our society. Energy demand increases every year and generation of green energy is essential to country growth that minimizes environmental impacts. Although there are many types of systems that can be used to generate green energy, sustainable infrastructures are needed to achieve sustainable development. Archimedean screw turbines (ASTs) are low-head hydropower generators suitable for industrial applications.

Low-head hydropower develops hydroelectric power using low heads, usually less than 20 m high. It is widely known as a renewable energy source. This highly efficient hydropower is inexpensive to manufacture and maintain. It reduces greenhouse gas emissions with minimal environmental impact. Microhydropower plants have been designed and studied in depth because of their advantages over traditional fossil fuels [1].

The first device used to generate electricity from running water was the waterwheel in the Fox River, Wisconsin, in 1882 [1]. Currently, hydropower devices which generate less than 100 kW are known as microhydropower generators. The common range of energy is from 5 kW to 100 kW but there are a great number of locations where small devices generating less than 5 kW could be installed [2].

There are many types of low-head devices for low-head hydropower applications, such as traditional waterwheels, Kaplan, Francis and Pelton turbines and ASTs. The AST was patented in 1991 by Karl-August Radlik [3]. It has gained popularity recently due to important advantages over alternatives, such as its suitability for low heads, the by "the simple construction process"needed for its installation and its reduced environmental impact [4–7]. Although the Archimedes screw was developed for pumping water from lower to higher levels, the use of this device in recent years focuses on the generation of sustainable energy. The AST is efficient in locations with low head and low flow [5]. The most relevant advantages of the AST are its low cost of manufacturing, installation and maintenance and its low environmental impact. In addition, the construction process needed is very simple and fish friendly [4–7]. Due to the low rotational speed and the atmospheric pressure, most fish species are able to pass through the turbine with no injuries. Although the benefits of ASTs make them a sustainable facility to generate green energy, adequate design methodologies must be followed.

Most previous studies model and test ASTs in small-scale prototypes in the laboratory. In 2012 Lashofer et al. published an overview of the AST technology including design and operation. An in-depth study of 74 ASTs installed in 71 locations across Europe was published in the same year [3]. The main parameters of AST geometry, the speed rotation, the civil construction time and the most relevant operating issues were presented. An important contribution of this report [3] was the provided AST design values established by manufacturers and operators from functioning AST plants in Europe. The plant efficiency was obtained as a function of load ratio, which is the actual flow at the time of measurement over the design flow. The results of the report show very similar efficiency patterns for all the plants studied, with peak efficiencies around 75%. Other studies find similar efficiencies of between 60% and 80% [2,4,8]. Most of the references agree that leakage losses are one of the most influential parameters in the efficiency of turbines [2]. Leakage losses are mainly due to the gap between the flights of the screw and the trough, which allows the screw to rotate.

Although there are many AST facilities around the world which have been tested for efficiency, there is a lack of understanding about how ASTs work and how to design them more efficiently. There are some authors that explain the AST as a mechanism similar to a waterwheel [8]. This similarity has not been analytically proved. Most of the studies are based on experimental experiences [3,5–7,9,10]. However, to understand the technology and optimize its efficiency, theoretical hypotheses concerning AST performance are needed. The losses in this kind of turbine and their influence on efficiency must be studied in depth.

In this study, the performance of ASTs is analyzed using an analogy between ASTs and bucket elevators. Based on this analogy, AST performance is explained, including the fundamentals required to obtain high efficiency, approximately 80%, in real facilities. Previously published efficiency curves of ASTs are discussed. A new methodology to design ASTs based on the leakage losses curve is presented. The authors establish that leakage losses follow a linear trend, which is a function of the ratio of flow (actual flow over maximum flow). This methodology was tested in two prototypes developed and installed in the north of Spain. Key parameters to design efficient ASTs are revealed and the sustainability of this technology is improved. A discussion comparing the efficiencies obtained in this work and the most relevant contributions from the literature is presented. Finally, conclusions are drawn that contribute to the development of sustainable facilities to generate green energy.

2. Materials and Methods

2.1. Description of the Archimedes Screw Turbine Technology

In this work, an analogy between ASTs and bucket elevators is used to identify the most influential parameter on total efficiency. The total efficiency of the AST is divided into hydraulic, mechanical and electrical efficiencies. In this study, the mechanical and electrical efficiencies were estimated as 90%. The present analysis only aims at improving the hydraulic efficiency of ASTs, which can improve the total efficiency by about 10%. Figure 1 shows the mechanisms of ASTs and bucket elevators, including their main design parameters, which are as follows:

- Z is the height of the head;
- Q, in m^3/s, is the total flow passing through the hydraulic machine, the AST or the bucket elevator;
- Q_{leak}, in m^3/s, is the leakage loss flow due to leakage between two consecutive levels;
- $Q_{overflow}$, in m^3/s, is the overflow loss when the design flow is exceeded. These losses appear in ASTs although they are not included in Figure 1a.

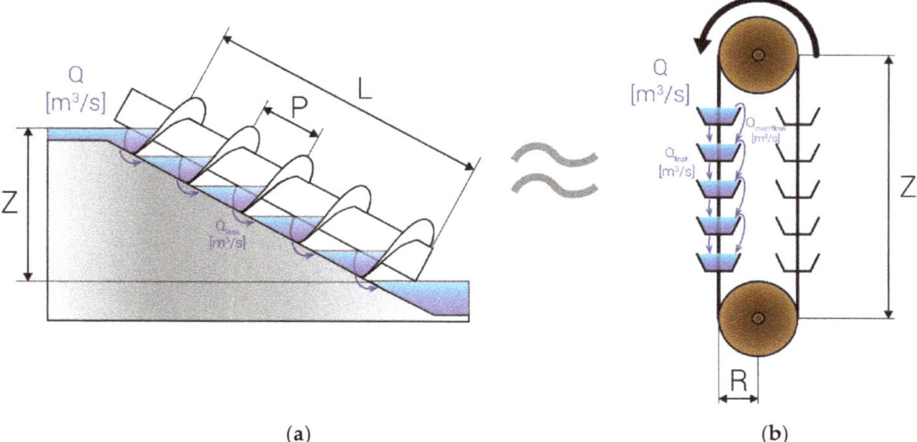

(**a**)　　　　　　　　　　　(**b**)

Figure 1. Analogy between Archimedean screw turbine (AST) and bucket elevator. (**a**) AST; (**b**) bucket elevator.

The design parameters of ASTs, shown in Figure 1a, are the following:

- P, in m, is the pitch of the screw;
- L, in m, is the length of the screw.

The design parameter of the bucket elevator, shown in Figure 1b, is R, the radius of the pulley used to move the belt, measured in m.

For an ideal machine where the maximum flow, Q, is the design flow, the efficiency will be 1. However, in real machines variable flows and losses flows must be taken into account to determine the real efficiency. Three operation conditions were used to study AST.

(a)　AST operates at the design flow, Qmax.:

When a bucket elevator is operating under design flow conditions, the buckets are full and the efficiency is mainly determined by the losses. Leakage losses in the buckets of the elevator are due to imperfections, holes or defects that are avoided during its manufacturing process and controlled during its operation. However, leakage losses in ASTs are mainly due to the gap between the flights of the screw and the trough. These losses are unavoidable in order to ensure the good performance of the AST. For this reason, leakage flow was analyzed in this study and its influence on the efficiency of the mechanism was determined. The unavoidable leakages were determined in this work using a small-scale prototype in experimental tests for different flow levels.

(b)　AST operates below the design flow:

If the total flow is less than the design flow and the rotation velocity is the same as the velocity of the design flow, the efficiency is lower. This is easy to understand in the bucket elevator because below the design flow the capacity of the buckets is lower and the torque of the mechanism is also lower. This reduction of torque leads to a lower efficiency and a lower generation of energy. This effect also takes place in ASTs. In this study, the influence of the volume of water on the torque, and consequently on the efficiency, is studied.

(c)   AST operates above the design flow, overflow:

The leakage due to overflow refers to the water losses due to overfilling when the total flow is exceeded. In a bucket elevator overflow occurs due to overfilling of the buckets. In ASTs these leakages also appear although their effect on the efficiency is lower than the effect due to leakage between the flights and the trough.

The hypothesis of this work defines leakage loss as a linear trend as follows. Unavoidable leakages were experimentally obtained with the AST stopped. Then, the increase of leakage losses was determined as a function of the rotation speed. This is presented in Equation (1).

$$\frac{Q_{leak}}{Q_{max}} = L_s + L_s \frac{\Omega}{60} \frac{Q}{Q_{max}} \tag{1}$$

where:

Q_{leak} is the loss due to leakage (flow in m^3/s);
Q_{max} is the maximum flow (m^3/s);
L_s is the static leakage loss (%);
Ω is the rotation speed of the screw (rpm);
Q is the actual flow through the AST (m^3/s).

This hypothesis was checked against efficiency curves published by other authors and manufacturers [11–13]. In 2015, Charisiadis published a review of the Archimedean screw as a low-head generator considered as an eco-friendly technology. ASTs and their benefits were also described in a booklet that collected data from previous works. In this publication, an AST efficiency curve was included and compared with traditional generators such as waterwheels, propellers and Kaplan and Francis turbines [11]. Another study compared different types of hydropower turbines [12]. The authors explained why AST efficiency under real conditions is lower than that claimed by the manufacturers. They also included efficiency curves of ASTs. The last efficiency curve included in this work is from the manufacturer Roncuzzi Renewable Energy [13]. In its publication, details about the civil work, the benefits of hydropower screws and performance were included.

The AST is a volumetric turbine and thus based on volumetric efficiency (see Equation (2)). The authors obtained the leakage losses curve from the efficiency curve of the AST.

$$\eta_v = 1 - \frac{Q_{leak}/Q_{max}}{Q/Q_{max}} \tag{2}$$

Using this hypothesis, and considering the information previously published, the authors determined the leakage losses curve of the AST from the references [11–13].

The real leakage losses can be obtained in ASTs which have already been installed and thus the adjustment of the linear trendline proposed in this work can be determined.

This hypothesis provides a new methodology to design efficient ASTs. The key points of the hypothesis used to design high-efficiency ASTs are the following:

1.   The AST must be understood as a bucket elevator and designed to take into account the filling level.
2.   The AST must be designed to ensure as low a percentage of leakage losses as possible.

In order to prove this hypothesis, numerical and experimental analyses were developed.

Firstly, the filling level of the buckets was studied through numerical models using the finite element method (FEM). In addition, two experimental prototypes were designed following this methodology based on linear leakage losses. Experimental data from the prototypes reveal the efficiency of this kind of AST and its leakage losses.

2.2. Study of Theoretical Performance of the AST through Numerical Models using the Finite Element Method

The complex geometry of the flights of a screw makes its analytical study very difficult. Numerical models using FEM are a powerful tool to study those complex geometries and the filling level and its effect on the efficiency of the turbine.

The numerical models presented in this work analyze the effect of the filling of the buckets on the torque. The torque is directly related to the efficiency of the AST. The higher the torque is, the more power is generated. The analysis was done considering the design flow.

The geometrical model of the AST is shown in Figure 2a. To study the effect of the filling of the buckets, only one bucket was studied. The numerical model is shown in Figure 2b. The type of finite structural element used to simulate the flights was SHELL93 [14]. This element is suitable for modeling curved shells. It has six degrees of freedom at each node: translations in x, y and z and rotations about x, y and z.

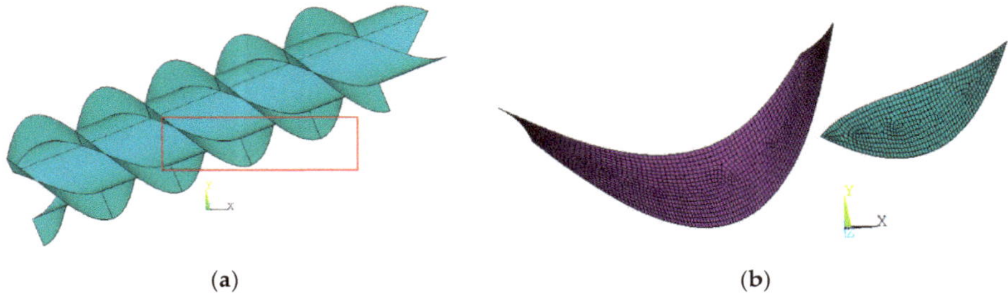

(a)   (b)

Figure 2. Finite element method (FEM) model. (**a**) Geometry of the AST; (**b**) numerical model of the bucket.

The numerically analyzed characteristics of the AST are detailed in Table 1.

Table 1. Values of the main parameters of the AST numerically studied.

Parameter (Units)	
Slope (°)	22
Outer diameter (m)	0.564
Inner diameter (m)	0.302
Pitch (m)	0.972
Length of the screw (m)	3.203
Head (m)	1.20
Gap between flights and trough (mm)	3

Four numerical models were developed to study the following levels of filling: 41%, 52%, 69% and 85%. The boundary conditions of the numerical model are shown in Figure 3.

Node A is in the axis of the screw. Displacements and rotations at node A were avoided. The filling of the bucket was modeled by applying the fluid pressure normal to the surface. The surface was obtained from the geometrical model and the value of the normal pressure depends on the filling level.

(a) (b)

(c) (d)

Figure 3. Boundary conditions of FEM model. (**a**) 49% full; (**b**) 52% full; (**c**) 69% full; (**d**) 85% full.

Although the theoretical performance of an ideal AST would provide an efficiency of 100%, a real mechanism has losses due to leakage, overflow and other factors. To study the real efficiency of the AST two experimental prototypes were manufactured and installed in the north of Spain.

2.3. Real Performance of the AST Using Experimental Prototypes

Two prototypes were developed and installed in two different locations. A small-scale prototype was installed in the Atlantic Botanic Garden of Gijón (Spain) in 2018 [15] (see Figure 4a) and a real-scale AST was installed in Barreda hydroelectric plant in Torrelavega (Spain) [16] (see Figure 4b).

The experimental setup recalculates the flow to measure the flow, torque and rotational speed of the AST prototypes. The system is shown in Figure 5.

(**a**)   (**b**)

Figure 4. AST experimental prototypes. (**a**) Atlantic Botanic Garden of Gijón (Spain); (**b**) Barreda hydroelectric plant in Torrelavega (Spain).

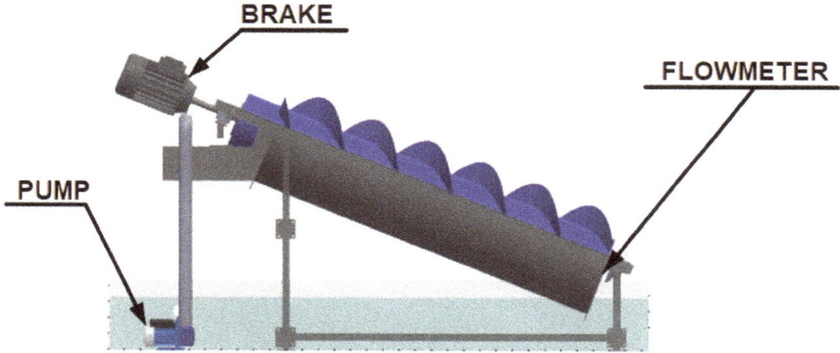

Figure 5. Experimental setup design of the recirculating system.

2.3.1. Small-Scale Prototype

The AST installed in Gijon was an experimental prototype designed for research purposes in the University of Oviedo. It was situated in a small lake and used to generate electricity to supply energy for an ultrasound system to avoid eutrophication issues. The head was very small, only 1.2 m high, and the design flow was 1.20 m^3/s. However, the AST prototype generated 1.2 kW and helped to supply the ultrasound control with energy efficiently and in a fish friendly and sustainable way.

The screw and the trough were made of steel and accurately manufactured. The gap between the flights and the trough was 3 mm. The maximum rotation speed was 73 rpm. The hypothesis presented in this work, Equation (1), was used to minimize leakage losses and obtain an efficient AST.

This small-scale prototype was tested under different flows to analyze the efficiency of the AST under real conditions.

The main design parameters of the small-scale AST are shown in Table 2.

Table 2. Values of the main parameters of the small-scale AST installed in the Atlantic Botanic Garden.

Parameter (Units)	
Slope (°)	22
Outer diameter (m)	0.564
Inner diameter (m)	0.302
Pitch (m)	0.972
Length of the screw (m)	3.203
Head (m)	1.20
Gap between flights and trough (mm)	3
Rotation speed (Ω)	73.239

2.3.2. Real-Scale Prototype

A real-scale prototype was also designed based on the leakage losses and their influence on the efficiency. The prototype was installed in a low head, less than 2 m high, of the Saja River. Two ASTs, able to generate 35 kW/turbine, were installed in the facility to generate 70 kW. The design flow was 2.50 m³/s per turbine. The maximum rotation speed was 50 rpm although there was a variable speed control.

The screw was made of steel and the trough was a simple civil work made of concrete.

The main parameters of the real-scale AST are shown in Table 3.

Table 3. Values of the main parameters of the real-scale AST installed in the Barreda hydroelectric plant.

Parameter (Units)	
Slope (°)	22
Outer diameter (m)	2.278
Inner diameter (m)	1.273
Pitch (m)	4.540
Length of the screw (m)	4.179
Head (m)	1.94
Gap between flights and trough (mm)	3
Rotation speed (Ω)	variable

3. Results

3.1. Numerical Results

Force reactions obtained in node A provided the torque of the screw for the four levels of filling studied. Figure 6 shows these results including the linear trendline. The adjustment of the trendline based on R-squared values was very good and showed the expected linear trend. For the design flow, the maximum theoretical torque was obtained when buckets were full. This effect, which is clear in a bucket elevator, is not so obvious in the AST design. In a bucket elevator, the torque is the weight of the bucket times the radious of the belt pulley. However, the concept of buckets and of filling in ASTs has not been clearly explained in previous research.

The results of this numerical analysis prove that the torque of the AST was directly proportional to the filling volume of the buckets. Filling volume and flow were also directly related by the rotation speed of the screw. Therefore, to obtain maximum efficiency the rotation speed of the screw plays a very important role.

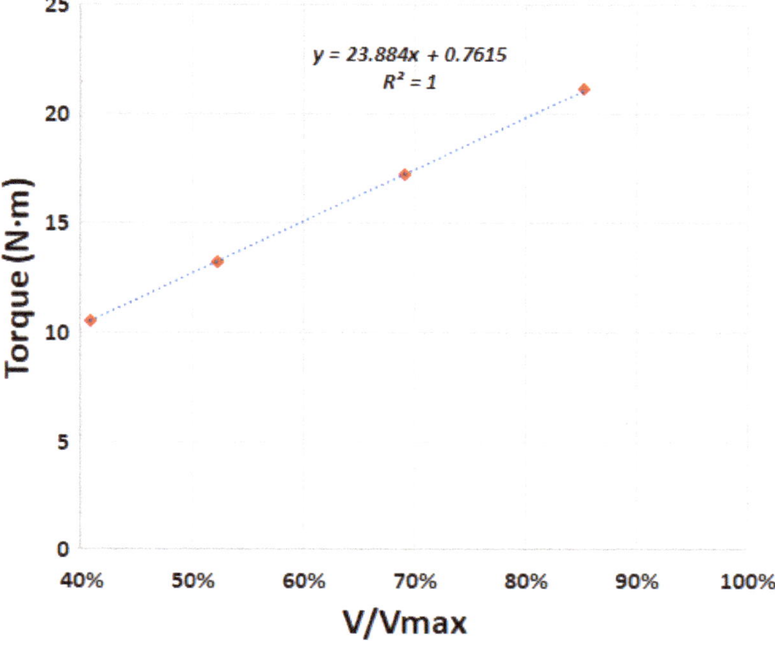

Figure 6. Torque obtained for four filling levels.

3.2. Experimental Results

3.2.1. Small-Scale Prototype Results

Tests done on the small-scale prototype demonstrated the efficiency of an AST for a very low head with different flows. The system was designed to work with the buckets 85% full. This value was identified as the optimum filling level in the experimental tests. Above that value, overflow occurs, as was seen in the experimental tests and in the decrease in efficiency.

Figure 7 shows the efficiency curve of this specific prototype based on the relation between the real flow measured and the maximum flow.

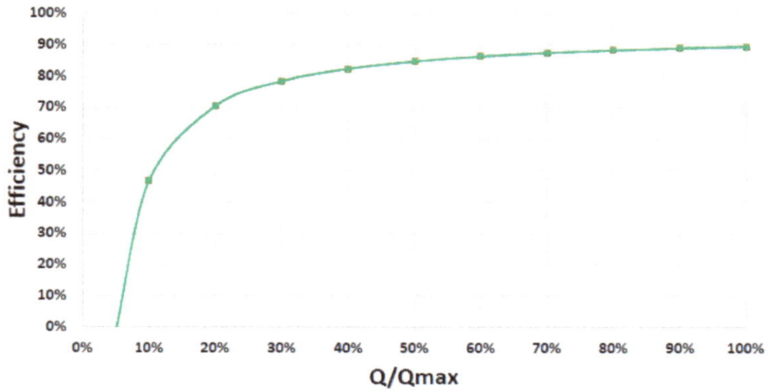

Figure 7. Efficiency curve of the small-scale AST prototype in the Atlantic Botanic Garden (Gijón, Spain).

Leakage losses were also obtained from the small-scale prototype. Figure 8 shows the percentage of leakage losses in the small-scale prototype. Static leakage losses were measured with the screw stopped. It was demonstrated that 5% losses were unavoidable. The leakage losses increased slowly, confirming the hypothesis established about the design of the AST.

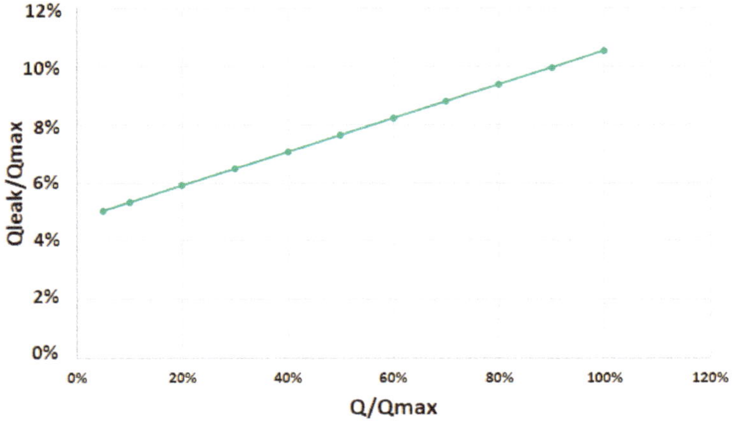

Figure 8. Percentage of leakage losses over the percentage of flow through the screw in the small-scale prototype.

3.2.2. Real-Scale Prototype Results

Results obtained for the real-scale prototype from Barreda hydroelectric power plant demonstrated high efficiencies, usually above 80%. This experimental prototype was designed to work with a bucket 85% full. In this prototype, the filling level was ensured using an active speed control. This active speed control checks the torque given by the AST every 10 min. The control varies the speed using a converter and an algorithm is provides five torque/speed combinations. All the combinations are immediately compared and the control chooses the highest power available and sets the speed to obtain that power as the nominal speed.

The efficiency was obtained as a function of the flow ratio: actual flow over the maximum flow. Figure 9 shows the efficiency curve of the real-scale prototype installed in Barreda hydroelectric plant.

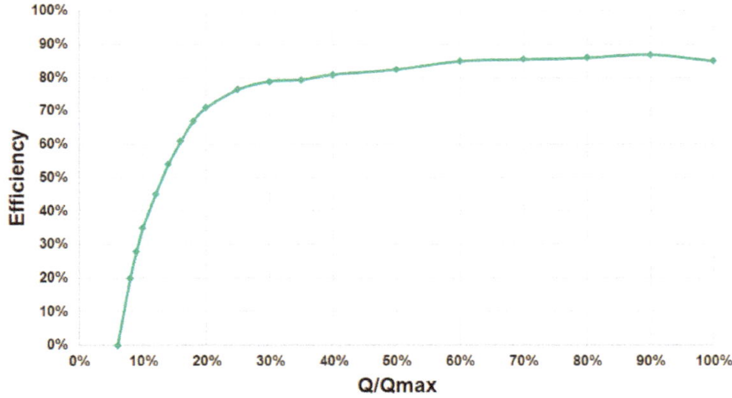

Figure 9. Efficiency curve of the real-scale AST prototype in Barreda hydroelectric plant (Torrelavega, Spain).

Leakage losses were also obtained for this real prototype. The real-scale prototype showed similar behavior to the small-scale prototype in experimental tests. Figure 10 shows the percentage of leakage losses in the real-scale prototype. In this turbine, static leakage losses were around 6%. Although the leakage losses for low flows were higher than those in small-scale prototypes, the increase of losses was smaller than in the small-scale prototype. For the small-scale prototype, the leakage losses at the design flow were around 10% (see Figure 8). However, the real-scale prototype had leakage losses below 9% (see Figure 10).

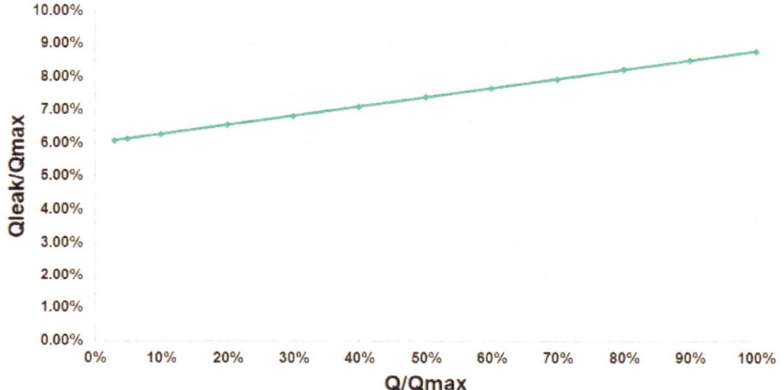

Figure 10. Percentage of leakage losses over the percentage of flow through the screw in the real-scale prototype.

4. Discussion

Numerical studies were used to explain how to design efficient ASTs. These numerical studies led us to understand that ASTs also have buckets and that their filling level is relevant to their design. An analogy between ASTs and bucket elevators was used to understand this concept.

The efficiency curves of the prototypes studied in this work were compared with previously published curves from previously installed ASTs [11–13]. Figure 11 shows a comparison of efficiency curves for five ASTs already installed. The ASTs designed in this work provided high efficiencies even for low flow percentages. For a flow ratio around 20%, the efficiency improved by between 5% and 10% in comparison with other ASTs. This improvement in hydraulic efficiency provides a significant increase in the total efficiency of these turbines.

The percentages of leakage losses for several ASTs were also studied. Figure 12 shows a comparison of the percentages of leakage losses as a function of the percentages of flow running through the turbines. The graph shows very similar static losses for all the ASTs, ranging from 5% to 8%. However, ASTs usually operate under higher flow percentages, above 50%, because their efficiency is higher. For a percentage of flow above 50%, the leakage losses of other designs [11–13] increased up to 16%. The ASTs designed in this work caused maximum losses of 10% at the maximum flow. This was due to the design based on the filling level of the AST buckets, which must be around 85% full to reduce overflow leakages. This consideration, as well as the speed control, significantly reduced the leakage losses and consequently improved the efficiency of the turbine.

Figure 12 shows a comparison of leakage loss percentages between previous ASTs and the ASTs designed and installed in this study. The trendlines are in good agreement with the actual data calculated. Figure 12 reveals that static leakage losses were almost the same in all the ASTs studied, around 5% of leakage losses. This value depends on the gap between the flights and the trough of the AST.

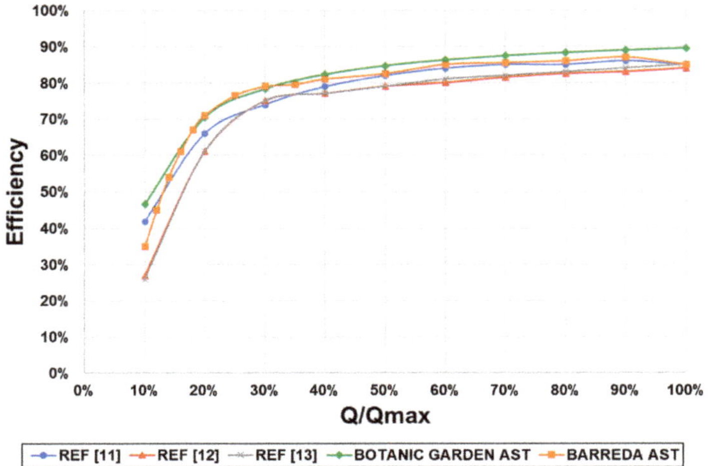

Figure 11. Comparison of previously published AST efficiency curves [11–13] and the two prototypes of this work.

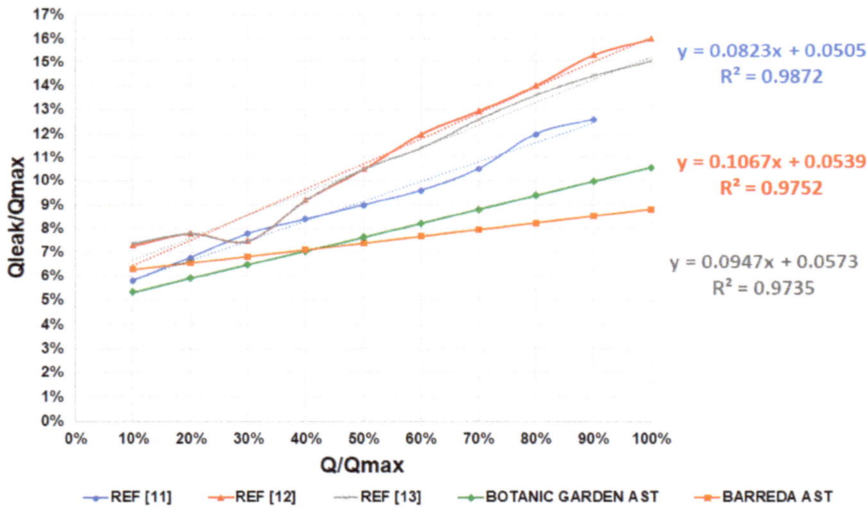

Figure 12. Comparison of the percentages of leakage losses for ASTs from three references [11–13] and the two experimental prototypes from this work.

In this study, it was also proved that the reduction of leakage losses in the design of an AST is an important factor for efficiency. The higher the losses due to leakage, the lower the efficiency is.

To minimize the slope of the leakage losses curve, the AST must be designed to work with the buckets at a filling level of around 85%. In this work, this was obtained using an active speed control which varied the speed as a function of the flow ratio. As shown in Figure 12, the lower slopes of the leakage losses curves were given by the prototypes developed in this work, which present the highest efficiency for any flow ratio in Figure 11.

A relevant contribution of this study is the significant improvement in efficiency for low flow ratios ranging between 15% and 40%.

5. Conclusions

The main conclusions of this work are listed as follows:

- Although buckets of the AST were not identified easily, the analogy with the bucket elevator presented in this work helps in understanding this concept. This analogy was a relevant contribution because the capacity of the buckets of an AST is one of the keys for an efficient design.
- Although the AST is a volumetric mechanism which works under atmospheric pressure, its efficiency strongly depends on the filling of the buckets. 100% filling would provide the highest efficiency in ideal ASTs. However, because of the movement of the screw, overflow losses occur above 85% filling and the efficiency of the turbine decreases.
- To ensure 85% filling of the buckets, an active speed control was suggested to ensure that the rotation speed decreased with the decrease of flow.
- Leakage losses, which are one of the most influential parameters on the efficiency of an AST, can be defined through the static leakage losses, the rotation speed and the design flow. This hypothesis was checked for other ASTs and for the two prototypes studied in this work.

Finally, this work demonstrates how an efficient design of this kind of facility is needed to ensure sustainable generation of energy. Green energy is needed to meet the goals of the 2030 Climate Target Plan but efficient methodologies for the generation of green energy are needed to ensure the sustainability of facilities and infrastructures.

Author Contributions: Conceptualization, M.A.-M. and J.L.S.S.; methodology, J.L.S.S. and J.J.d.C.D.; validation, J.L.S.S., J.J.d.C.D. and J.E.M.-M.; investigation, M.A.-M. and J.E.M.-M.; resources, J.J.d.C.D. and M.A.-M.; data curation, J.L.S.S.; writing—original draft preparation, M.A.-M.; writing—review and editing, J.E.M.-M.; visualization, M.A.-M.; supervision, J.J.d.C.D.; project administration, J.L.S.S.; funding acquisition, M.A.-M. and J.J.d.C.D. All authors have read and agreed to the published version of the manuscript.

Funding: This research was funded by "Agencia Estatal de Investigación" (ESBH project RTC-2017-6574-3) FEDER funds, and Gobierno del Principado de Asturias and FICYT (GRUPIN Research Project IDI/2018/000221).

Acknowledgments: The authors of the research presented in this paper acknowledge the financial support provided by the Ministry of Science, Innovation and Universities through the ESBH project RTC-2017-6574-3 and "Gobierno del Principado de Asturias" and FICYT under Research Project GRUPIN-IDI/2018/000221, which was co-financed with FEDER funds. Furthermore, the authors also thank SinFin Energy for all the information provided from Barreda hydroelectricity plant. Finally, thanks to Swanson Analysis Inc. for the use of ANSYS University Research software.

Conflicts of Interest: The authors declare no conflict of interest.

References

1. Jawahar, C.; Michael, P.A. A review on turbines for micro hydro power plant. *Renew. Sustain. Energy Rev.* **2017**, *72*, 882–887. [CrossRef]
2. Hoghooghi, H.; Durali, M.; Kashef, A. A new low-cost swirler for axial micro hydro turbines of low head potential. *Renew. Energy* **2018**, *128*, 375–390. [CrossRef]
3. Lashofer, A.; Hawle, W.; Pelikan, B.; Kampel, I.; Kaltenberger, F. State of technology and design guidelines for the Archimedes screw turbine. In Proceedings of the Hydro 2012—Innovative Approaches to Global Challenges, Bilbao, Spain, 29–31 October 2012.
4. Shahverdi, K.; Loni, R.; Ghobadian, B.; Gohari, S.; Marofi, S.; Bellos, E. Numerical Optimization Study of Archimedes Screw Turbine (AST): A case study. *Renew. Energy* **2020**, *145*, 2130–2143. [CrossRef]
5. Kozyn, A.; Lubitz, W.D. A power loss model for Archimedes screw generators. *Renew. Energy* **2017**, *108*, 260–273. [CrossRef]
6. Lavrič, H.; Rihar, A.; Fišer, R. Simulation of electrical energy production in Archimedes screw-based ultra-low head small hydropower plant considering environment protection conditions and technical limitations. *Energy* **2018**, *164*, 87–98. [CrossRef]

7. Dellinger, G.; Simmons, S.; Lubitz, W.D.; Garambois, P.-A.; Dellinger, N. Effect of slope and number of blades on Archimedes screw generator power output. *Renew. Energy* **2019**, *136*, 896–908. [CrossRef]

8. YoosefDoost, A.; Lubitz, W.D. Archimedes Screw Turbines: A Sustainable Development Solution for Green and Renewable Energy Generation—A Review of Potential and Design Procedures. *Sustainability* **2020**, *12*, 7352. [CrossRef]

9. Lubitz, W.D.; Lyons, M.; Simmons, S. Performance Model of Archimedes Screw Hydro Turbines with Variable Fill Level. *J. Hydraul. Eng.* **2014**, *140*, 04014050. [CrossRef]

10. Rorres, C. The Turn of the Screw: Optimal Design of an Archimedes Screw. *J. Hydraul. Eng.* **2000**, *126*, 72–80. [CrossRef]

11. Charisiadis, C. *An Introductory Presentation to the "Archimedean Screw" as a Low Head Hydropower Generator*; Leibniz Univerity Hannover: Hannover, Germany, 2015.

12. Cappi, A. *Hydropower Turbine Types Comparison for Ebley Mill, Stroud*; Water 21: Bristol, UK, 2012.

13. Hidrodynamic Screws PAE. Available online: http://www.roncuzzi.com/pae/download.asp (accessed on 28 October 2020).

14. Ansys®. *Academic Research Mechanical*; ANSYS Inc.: Canonsburg, PA, USA, 2011.

15. Jardín Botánico Gijón. Available online: http://www.sinfinenergy.com/en/projects/jardin-botanico-gijon/ (accessed on 24 October 2020).

16. Solvay. Sinfin Energy n.d. Available online: http://www.sinfinenergy.com/en/projects/solvay/ (accessed on 24 October 2020).

Publisher's Note: MDPI stays neutral with regard to jurisdictional claims in published maps and institutional affiliations.

International Journal of
***Environmental Research
and Public Health***

Article

Exploring Environmentally Friendly Biopolymer Material Effect on Soil Tensile and Compressive Behavior

Chunhui Chen [1], Zesen Peng [2], JiaYu Gu [1], Yaxiong Peng [3], Xiaoyang Huang [4] and Li Wu [2,*]

[1] Three Gorges Research Center for Geo-Hazards of Ministry of Education, China University of Geosciences, Wuhan 430074, China; chenchunhui@cug.edu.cn (C.C.); Jia_yuGu@163.com (J.G.)
[2] Faculty of Engineering, China University of Geosciences, Wuhan 430074, China; zxc455678125@163.com
[3] Hunan Provincial Key Laboratory of Geotechnical Engineering for Stability Control and Health Monitoring, Hunan University of Science and Technology, Xiangtan 411100, China; 1020172@hnust.edu.cn
[4] Centre for Doctoral Training in Catalysis, Cardiff Catalysis Institute, School of Chemistry, Cardiff University, Park Place, Cardiff CF10 3AT, UK; HuangX17@cardiff.ac.uk
*   Correspondence: lwu@cug.edu.cn

Received: 29 October 2020; Accepted: 1 December 2020; Published: 3 December 2020

Abstract: The study of the high-performance of biopolymers and current eco-friendly have recently emerged. However, the micro-behavior and underlying mechanisms during the test are still unclear. In this study, we conducted experimental and numerical tests in parallel to investigate the impact of different xanthan gum biopolymer contents sand. Then, a numerical simulation of the direct tensile test under different tensile positions was carried out. The micro-characteristics of the biopolymer-treated sand were captured and analyzed by numerical simulations. The results indicate that the biopolymer can substantially increase the uniaxial compressive strength and tensile strength of the soil. The analysis of the microparameters demonstrates the increase in the contact bond parameter values with different biopolymer contents, and stronger bonding strength is provided with a higher biopolymer content from the microscale. The contact force and crack development during the test were visualized in the paper. In addition, a regression model for predicting the direct tensile strength under different tensile positions was established. The numerical simulation results explained the mechanical and fracture behavior of xanthan gum biopolymer stabilized sand under uniaxial compression, which provides a better understanding of the biopolymer strengthening effect.

Keywords: biopolymer; numerical simulation; micro-behavior; green technology

---

1. Introduction

Traditional cement materials have been employed as stabilizing agents in civil engineering for a long time. Nevertheless, the extensive production and application of these materials have caused serious impacts on the environment, including solid waste, soil contamination, carbon emissions, dust, and water pollution [1–5]. Due to environmental concerned, the exploration of eco-friendly biomixture materials has recently been introduced as a potential replacement of cemented materials. A biopolymer is a high-performance and current eco-friendly material from microorganisms obtained by fermentation. The utilization of biopolymers in civil engineering is a sustainable technology because biopolymers can be used as organic additives in traditional cement materials. According to their composition unit and structure, most biopolymers are polysaccharide polymers. They are high molecular weight polymers formed by the dehydration of multiple monosaccharide molecules. Biopolymers have a profound influence on the soil in terms of the hydroconductivity [6], strength [7], and durability [8] by conducting geotechnical tests [9]. Biopolymers mostly contain hydrophilic groups, and the whole molecule has strong hydrophilicity, which leads to the strong viscosity of aqueous solutions. They can reduce the hydro-conductivity of soil and be used as candidate materials for temporary seepage barriers [10]. The interparticle cohesion provided by the biopolymer reduced soil loss and erosion retain water against evaporation [11]. Meanwhile, the addition of biopolymers can remarkably increase the soil shear strength and compressive strength. Moreover, the use of high content biopolymers can increase the soil strength, which is comparable to concrete to some extent. This strengthening effect of biopolymers is influenced by the biopolymer concentration and type, curing time, dehydration condition, and soil type [12–15]. Although geotechnical tests have investigated the mechanical behavior of biopolymer-treated soil on a macro scale and explained their possible strengthening mechanism, the interparticle interaction between the soil and failure mechanism during the test remains unknown.

The discrete element method (DEM) is currently a good tool for solving geotechnical problems. The DEM exhibits some advantages in solving discontinuous problems by modeling particle interactions with granular materials, as proposed by Cundall and Strack [16]. The crushable soil/rock materials are assembled by spheres with different contacts. All the particles are rigid bodies while interaction rules are embodied at the particle contact [17]. For discrete sand materials, PFC software can record individual particle displacement, speed, and rotation to analyze the micromovement of sand [18]. With outer static or dynamic loading, deformation occurs at the contact, and PFC can capture the internal micro behavior process of a material. Potyondy and Cundall [19] first proposed the linear parallel bond model to describe the contact of cemented materials. The samples were assembled by rigid spherical particles and jointed by the bond contact model. After applying external forces, the bond contacts become broken and can no longer provide adhesion force [20]. The biopolymer can be regarded as a cemented material that increases soil strength, and the soil particle interaction provided by the biopolymer in particle flow code (PFC) can be quantified by contact model parameters [21]. Considering the computational capacity, as the DEM model often contains thousands of small granular materials, it is appropriate to use PFC to simulate laboratory tests at a small scale. Previous studies used PFC to observe and analyze the specimen micro characteristics with different geotechnical tests. By modeling the direct shear test [22], triaxial test [23], and compressive test [24], the dynamical microbehaviors were monitored and captured to explain the corresponding experimental phenomena in terms of the micro characteristics.

However, biopolymer treated soil has shown improved properties and has been explored at the macro scale. However, the micro-behavior and the in-depth analysis of their characteristics are still unclear. The purpose of this study is to use both experimental and numerical simulations to investigate biopolymer treated soil with respect to macro and microgeotechnical behaviors. We conducted a series of uniaxial compressive tests for soil treated with different biopolymer contents from a macro perspective. For the corresponding numerical simulation, the linear contact bond model was proposed and material parameters were calibrated. Based on these data, direct tensile tests were simulated with different tensile positions which were impossible to carry out in the laboratory. Accordingly, we analyzed the micro parameters, internal force, and crack propagation in detail which were hard to observe via laboratory tests. Finally, the relationship between the tensile strength and compressive strength was analyzed.

2. Materials and Methods

2.1. Material

Natural silica sand with a specific gravity of 2.65 was employed for the uniaxial compressive test. The particle size distribution is presented in Figure 1. Commercial microbial exopolysaccharide xanthan gum was used in this research. Xanthan gum is a xanthomonas campestris fermented high molecular weight polysaccharide. Dry xanthan gum is a white powder, that forms a viscous solution when dissolved in water. Under different PH values, temperatures, and ionic strengths, xanthan gum gel still presents stable behavior [25]. It acts as a food thickener, drilling lubrication, and concrete viscosity modifier, and has been used in soil stabilization [26].

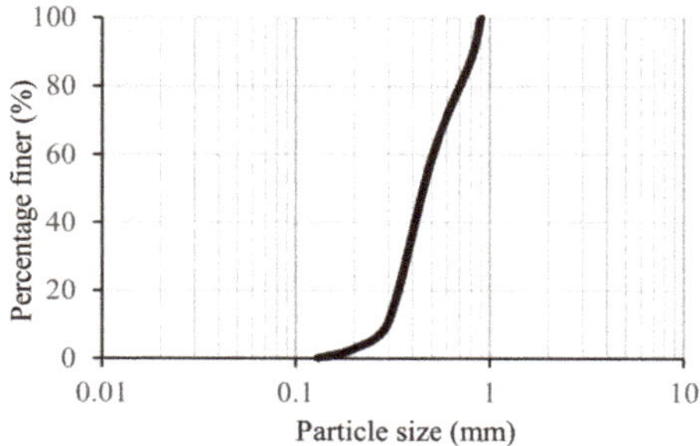

Figure 1. Particle size distribution.

2.2. Uniaxial Compressive Test

Due to xanthan gum's strong hydrophilic property, if poured into the water without proper stirring, the outer sphere will absorb water which then prevents water from permeating into the inter sphere, forming a white clump. Hence, magnetic stirring was used to uniformly mix xanthan gum powder with deionized water. A wide range of biopolymer contents (0.2%, 0.5%, 1%, 1.5% and 2% to the mass of sand) were mixed with 9.6% water, which was the optimum water content of the soil. Then, xanthan gum gel was blended with the sand via blender proper mixing for 10 min and poured into cubic molds with dimensions of 50 mm × 50 mm × 50 mm, as shown in Figure 2. The dry density was set to 1.63 g/cm^3, and the specimen was compacted layer by layer with the hammer layer by layer. Finally, all the samples were created in triplicate and preserved in a 30 °C oven until a constant weight was reached to represent dry conditions. The uniaxial compressive tests were conducted at 1%/min strain rate. All the tests were conducted under the guidance of Chinese standard GB/T 50123-1999. All tested were carried out in triplicate to minimize experimental error.

Figure 2. Preparation samples for experimental test ((**a**): sample mold; (**b**): sample dimension).

2.3. Scanning Electron Microscope (SEM)

To provide information on the biopolymer and sand interaction for the subsequent numerical simulation model, a scanning electron microscope was adopted. All the SEM images in this study were obtained by TM3030 Tabletop Microscope from Hitachi High-Tech. Samples were loaded on conductive carbon tape (SPI SUPPLIES). The measurement was operated in a vacuum (3–5 Pa) under an acceleration voltage of 15 kV with signals of backscattering electron (BSE). Relevant images and results are displayed in Figure 3.

Figure 3. SEM image of biopolymer treated sand and clean sand ((**a**): clean sand with 300 magnification; (**b**): biopolymer treated sand with 150 magnification).

3. Numerical Investigation

The experimental tests provided preliminary outcomes of the geomechanical properties of xanthan gum treated soil. The results indicated that the addition of biopolymer into the soil could largely increase the soil strength which is illustrated in detail in the following sections. To enhance the understanding of biopolymer stabilization soil towards microstructural behavior, particle flow code software was used to analyze the underlying mechanism of the interaction between the treated sands. By simulating the tests, the following processes and problems should be solved and executed:

(1) Contact model. The rigid particles interact with each other at particle surface contacts. Contact modes are assigned at the contacts to develop internal forces for various contact mechanics. The SEM images in Figure 3 present the differences between the clean sand and xanthan gum biopolymer-treated sand. The clean sand has irregular and isolated particles (Figure 3a) while the biopolymer-treated sand is covered and connected by the biopolymer (Figure 3b). The formed hydrogen bonding between the sand particles and xanthan gum provides this bonding connection [27]. In the PFC component, the linear parallel bond exhibits elastic interactions between particles that transmit a normal force, shear force, and moment [17]. When particles are bonded together, they delivered resistance to the particle rotation and therefore exhibit linear elastic to present bond properties. When the bond is broken, it acts as a linear model that cannot resist rotation and tension. Details can be found in the research by Potyondy and Cundall [19]. Thus, the linear parallel bond model can simulate the bonded and unbonded state of the sample particles in the test.

(2) Biopolymer bonding. In the SEM images, biopolymers were distributed in the soil and connected the soil particles. Figure 4 can reproduce this condition by adding a linear parallel contact bond between the sand particles. Contact parameters were embodied in all the particle contacts. Thus, all the particle contacts were set as linear parallel bonds. The different contents of the biopolymer were represented by the different contact parameters.

(3) Sample generation. Previous studies have pointed out that particle sizes and numbers are important factors that affect the simulation time [28]. Considering the computational efficiency and realistic particle size distribution, the particle radius is scaled up 2 times. This is a common way to reduce simulation time with fewer particle balls [29]. The specimen was generated with dimensions of 50 mm × 50 mm in accordance with the experimental sample, as presented in Figure 4. After multiple cycling calculations, the balls reached the equilibrium state in the subsequent test.

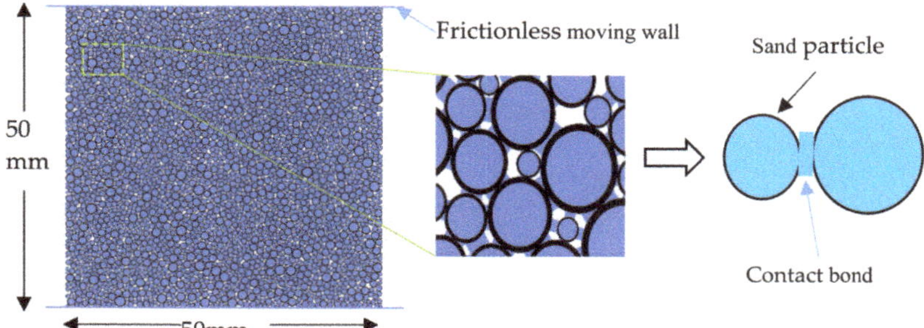

Figure 4. Characteristics of the numerical specimen (all sand particles contacts were linear parallel bond contact which was plot as blue line).

(4) Calibrate material behavior. The PFC parameter calibration is a process that reproduces the macro experimental behavior by selecting corresponding numerical parameters to match the experimental data. Due to the complicated interaction in realistic geo-mechanics and simplification in the simulation model, it is difficult to expect these parameters to be precisely recorded and matched with the real conditions. The parameters are given to the PFC components including the ball, wall and contact. As it is difficult to calculate the input parameters through the macroscopic response of the specimen, the most commonly used method of "trial and error" is adapted to calibrate the parameters [30]. Table 1 summarizes the biopolymer binding model parameters.

Table 1. Bond parameters of contact model.

Parameter	Symbol	Biopolymer Content				
		0.2%	0.5%	1%	1.5%	2%
		Value				
Sand particle density (kg/m^3)	ρ_s	2600	2600	2600	2600	2600
Bond tensile stress (Pa)	pb_ten	1×10^5	5.8×10^5	6.1×10^5	7.4×10^5	8×10^5
Bond cohesion (Pa)	pb_coh	5×10^4	2×10^5	3.5×10^5	4.6×10^5	6×10^5
Bond normal-to-shear stiffness ratio	kratio	2	2	2	2	2
Friction coefficient	μ	0.34	0.45	0.61	0.64	0.68
Bond effective modulus (Pa)	emod	9.8×10^6	1×10^7	1.3×10^7	1.4×10^7	1.5×10^7

(5) Remove floaters. When generating ball particles in PFC software, it is evident that there are no contacts around some particles. These floating particles called floaters are isolated and are not connected with the other particles from the initial point of the test. However, with the deformation of the samples during the test, these floaters may contact other particles and influence their properties. Thus, multiple cycles are set to remove the floaters.

(6) Simulating the uniaxial compressive test. Frictionless walls were applied as confined caps at the top and bottom of the sample in the simulation model. The compressive rate was identical to the experimental test. To analyze the micro-behavior of the samples, the stress-strain, internal force, and crack images were tracked and captured during the simulation. The compressive strength was recorded as the average force value on the top and bottom walls during the test.

(7) Simulating the direct tensile test. Unlike the uniaxial compressive test, loading walls were removed in the direct tensile test. A measurement circle was embedded in the specimen to record the axial tensile strength through the built-in FISH language. As the parameters were successfully calibrated to match the experimental test results for the uniaxial compressive test, the direct tensile tests were conducted by using identical parameters to those of the uniaxial compressive test. The particles were divided into an upper group (1~25 mm; 5~25 mm; 10~25 mm;

15~25 mm; 20~25 mm; 24~25 mm;) and a lower group (−1~−25 mm; −5~−25 mm; −10~−25 mm; −15~−25 mm; −20~−25 mm; −24~−25 mm;) according to their position. Then these two groups were given an opposite movement velocity to simulate the direct tensile test. The tensile tests were set for specimens a~f according to their different tensile positions. The schematic diagram is presented in Figure 5.

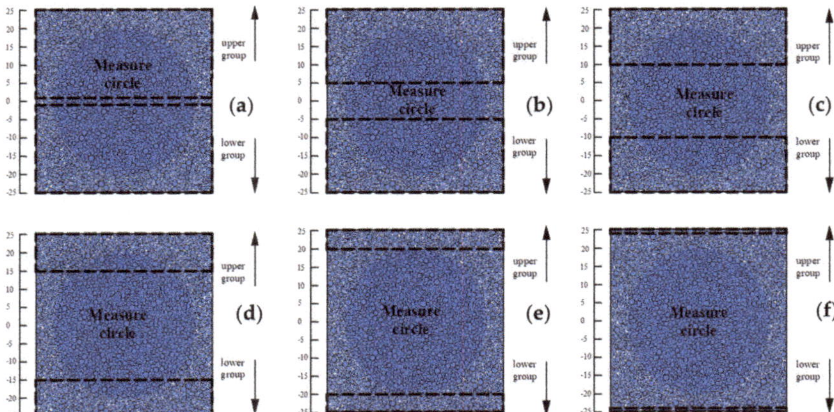

Figure 5. Schematic diagram of specimen tensile movement position ((**a**): tensile position 1~25 mm and −1~−25 mm; (**b**): tensile position 5~25 mm and −5~−25 mm; (**c**): tensile position 10~25 mm and −10~−25 mm; (**d**): tensile position 15~25 mm and −15~−25 mm; (**e**): tensile position 20~25 mm and −20~−25 mm; (**f**): tensile position 24~25 mm and −24~−25 mm;).

4. Results and Discussion

4.1. Comparison of Experimental and Numerical Uniaxial Compressive Test Results

Figure 6 lists the average value of the stress-strain curve of the experimental test results and simulation results of biopolymer treated sand at each content (0.2%, 0.5%, 1%, 1.5%, and 2%). The uniaxial compressive strength of the samples increased with increasing biopolymer content. Compared to cohesionless sand, for which it is difficult to obtain its uniaxial compressive test, there is a similar increasing trend, and the biopolymer has a notable effect on increasing the compressive strength. Simulation results can match the experimental test up to the peak strength. However, after reaching the peak strength, the simulation curves fall off rapidly. This may be caused by the absence of strain localization and the irregularity of the broken fragment. The uniaxial compressive strength of the tested samples increases from 131 kPa (0.2% content) to 1412 kPa (2% content). When the xanthan gum biopolymer is mixed with water, it formed a viscous hydrogel. By properly mixing the xanthan gum biopolymer with sand, these viscous gels are in contact with individual sand particles [31]. There are attachments between the soil particles and biopolymer solution followed by drying to form the adhesive connection [32]. In this case, particles are firmly bonded by the biopolymer to resist outer forces due to increased strength. From the micro parameter perspective, the bond tensile stress and cohesion value of the numerical bond contact in Table 1 improve with increasing biopolymer content, confirming that a higher biopolymer content can directly increase the bonding strength of sand. This eventually leads to an increase in compressive strength. The experimental and numerical results with each xanthan content indicate that the calibrated material behavior and contact model effectively simulate the geotechnical behavior of the samples. It is appropriate to use these numerical parameters to test the specimen tensile strength and analyze the specimen interior micro response and particle behavior.

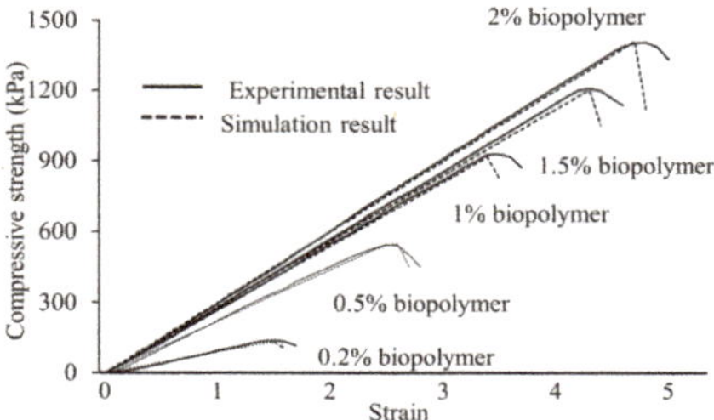

Figure 6. Stress-strain curve of biopolymer treated soil under uniaxial compressive test.

4.2. Tensile Strength

In the SEM image and compressive test simulation, it can be found that the biopolymer provided the most cohesion force to bind the sand particles together. The compression and tension contact force chain development also confirmed the contribution of the biopolymer on the soil cohesion property. Details can be found in the next section. The soil tensile strength directly reflects the mutual attraction of the soil particles which is an important indicator of soil cohesion characteristics. Compared to the compressive strength and shear strength, the soil tensile strength is relatively small and easy to neglect. However, failure modes due to foundation settlement, earth dams and slope are closely related to tensile cracks [33]. Thus, it is necessary to investigate the biopolymer effect on the soil tensile strength of the soil.

To test the tensile strength of the soil, the method can be categorized as the indirect test method and direct test method. The direct test method commonly uses a self-developed apparatus to pull the soil sample and directly test its tensile strength [34]. The main problem with this method is determining how to fix the specimen during the test. In addition, stress concentrations may occur at the specimen fixation position. Indirect tests use other loading and measurement methods to determine the relevant parameters and calculate the soil tensile strength. This is influenced by the specimen size and loading conditions [35]. Only the peak value rather than a series of stress strength data can be obtained by the indirect test. Furthermore, the calculated result cannot precisely match the realistic tensile strength. Both the indirect test and direct test methods cannot observe the influence of the tensile position on tensile strength.

To analyze the interior force performance, the different biopolymer-treated sample force contours under different tensile positions are presented in Figure 8. The interior forces were illustrated with different colored lines, and the width of the force chain lines was proportional to the force magnitude. From Figure 8, we can conclude that at the same tensile position, the interior force increased with increasing biopolymer content, which can explain the increase in the tensile strength in Figure 7. Chen, Wu and Harbottle [36] found that biopolymer forms a thin film between the sand particles and that this polymer film directly connects the sand particles. This tensile force increased with increasing biopolymer content, which led to an increase in the interior force and tensile strength. When the biopolymer-treated sand was under tension at different positions, there were two force interfaces at each tensile position. The magnitudes of the interior forces were quite different and can be classified as the outer part and inner part according to their relative position to the tensile position. The outer part was located at the moving part of the specimen, where the forces were relatively small. Because all the particles at the outer part were given the same movement velocity, these balls did not have relative displacement. There were no extra interior forces at the moving part to control the deformation. Compared to the outer part, the inner part was located at the extended part of the specimen, which had a high value of forces. The inner part particles did exhibit have movement at first. Due to the contact bond between the particles, inner particles were dragged by the outer part particles from the opposite sides. Thus, the bond contact forces gradually increased to resist tensile deformation.

Figure 7 plots different biopolymer treated sand tensile positions and peak tensile strength relationships. The peak tensile strength increased with higher biopolymer content. At each content, the specimen peak tensile strength first increased and then decreased. Thus, the commonly used multiple regression analysis was applied to estimate the tensile strength. The predictive polynomial function and coefficient of determination (R^2) are listed in equation 1. The second-degree polynomial function is highly correlated (with the high values of R^2).

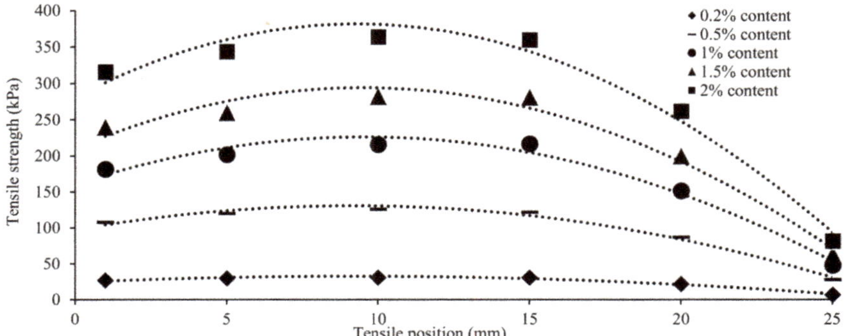

Figure 7. Specimens tensile strength according to tensile position and biopolymer content.

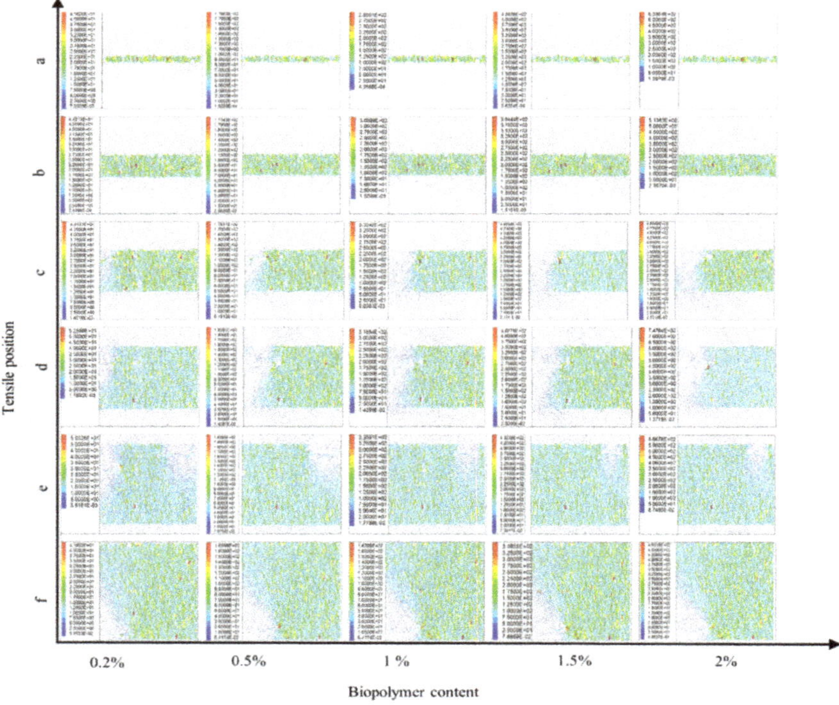

Figure 8. Different biopolymer treated sample force contour under different tensile position.

4.3. Internal Force and Crack Propagation Patterns

Sand particles are granular materials that convey forces via particle contacts in PFC simulations. During the test, the contact force changes rapidly with the specimen deformation [37]. The discrete element model can effectively visualize these force change networks by plotting force chains. The different contact force chains and fractures vary with the calculation time step. Figure 9 presents the contact force chain and fracture performance of 0.5% biopolymer-treated samples under the uniaxial compression test. The transmitted forces of the sand particles are visualized by green and blue lines: green represents tension forces, while blue represents compression forces. The width of these force chain lines is proportional to the force magnitude. When the specimens experience compressive loading, the aligned particles with internal forces formed the network of the force chain to resist the applied force. Lévy-Véhel [38] mentions that the force chain development is related to strength behavior. According to the development of fracture numbers, the biopolymer-treated sand under uniaxial compression can be described by three phases: the compact stage, crack development stage, and failure stage. At the compact stage (red background in Figure 9), the fracture number remains at zero. Because the contact force chain networks are almost intact, the compressive load is delivered by these networks and no fracture occurs (Figure 9 initial state). The magnitude and form of the internal force maintain the dynamic changes during compression. The slim compression force chain become bold and wide which demonstrates the constant increase in the compressive force provided by the biopolymer at this stage. Chang [39] demonstrated that the grain particle surface coated with xanthan gum biopolymer can enhance interparticle interactions, and the strength of the treated soils depends on the xanthan gum matrix. This cementing effect can be observed in this study and explained in more detail. Under compression, the tension force provided by the biopolymer presents a subhorizontal direction to resist the compressive dilatant deformation. With continuous compression, these forces eventually reach the threshold, they break and can no longer transmit the force which leads to the appearance of cracks. Thus, the sample reaches the crack development stage (yellow background in Figure 9). The overall fracture number increases gradually. The cracks occur (Figure 9b), develop afterward (Figure 9c), and finally penetrate through the sample (Figure 9d). After the compressive strength of the sample reaches its peak strength, the overall fracture number decreases substantially. Crushing is initiated, and relative occurs between the soil particles which results in fragment separation. This phase is called the failure stage (blue background in Figure 9). The break of the contact bonding force is accompanied by the appearance of fractures [40]. This can be confirmed by the obvious failure band marked with circles and concomitant fracture in Figure 9d. The fractures are located in the same position and present a similar shape.

Figure 10 presents the contact force chain and fracture performance of 0.5% biopolymer treated samples under the tensile test. The tensile test position was selected as the upper group from 24~25 mm and the lower group from −24~−25 mm in Figure 5f. Compared to the uniaxial compression test, the development of the contact force chain and fracture performance of 0.5% biopolymer treated samples under tensile tests exhibited a slightly different performance. The crack number first remained at zero and then increased substantially in Figure 10. It was not until the tensile strength reached 90% of the peak tensile strength that the first fracture appeared. The specimen then started to break from this point and reached failure quickly. Thus, the biopolymer-treated sand under the tensile test could only be described by the tensile stage and failure stage. When the specimen was under compression, cracks started to appear and develop at any possible position. However, when the specimen was under tension, cracks developed along with the first crack position and penetrated through the sample, thus, the crack number of the specimen under tension was much less than that of the specimen under compression. The formation of cracks led to stress redistribution and formed a tensile-zone which eventually resulted in specimen tension failure.

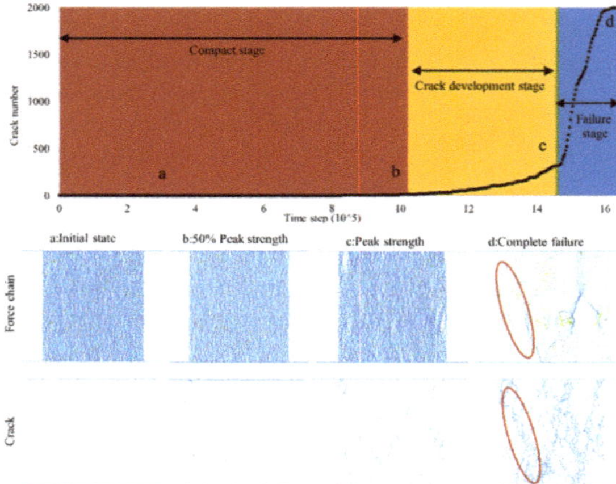

Figure 9. Force chain and crack development of 0.5% biopolymer treated sand under the uniaxial compressive test.

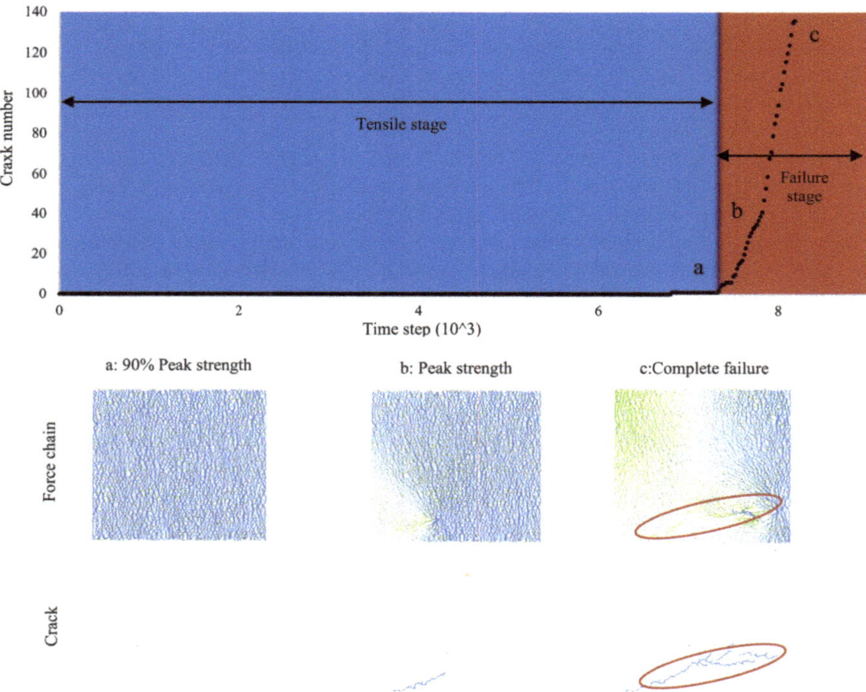

Figure 10. Force chain and crack development of 0.5% biopolymer treated sand under the tensile test.

4.4. Correlation of Compressive Verse Tensile Strength

Table 2 lists the biopolymer-treated soil tensile strength (qt), uniaxial compressive strength (qu) and the calculation of qt/qu. From the SEM images, it can be seen that the biopolymer occupied the pore space and acted as a bridge to connect sand particles, and biopolymer-treated sand can be regarded as a cemented soil. For cemented soil, the biopolymer content is the main factor influencing the soil strength. Although the biopolymer content had a substantial effect on the compressive strength and tensile strength of the soil, the ratio between the tensile strength and compressive strength can be regarded as a constant value that does not rely on the biopolymer content. Considering that tensile strength varies for different tensile positions, the qt/qu ratio remains a constant and presents a small deviation at each tensile position. This was similar to the results of N.C. Consoli found that the ratio of the lime cemented soil split tensile strength and unconfined compressive strength maintains a unique value of 0.16, independent of other factors [41].

Table 2. Biopolymer treated soil tensile strength (qt), uniaxial compressive strength (qu) and the calculation of qt/qu.

Biopolymer Content	qu (kPa)	Tensile Position											
		1~25 mm −1~−25 mm		5~25 mm −5~−25 mm		10~25 mm −10~−25 mm		15~25 mm −15~−25 mm		20~25 mm −20~−25 mm		24~25 mm −24~−25 mm	
		qt (kPa)	qt/qu	qt (kPa)	qt/qu	qt (kPa)	qt/qu	qt (kPa)	qt/qu	qt (kPa)	qt/qu	qt (kPa)	qt/qu
0.2%	131	27	0.206	30	0.229	31	0.237	31	0.237	22	0.168	7	0.053
0.5%	548	108	0.197	120	0.219	126	0.230	122	0.223	87	0.159	28	0.051
1%	920	182	0.198	202	0.220	216	0.235	217	0.236	152	0.165	48	0.052
1.5%	1204	240	0.199	260	0.216	282	0.234	281	0.233	200	0.166	63	0.052
2%	1412	316	0.224	344	0.244	364	0.258	360	0.255	262	0.186	82	0.058
Average value	-	-	0.205	-	0.225	-	0.239	-	0.237	-	0.169	-	0.053
Standard deviation	-	-	0.0100	-	0.0101	-	0.0098	-	0.0104	-	0.0090	-	0.0024

5. Conclusions

In this study, both experimental and numerical tests were conducted to investigate the impact of different contents of xanthan gum biopolymers on the uniaxial compression strength and direct tensile strength of sand. The linear contact bond method was proposed to represent the biopolymer binding effect, and the corresponding numerical parameters were calibrated. The micro characteristics of the biopolymer cementation effect were captured and analyzed through the DEM simulation. The following conclusions can be obtained:

(1) The experimental and numerical results with different xanthan contents indicate that the calibrated material behavior and contact model can effectively simulate the geotechnical behavior of the samples. They all indicate that the uniaxial compression strength and tensile strength in the xanthan gum biopolymer-treated sand increase with a higher biopolymer content.

(2) The bond tensile stress and cohesion value in the PFC numerical simulation model increased with increasing biopolymer content, illustrating that a higher biopolymer content can provide a stronger bond. This can also be confirmed by a decrease in the fracture number when the biopolymer content increases.

(3) There was a second-degree polynomial function relationship between the tensile position and tensile strength.

(4) According to the development of the contact force chain and crack propagation pattern in PFC, the behavior from biopolymer-treated sand under uniaxial compression can be classified into three stages: the compact stage, crack development stage, and failure stage, which can be regarded as the tensile stage and failure stage when the specimen is subjected to the tensile test.

(5) Although biopolymer content had a substantial effect on the compressive strength (qu) and tensile strength (qt) of sand, the ratio of qt/qu remained constant at each tensile position, independent of other factors.

Author Contributions: Formal analysis, Y.P.; investigation, J.G.; resources, X.H.; software, Z.P.; writing—review and editing, C.C. and L.W. All authors have read and agreed to the published version of the manuscript.

Funding: This work was supported by the National Natural Science Foundation of China (Grant Number 41402260) and by the Fundamental Research Funds for the Central Universities, China University of Geosciences (Wuhan) (No.2020019).

Conflicts of Interest: The authors declare no conflict of interest.

References

1. Van den Heede, P.; de Belie, N. Environmental impact and life cycle assessment (LCA) of traditional and 'green'concretes: Literature review and theoretical calculations. *Cem. Concr. Compos.* **2012**, *34*, 431–442. [CrossRef]
2. Salas, D.A.; Ramirez, A.D.; Rodríguez, C.R.; Petroche, D.M.; Boero, A.J.; Duque-Rivera, J. Environmental impacts, life cycle assessment and potential improvement measures for cement production: A literature review. *J. Clean. Prod.* **2016**, *113*, 114–122. [CrossRef]
3. Miller, S.A.; Moore, F.C. Climate and health damages from global concrete production. *Nat. Clim. Chang.* **2020**, *10*, 439–443. [CrossRef]
4. Vimercati, L.; Cavone, D.; Delfino, M.C.; Caputi, A.; de Maria, L.; Sponselli, S.; Corrado, V.; Ferri, G.M.; Serio, G. Asbestos air pollution: Description of a mesothelioma cluster due to residential exposure from an asbestos cement factory. *Int. J. Environ. Res. Public Health* **2020**, *17*, 2636. [CrossRef]
5. Zhou, Z.W.; Alcalá, J.; Yepes, V. Bridge carbon emissions and driving factors based on a life-cycle assessment case study: Cable-stayed bridge over Hun He river in Liaoning, China. *Int. J. Environ. Res. Public Health* **2020**, *17*, 5953. [CrossRef]
6. Gates, W.P.; Bouazza, A.; Ranjith, P.G. Hydraulic conductivity of biopolymer-treated silty sand. *Géotechnique* **2009**, *59*, 71–72.
7. Muguda, S.; Booth, S.J.; Hughes, P.N.; Augarde, C.E.; Perlot, C.; Bruno, A.W.; Gallipoli, D. Mechanical properties of biopolymer-stabilised soil-based construction materials. *Géotechnique Lett.* **2017**, *7*, 309–314. [CrossRef]
8. Chen, C.; Wu, L.; Harbottle, M. Influence of biopolymer gel-coated fibres on sand reinforcement as a model of plant root behavior. *Plant Soil* **2019**, *438*, 361–375. [CrossRef]
9. Soldo, A. Biopolymers for Enhancing the Engineering Properties of Soil. Ph.D. Thesis, Auburn University, Auburn, AL, USA, 2020.
10. Pacheco-Torgal, F.; Ivanov, V.; Karak, N.; Jonkers, H. *Biopolymers and Biotech Admixtures for Eco-Efficient Construction Materials*; Woodhead Publishing: Cambridge, UK, 2016.
11. Chang, I.; Prasidhi, A.K.; Im, J.; Shin, H.-D.; Cho, G.-C. Soil treatment using microbial biopolymers for anti-desertification purposes. *Geoderma* **2015**, *253*, 39–47. [CrossRef]
12. Soldo, A.; Miletić, M.; Auad, M.L. Biopolymers as a sustainable solution for the enhancement of soil mechanical properties. *Sci. Rep.* **2020**, *10*, 1–13. [CrossRef]
13. Chen, C.; Wu, L.; Perdjon, M.; Huang, X.; Peng, Y. The drying effect on xanthan gum biopolymer treated sandy soil shear strength. *Constr. Build. Mater.* **2019**, *197*, 271–279. [CrossRef]
14. Dehghan, H.; Tabarsa, A.; Latifi, N.; Bagheri, Y. Use of xanthan and guar gums in soil strengthening. *Clean Technol. Environ. Policy* **2019**, *21*, 155–165. [CrossRef]
15. Chen, R.; Zhang, L.; Budhu, M. Biopolymer stabilization of mine tailings for dust control. *J. Geotech. Geoenvironmental Eng.* **2013**, *139*, 1802–1807. [CrossRef]
16. Cundall, P.A.; Strack, O.D. A discrete numerical model for granular assemblies. *Geotechnique* **1979**, *29*, 47–65. [CrossRef]
17. *Particle Flow code in Two Dimensions (PFC2D)*; Itasca Consulting Group, Inc.: Minneapolis, MN, USA, 2008.
18. Mak, J.; Chen, Y.; Sadek, M. Determining parameters of a discrete element model for soil–tool interaction. *Soil Tillage Res.* **2012**, *118*, 117–122. [CrossRef]
19. Potyondy, D.O.; Cundall, P. A bonded-particle model for rock. *Int. J. Rock Mech. Min. Sci.* **2004**, *41*, 1329–1364. [CrossRef]
20. Tang, Y.; Xu, G.; Lian, J.; Yan, Y.; Fu, D.; Sun, W. Research on simulation analysis method of microbial cemented sand based on discrete element method. *Adv. Mater. Sci. Eng.* **2019**, *2019*, 1–13. [CrossRef]
21. Chen, R.; Ding, X.; Ramey, D.; Lee, I.; Zhang, L. Experimental and numerical investigation into surface strength of mine tailings after biopolymer stabilization. *Acta Geotech.* **2016**, *11*, 1075–1085. [CrossRef]

22. Sadek, M.A.; Chen, Y.; Liu, J. Simulating shear behavior of a sandy soil under different soil conditions. *J. Terramechanics* **2011**, *48*, 451–458. [CrossRef]

23. Nandanwar, M.; Chen, Y. Modeling and measurements of triaxial tests for a sandy loam soil. *Can. Biosyst. Eng.* **2017**, *59*, 2.1–2.8. [CrossRef]

24. Pourmand, S.; Chakeri, H.; Sharghi, M.; Ozcelik, Y. Investigation of soil conditioning tests with three-dimensional numerical modeling. *Geotech. Geol. Eng.* **2018**, *36*, 2869–2879. [CrossRef]

25. García-Ochoa, F.; Santos, V.E.; Casas, J.A.; Gómez, E. Xanthan gum: Production, recovery, and properties. *Biotechnol. Adv.* **2000**, *18*, 549–579. [CrossRef]

26. Plank, J. Applications of biopolymers in construction engineering. In *Biopolymers Online*; Wiley-VCH VerlagGmbH & Co. KGaA: Weinheim, Germany, 2005.

27. Lee, S.; Chang, I.; Chung, M.-K.; Kim, Y.; Kee, J. Geotechnical shear behavior of Xanthan Gum biopolymer treated sand from direct shear testing. *Geomech. Eng.* **2017**, *12*, 831–847. [CrossRef]

28. Mehranpour, M.H.; Kulatilake, P.H.S.W. Improvements for the smooth joint contact model of the particle flow code and its applications. *Comput. Geotech.* **2017**, *87*, 163–177. [CrossRef]

29. Feng, K.; Montoya, B.; Evans, T. Discrete element method simulations of bio-cemented sands. *Comput. Geotech.* **2017**, *85*, 139–150. [CrossRef]

30. Cao, R.H.; Cao, P.; Lin, H.; Ma, G.W.; Fan, X.; Xiong, X.G. Mechanical behavior of an opening in a jointed rock-like specimen under uniaxial loading: Experimental studies and particle mechanics approach. *Arch. Civ. Mech. Eng.* **2018**, *18*, 198–214. [CrossRef]

31. Chang, I.; Im, J.; Cho, G.-C. Introduction of microbial biopolymers in soil treatment for future environmentally-friendly and sustainable geotechnical engineering. *Sustainability* **2016**, *8*, 251. [CrossRef]

32. Patel, A.K.; Mathias, J.-D.; Michaud, P. Polysaccharides as adhesives. *Rev. Adhes. Adhes.* **2013**, *1*, 312–345. [CrossRef]

33. Lu, N.; Wu, B.; Tan, C.P. Tensile strength characteristics of unsaturated sands. *J. Geotech. Geoenvironmental Eng.* **2007**, *133*, 144–154. [CrossRef]

34. Hamzah, M.O.; Yee, T.S.; Golchin, B.; Voskuilen, J. Use of imaging technique and direct tensile test to evaluate moisture damage properties of warm mix asphalt using response surface method. *Constr. Build. Mater.* **2017**, *132*, 323–334. [CrossRef]

35. Carmona, S. Effect of specimen size and loading conditions on indirect tensile test results. *Mater. Constr.* **2009**, *59*, 7–18. [CrossRef]

36. Chen, C.; Wu, L.; Harbottle, M. Exploring the effect of biopolymers in near-surface soils using xanthan gum-modified sand under shear. *Can. Geotech. J.* **2020**, *57*, 1109–1118. [CrossRef]

37. Peters, J.F.; Muthuswamy, M.; Wibowo, J.; Tordesillas, A. Characterization of force chains in granular material. *Phys. Rev. E Stat. Nonlinear Soft Matter Phys.* **2005**, *72*, 041307. [CrossRef] [PubMed]

38. Lévy-Véhel, J. *Fractals in Engineering*; Springer: Berlin/Heidelberg, Germany, 2005.

39. Chang, I.; Jooyoung, I.; Prasidhi, A.K.; Cho, G.-C. Effects of Xanthan gum biopolymer on soil strengthening. *Constr. Build. Mater.* **2015**, *74*, 65–72. [CrossRef]

40. Mitchell, J.K.; Soga, K. *Fundamentals of Soil Behavior*; John Wiley & Sons: Hoboken, NJ, USA, 2005.

41. Consoli, N.C.; Johann, A.D.R.; Gauer, E.A.; Santos, V.R.d.; Moretto, R.L.; Corte, M.B. Key parameters for tensile and compressive strength of silt–lime mixtures. *Géotechnique Lett.* **2012**, *2*, 81–85. [CrossRef]

Publisher's Note: MDPI stays neutral with regard to jurisdictional claims in published maps and institutional affiliations.

International Journal of
Environmental Research and Public Health

Article

Techno-Cultural Factors Affecting Policy Decision-Making: A Social Network Analysis of South Korea's Local Spatial Planning Policy

Eun Soo Park [1] and An Yong Lee [2,*]

[1]   Department of Architecture, Sahmyook University, Seoul 01795, Korea; espark@syu.ac.kr
[2]   Department of Smart Factory, Korea Polytechnics, Incheon 21417, Korea
*   Correspondence: adragon@kopo.ac.kr; Tel.: +82-32-510-2292

Received: 28 September 2020; Accepted: 19 November 2020; Published: 25 November 2020

Abstract: Increasing interest in various local construction forms necessitate examining its link to human life. Construction culture should be adapted and applied to the contemporary context to create a harmonious coexistence with diverse local cultures and to strengthen regional sustainability, avoiding the rigid, one-dimensional local construction development. Thus, this study aims to analyze the factors of influence needed for policy decision-making at the local spatial planning stage, with regional technologies and cultural contents from a convergent perspective taken into consideration. This study derived tangible and intangible policy decision-making factors during the spatial planning stage using text mining analysis. Additionally, social network analysis was also used to seek multi-angle correlations among factors. Through big data analytics, 16 key decision-making contents in the spatial planning stage were derived, with 'regional development, urban policy' as most influential. Such a result indicates the need for regional and urban policy engagement with strategic development from a holistic perspective—in view of socio-cultural relations and forms of change—and local perceptions of spatial value and significance affecting decision-making in the local spatial planning stage (LSPS). Understanding the decision-making process in the spatial planning stage requires a holistic approach with both visible technological factors (structure, form, and construction method) and invisible cultural factors (ways of life projected during space formation, zeitgeist, religion, learning, and art) included.

Keywords: local spatial planning stage; policy decision-making; social network analysis; convergence analysis; text mining analysis

1. Introduction

Along with rapid changes in the Information Age, the culture of each region representing a given era is becoming even more segmented and complex. Increasing cultural diversity gradually turns regional characteristics, factors, and influence, (for example, invisible factors, such as local spirit, ideas, customs and traditions, and regional characteristics of economic, policy, geography, and environmental factors), which used to be understood from a fragmentary perspective in the past, more difficult in terms of any single field. Likewise, predicting and responding to future changes is also becoming even more challenging. Advances in information technology for knowledge delivery have led to the creation or destruction of diverse cultural content forms, many of which cannot be explained from a single viewpoint.

Therefore, the regional development sector also needs to consider the diversity of local culture in the establishment of local construction policies by adopting a broader perspective and approach to the preservation, development, and management of historical or cultural assets, including natural resources, unique to each region. Currently, Government policies in place on regional development operate from a technological-based regional development context. According to the current Ministry of Land, Infrastructure and Transport Act, 46 factors for impact checklists are provided for decision-making in the local spatial planning stage. Among these factors, only two cultural factors are considered (the review of the humanities and environment, the inclusion of cultural heritage preservation values). As a cultural and regional development framework, this would be a good model of balanced regional development that can draw on the regional identity and self-esteem of local residents, as well as maximize the use of traditional contents and local traits that are passed on in each region. Accordingly, regional construction may draw out unique innovations and development of diverse cultural contents if policy measures at the government level are taken into consideration from multiple angles in the local spatial planning stage (LSPS).

In particular, although specific details needed to establish a local spatial plan will differ according to the scale and requirements of construction in each region, various tangible and intangible factors generally connected with technological factors relevant to construction, such as the characteristics of each region, direction, and indicators set for the plan, adoption of a spatial structure, conservation and management of the environment, infrastructure, as well as economic, social, and cultural growth exist [1]. Various intangible elements—historical facts, political background, cultural trends, and climate conditions of a given region—pervade that region in different forms as part of the local life and culture.

Julia and Michael [2] have investigated the methods of analyzing the contents of oral traditions, tales, etc., handed down from the local private sector, local participation and observation, study of official organizations, community research designs, and field surveys of humans and their culture.

In 2006, the Korea Rural Development Administration (RDA) [3] conducted research on amenities in urban and rural areas. This study developed the discovery and inventory of local amenity resources, derived important resources and classification systems, and conducted the valuation of resources. Based on this, the factors and characteristics of vitalization of local resources were analyzed to develop regional planning guidelines and technologies and to propose policy proposals and commercialization measures.

For the purposes of this study, region type refers to a special space where particular traditions and cultures have taken shape and are passed on, and the flow of events in accordance with the local form of culture comprise the continuous production of time. In this context, 'region' signifies a living space where humans seek to lead their daily lives imbued with happiness [4]. Additionally, humans are members of local communities that produce the cultures adapted to various conditions and environments and subsequently passed them on to create their future. This is similar to Carl Sauer's theoretical approach. In 1925, Carl Sauer defined a cultural landscape as a natural landscape that had been modified by a cultural group. The cultural landscape will change itself according to the conditions of a given culture and over time, experience development, go through stages or reach almost the last stage of the development cycle [5].

In line with the abovementioned culture flow, construction policies in each region have also coexisted in relation to various factors. In recent regional development, the aesthetic and cultural elements have a great influence on the urban image and cultural industry. It is necessary to define the harmony between buildings and surrounding spaces and the consideration of urban landscape as public elements. It is also necessary to implement construction policies that emphasize the importance of the quality of local development in the environment and culture. Regions thus become points of contact for the symbiosis of the past and present taking place in the space that humans maintain. The present study focuses on identifying such points of contact through cultural factors that can be used to prepare for the future by looking back on the past and reflecting on the present. Technological and cultural factors indigenous to a region and relevant to the establishment of local construction policies are thus analyzed to examine the diverse policy-related local factors for promoting the coexistence of technology and culture relevant to decision-making in LSPS.

Therefore, the purpose of this study is to examine the previously mentioned cultural factors of the local space as well as the technical development conditions previously reflected in the LSPS decision-making process. Additionally, this study analyzes which of the cultural factors analyzed has an effect on the technical factors. Through this, the research can contribute to the creation of regional development as a future-oriented space element into a living space where humans seek to lead their daily lives imbued with happiness, rather than just infrastructure development.

As essential data that shed light on the correlations relevant for decision-making, the study results can be utilized to derive policy decision-making indicators through the application of data analysis methodology and the convergence perspective.

2. Literature Review: Construction Planning Stage and Local Spatial Planning Stage

In general, construction planning is a perfunctorily part of the predesign stage and is comprised of tasks such as setting the course of the design and conducting a feasibility study for the construction project. In reality, however, it can be designated as a comprehensive coordination task for the entire design process, from project conceptualization to actual construction work [6]. In particular, the construction planning stage is a very important period that affects the calculation of about 80% of the total project cost. Even small-scale construction projects can cause defects in actual construction if they do not have a clear concept in the planning stage [7]. Therefore, construction planning is not confined to the predesign stage but extends from the employer's project proposal stage to the stage just before project implementation. Likewise, the planning task comprehends in its scope the preliminary spatial plan, working design, and actual spatial plan [8].

The main task of the planning stage can be the establishment of the basic course of the project at a certain level, similar to the concept establishment process in computer programming [9]. In other words, it may be referred to as the stage for carrying out tasks such as decision-making by the employer or project participants, conducting a feasibility study for the project, specifying the requirements for the target object, and initial space planning [10]. It is also the stage wherein initial project information is generated with the project launch, providing basic information such as the project concept, aim, intended use, and scale. Generally, such information includes site environment, marketability, project-related requirements (scale, shape, space, and others), and initial space plan, which are transformed into a construction model by going through the stages of concept design, working design, actual construction, and maintenance. In a broader sense, the planning stage makes up the design establishment process.

2.1. Classification of the Characteristics of the Local Spatial Planning Stage and Decision-Making Targets

Accurate decision-making in the spatial planning stage may vary greatly depending on the application characteristics of the planning stage. Therefore, it is necessary to consider the timing and purpose of the application that reflects the various characteristics of the planning phase [11]. While the spatial planning stage has the same overall analytic framework across various cases, there are differences in the core analytic tasks based on the project's development domain, promotion method, and type [12]:

- Scale of the project development domain. Relevant regulations established by the Ministry of Land, Infrastructure, and Transport (MOLIT), distinguish domains according to the scale such as metropolitan, city, county, district, and others.
- Project promotion method and developer such as private and government-funded investment.
- Project type (distinguished by work type and function), including size and classification of the social infrastructure.

These variations thus imply that planning stage tasks have different characteristics with regard to applied items and component technologies according to the project's target object [13]. The spatial planning stage program may be referred to as a process involving a systematic interpretation of the organizational, group, and individual aims and duties derived from various elements such as activity, human, and equipment relationships through relevant programming.

2.2. Decision-Making Policy System

As the regions covered in this study are social living spaces, decision-making in the spatial planning stage is understood in terms of valuable spaces where cultural preservation and technological development are harmonized during the promotion of local construction. Although specific details required for decision-making in the LSPS will differ according to the construction scale and requirements, various factors are generally connected with technological factors relevant to construction such as the characteristics of each region, direction, and indicators set for the plan, adoption of a spatial structure, conservation, and management of the environment and infrastructure, as well as the economic, social, and cultural growth.

In this study, the selected local spatial plans were divided into regional plans and sectoral plans, following the Framework Act on the National Land. Figure 1 shows how the said Act is divided into sublevel plans.

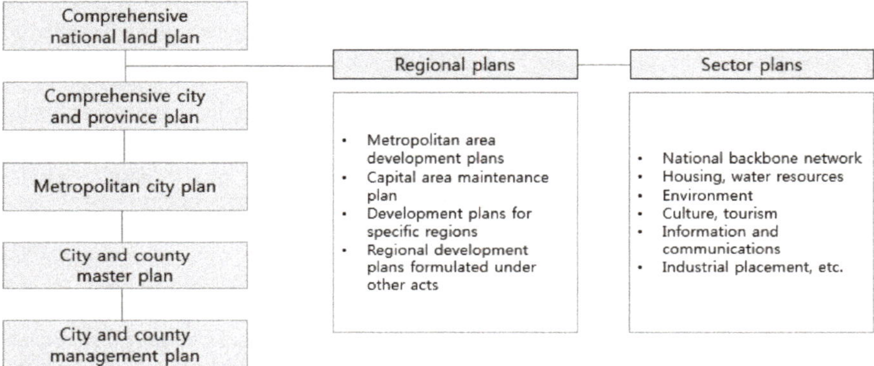

Figure 1. National land space planning system.

2.3. Decision-Making Agents

According to enforcement procedures for the relevant laws [14], the key policy-makers in the national land space planning system are the heads of local governments, and the feasibility study for a given plan and deliberation thereof are generally conducted through an urban planning committee, which is generally comprised of experts and representatives of residents. Such committees include the Central Urban Planning Committee, which deliberates on large-scale urban development projects, as well as their local counterparts that deliberate on major projects involving local governments in the national land-space planning system [14]. Figure 2 shows how these committees generally implement their master plan.

These committees aim to prevent the rigid operation of urban planning that regulates private property rights for the public's benefit, as well as ensure flexibility appropriate to local circumstances through the management of deliberation and counseling on proposed urban plans. Recently, however, questions regarding the professionalism and fairness of the local urban planning committee members in their evaluation arose, based on the result of the survey on the operation of the urban planning committee and architecture committee of local governments [15]. The main issues involved are subjective judgments based on qualitative factors, insufficient data on licensing criteria, lack of review and review checklists, and excessive documents that are not relevant. Moreover, the lack of consistency in the deliberation standards employed in each local government has led to confusion in local land use. Additionally, there are no consistent and explicit criteria to adjudicate among the contradictory opinions of different committees. In this study, organizations in charge of the final deliberations in spatial planning, such as the local urban planning committees, are thus selected as target agents.

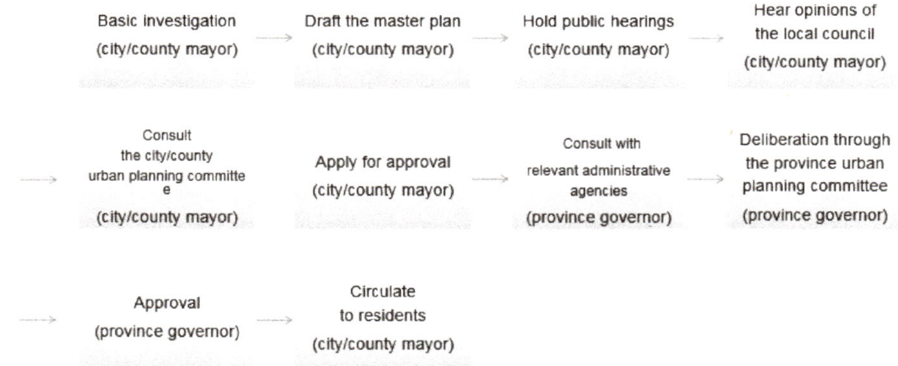

Figure 2. City/county master plan implementation procedure.

3. Materials and Methods

First, key search terms were selected and analyzed through consultations via interviews and e-mail queries with two practical experts in construction work, two experts in construction management, and two experts in the field of spatial culture from 1 May to 15 May 2020. Research feedback was received through data revision and supplementation. The selected search terms were used to collect data from a total of 1705 research papers in Korean academic journals from 1970 to 2020 [16]. Then, data from Korean MOLIT (Ministry of Land, Infrastructure, and Transport, the Korean government) news articles and KERIS (the Korea Education and Research Information Service) research papers were collected, and the top 50 keywords were derived and subsequently categorized. The Delphi method was used to derive the major factors affecting the research contents, which were subsequently placed in a matrix diagram. The Delphi method of this study is a decision-making methodology that effectively collects expert opinions through several surveys and interviews. This study conducted three rounds of open expert surveys to advance the convergence opinions of technology and cultural elements. The three rounds of Delphi methods have resulted in the final 16 cultural and technological factors being derived through examination, supplementation, and review. Expert opinions on results to derive more objective information on factors influencing decision-making in the LSPS were also collected. NetMiner 4.0 (Cyram, Gyeonggido, South Korea) was then used for the social network analysis of key contents to academic research data. Subsequent sections explain in more detail how the study proceeded.

3.1. Text Mining Analysis

Big data refers to the construction of a data system using hardware such as storage media or servers to process huge amounts of data. On the contrary, this study intends to focus on the analysis of stored data to produce valuable data through text mining analysis, which extracts valuable information from texts comprising unstructured data. It is an effective method for structuring data found in various texts and documents.

In this study, R programming (release 3.6.3, The R Foundation, Vienna, Austria), a programming language and software environment for statistical computing, was used to extract important information from unstructured texts of academic papers and news articles, as well as analyze related contents [17]. Figure 3 shows the text mining framework adopted in this study.

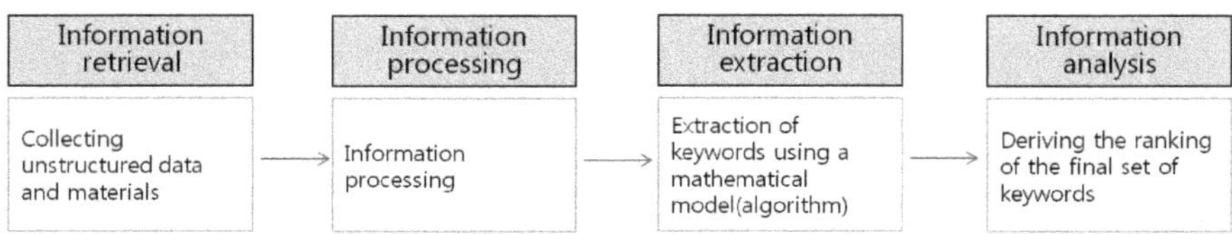

Figure 3. Text mining analysis.

For keyword derivation, words found in the collected data are first extracted, stop words are removed, and synonyms are then replaced by representative words. Correlations among these words were subsequently analyzed. In this study, the programming language and software suite R (version 3.6.3) was used to collect, organize, and analyze data. KoNLP (Korean Natural Language Process toolkit, Heewon Jeon, Seoul, South Korea) and word cloud packages for R were used for data analysis. Derived techno-cultural factors from these keywords and research data on decision-making in the LSPS were then analyzed.

Keyword extraction used in this study utilized R programming package, and a dictionary provided by KoNLP was used for word extraction. The process of text mining analysis was performed in the order of noun extraction, frequency table data frame creation, limiting word setting, and word-specific extraction. The source code used for text mining is stated in Appendix B.

3.1.1. Collection and Extraction of Data Related to Decision-Making in LSPS

The scope of this study region is a special space for places inherited by the formation of traditions and cultures for humans. This is a continuous result of time made by the flow of events in accordance with the cultural types in the region. Regions mean a place of life where humans want to enjoy happy daily lives. Here, humans are members of the community who create culture according to various conditions and circumstances and inherit it to create the future. Regional development policies have also been co-existing in this cultural trend due to various influences. Therefore, this study focuses on looking back on the past and finding a point of contact as a cultural influence factor that reflects the present.

Search terms selected for this study were 'construction', 'civil works', 'national land', and 'city', along with the basic terms 'culture' and 'region'. Table 1 shows how these terns were categorized and the corresponding number of extracted words. These data were collected through the Research Information Sharing Service (RISS), provided by the Korea Education and Research Information Service (KERIS).

Table 1. Collected Research Papers.

Search Category	Number of Academic Papers	Total No. of Extracted Words
Construction + Culture + Region	471	78,112
Civilworks + Culture + Region	34	4490
National land + Culture + Region	239	34,778
City + Culture + Region	961	112,166

Press releases from the Urban Policy Division of the Ministry of Land, Infrastructure, and Transport (MOLIT) were utilized as additional data sources. The division is in charge of all work related to regional national land development. The said press releases provide clear statements of the government policies, indicating the most important issues in local spatial planning in the course of national land development. Therefore, they contain important keywords related to local construction policies and national policy as well as technological, legal, and institutional terminology that can be used as basic data. Accordingly, 144 relevant press releases from the Urban Policy Division posted on the MOLIT website, Government 3.0, from 21 February 2006 to 20 February 2020 were used as a basis for information on factors relevant to decision-making in the spatial planning stage.

The data from this study were collected in two methods. First, information data of academic papers were collected in XML (Extensible Markup Language) format through the API (Application Programming Interface) service provided by RISS. Second, the press release data of the government portal bulletin board of the Ministry of Land, Infrastructure, and Transport (MOLIT) was collected through R programming-based web crawling.

3.1.2. Keyword Derivation and Classification through Text Mining Analysis

Text mining was used to analyze keywords from the collected 1705 research papers and 144 press releases. Using software suite R to extract the synonyms and stop words, at least 30 of the most frequently repeated keywords were derived in each search category (Table 2), thus arriving at a more accurate assessment of the influence of the associated factors.

Table 2. Top 30 Repeated Keywords.

Search Category (Papers)	Top 30 Repeated Keywords
Construction + Culture + Region (471 papers)	region, society, culture, construction, city, economy, problem, growth, relationship, state, nation, policy, history, persons with disabilities, politics, development, space, change, education, China, government, tradition, unification, promotion, diversity, world, strategy, security, the people, citizens
Civil works + Culture + Region (34 papers)	region, culture, informatization, scenic view, change, development, space, railroad, history, visualization, heritage, construction, bridge, place name, ground, travel, nature, characteristics, distinctive character, building, technology, building information modeling, value, element, location, computer, civil works, economy, ancient burial mound, planning
National Land + Culture + Region (239 papers)	culture, development, region, economy, society, environment, city, growth, scenic view, state, government, innovation, national land, relationship, corporation, analysis, cluster, world, space, technology, policy, change, assessment, the people, project, region, tourism, management, plan, industry
City + Culture + Region (961 papers)	city, region, culture, space, society, design, development, environment, history, tradition, diversity, economy, town, growth, scenic view, change, industry, public, identity, purpose, woman, policy, project, value, place, life, residents, festival, Busan, creation

Among the derived keywords, those expressing the same content, such as those involving the mixed-use of Korean and English or different terms having the same meaning, were merged together. Related keywords were thus grouped first into sets and sorted according to a classification scheme that reflects the opinions of experts, before the analysis [18]. Fifty of the most frequent keywords in the data collected from all target papers and articles were subsequently derived from investigating the cultural and technological factors relevant to decision-making in the LSPS.

Given the nature of text mining analysis, the higher-ranked keywords in Table 2 above comprise the most basic terminology for decision-making in the spatial planning stage. Keywords with a frequency of at least 300, such as culture, space, region, history, and policy, have cultural and technological connotations. In particular, words with cultural meaning—various factors associated with the region to be considered for decision-making in the spatial planning stage—or those that represent social phenomena, such as practical change and diversity, were derived as keywords related to policy considerations in the decision-making of the spatial planning stage.

Based on the 50 derived keywords, the classification of factors was carried out through the preceding study of extant literature and round one of the consultations with experts. Taking into account the diversity of decision-making in the spatial planning stage, the cultural and technological factors were subsequently classified.

For the classification of cultural factors, basic codes forming the cultural basis of decision-making in the spatial planning stage—local ethos, thought and religion—are shaped by the diverse cultural backgrounds and local history of a given region. Notably, these are relevant to the regional development background, urban design, planning, and policy decision-making among others.

Regarding the classification of technological factors, 10 factors relevant to decision-making in the LSPS were identified through consultation with experts, based on this study's aims and an analysis of the deliberative criteria on technologies given in the Framework Act on the National Land.

The cultural and technological factors relevant to decision-making in the LSPS may be regarded as the sum of mutually opposed concepts, but they can all be regarded as spatial contents created through various interrelationships based on local construction. For example, thought and tradition greatly influenced specific policies and technological advances. Additionally, information is provided to locals as specific systems and policies through a harmonious application of the relevant facilities and technologies. The concept of scenic view as a practical, aesthetic space, which is not found in extant discussions of cultural and technological factors, is also important; therefore, a separate keyword that can reflect this concept was introduced.

3.2. Delphi Research Design

The Delphi method is a decision-making technique for generating a consensus among experts through several rounds of the survey. This method is used when there are few materials to refer to, and expert opinions can be used as important information. Thus, it was deemed appropriate to re-examine the results derived in this study which supplement the data. As the current study's aims lie in analyzing the content, not just in the field of regional policy and construction but in the wide-ranging field of policy decision-making through the inclusion of a sociocultural approach, the Delphi technique would be effective as it allows for a more convergent approach that arrives at a consensus by aggregating diverse opinions from a group [19]. Here, the Delphi technique proceeded through individual meetings and three rounds of open-ended questionnaires and e-mail questionnaires sent to a panel of experts from construction, environment, work, design, and management to analyze relevant technological factors and to experts in spatial culture, spatial philosophy, and regional policy for cultural factors. See Table 3 for the detailed stages of the research design.

3.2.1. Derivation of Factor Indicators through Delphi Analysis

Table 4 shows the indicators derived through the Delphi survey. Here, 16 indicators of technological factors and 31 indicators of cultural factors were derived. The 16 indicators of techno-cultural factors (Table 5) were subsequently examined to derive the convergent factors.

Table 3. Delphi Research Design.

Stage		Content							
				Construction				Spatial	Regional
		Field of expertise	Environment	Work	Design	Management	Culture	Philosophy	Policy
Preliminary basic research	Task	Examining keywords and data for techno-cultural factor analysis							
	Experts								
		Number of participants	1	1		2	1	1	
Delphi research	Task	Examining the derivation of indicators for and assignment of weights to techno-cultural factors for social network analysis							
	Round 1	Delphi panel selection Classification of data and examination of indicators Open-ended questionnaire, interview, e-mail questionnaire	1	1	3	2	1	2	1
	Round 2	Supplementation of derived indicators from round 1 Open-ended questionnaire, interview, email questionnaire	1	2	1	2	1	1	2
	Round 3	Review of factor indicators and assignment of weights for data accuracy Open-ended questionnaire, interview, e-mail questionnaire	1	2	1	3	1	1	1

Table 4. Indicators derived through the Delphi Survey.

Round	Survey Content	Derived Indicators			Participating Experts
		Technological Factors	Cultural Factors	Techno-Cultural	
1st	Examination of factor indicators	30	45	17	11
2nd	Supplementation of factor indicators	18	33	17	10
3rd	Review of factor indicators/weight assignment	16	31	16	10

Table 5. Factor Indicators derived through three Rounds of Delphi Survey.

Technological Influence	Cultural Influence		Techno-Cultural Factors
regional thought, philosophy, and values in an era	region	construction	coherence with related plans
regional development, urban policy	culture	China	local environment
commercial economy, market formation	society	industry	natural environment
nature, surroundings, scenic view	space	globalization	future conditions
tradition, the character of an era	development	project feasibility	resident environment
remains, preservation, history	economy	strategy	infrastructure
coherence with higher-level plan	growth	value	impact on surrounding areas
consideration of environmental pollution, destruction	history	scenic view	site scale
future conditions	environment	government	finances, procurement
infrastructure, attached facilities	change	identity	public contribution
impact on surrounding areas	policy	town	suitability of development density
finances, procurement	design	education	placement of facilities
expansion of welfare, facilities	characteristics	residents	scenic view plan
design, planning	diversity	plan	environmental pollution
transport access	state	social perception	transport convenience
expansion of disaster-prevention, safety facilities	tradition		safety facilities

Mainly, the Delphi panel experts examined the validity of the preliminary factor indicators to check whether the research aims and direction of the analysis were adequately reflected in these indicators. Taking the experts' consensus into account, three Delphi survey rounds were conducted to secure an adequate level of validity for the factor indicators.

3.2.2. Matrix Construction and Weight Assignment: Correlation Analysis

After deriving the techno-cultural factor indicators, a matrix to determine the actual correlations among the factors was constructed. NetMiner 4.0 (Cyram, Gyeonggido, South Korea), a social network analysis program developed by Cyram, was used to analyze the matrix.

In keeping the data format of NetMiner 4.0, the matrix created in this study is comprised of a network or relational dataset that can be used to construct a meta-matrix. Network analysis can be expressed in the form of a matrix or edge list, and the matrix was selected for effective expression of the properties and relationships found in the dataset, along with the weight of influence [20].

Weights are assigned through a one-to-one arrangement of decision-making factors in the LSPS, each indicating the weight of the influence that a source factor has on a target factor. As for the matrix construction process, the derived factors were arrayed against the entire data, and their degree of influence in each instance was reviewed by the panel of experts. Weights were subsequently assigned to these degrees of influence in the course of a further review. Correlations were found as shown in Table 6.

Table 6. Correlations between Factors and Degree of Influence.

A [1] \ B [1]	Regional Thought, Philosophy, and Values in an Era ©	Regional Development, Urban Policy ©	Commercial Economy, Market Formation ©	Nature, Surroundings, Scenic View ©	Tradition, Character of an Era ©	Remains, Preservation, History ©	Coherence with Higher-Level Plan	Consideration of Environmental Pollution Destruction	Future Conditions	Infrastructure, Attached Facilities	Impact on Surrounding Areas	Finances, Procurement	Expansion of Welfare, Facilities	Design, Planning	Transport Access	Expansion of Disaster-Prevention, Safety Facilities
regional thought, philosophy, and values in an era ©	0.000	3.429	3.571	2.714	7.286	6.429	2.571	0.857	4.143	2.143	3.571	1.286	1.714	4.143	1.286	1.000
regional development, urban policy ©	2.714	0.000	6.857	5.857	2.143	4.571	4.143	6.429	4.000	6.429	6.571	3.000	3.714	4.143	6.286	4.000
commercial economy, market formation ©	2.286	6.714	0.000	2.857	2.143	1.571	2.571	3.429	4.857	4.571	3.143	5.143	3.714	4.000	5.286	4.429
nature, surroundings, scenic view ©	3.429	3.429	2.429	0.000	2.571	3.714	3.000	6.857	3.714	2.143	5.286	1.143	1.000	3.571	2.714	1.000
tradition, character of an era ©	7.143	3.143	2.857	2.429	0.000	6.714	1.571	2.429	2.143	1.571	3.571	1.000	1.143	2.429	1.714	1.000
remains, preservation, history ©	5.857	5.000	3.857	4.857	5.000	0.000	3.143	4.571	4.286	2.571	5.000	2.000	1.857	3.000	3.714	2.286
coherence with higher-level plan	1.000	5.714	6.286	3.429	3.000	3.429	0.000	3.571	5.429	6.429	5.143	4.000	3.714	3.429	3.714	3.143
consideration of environmental pollution, destruction	2.286	3.571	1.429	8.000	3.000	6.429	2.571	0.000	3.000	2.571	3.429	2.286	0.714	3.286	1.000	3.571
future conditions	2.714	6.429	4.857	3.000	0.857	3.571	3.857	3.286	0.000	4.143	5.429	3.857	5.429	4.571	5.857	4.000
infrastructure, attached facilities	1.429	3.571	2.571	1.714	0.857	0.857	4.429	2.714	3.143	0.000	5.143	3.000	4.143	2.857	5.000	2.714
impact on surrounding areas	1.143	4.857	4.143	4.571	1.857	2.286	2.714	4.000	3.429	4.000	0.000	3.857	2.429	2.857	4.000	1.143
finances, procurement	1.286	4.143	5.143	1.857	0.857	2.000	4.000	1.429	4.286	3.857	2.571	0.000	4.286	3.714	6.714	3.143
expansion of welfare, facilities	1.857	6.857	4.714	1.000	0.000	0.000	2.857	0.286	4.857	2.857	3.429	1.429	0.000	2.000	3.143	3.000
design, planning	2.857	2.857	3.143	4.000	2.286	3.143	3.143	3.143	3.857	2.571	3.143	1.714	1.143	0.000	3.000	3.000
transport access	0.000	6.857	8.000	2.286	0.429	1.143	3.000	0.000	5.000	4.429	3.857	4.571	0.000	1.000	0.000	0.714
expansion of disaster-prevention, safety facilities	0.000	3.143	2.857	1.571	0.000	1.143	2.857	3.857	5.000	3.143	0.857	0.000	0.714	3.143	0.000	0.000

[1] Directionality → (A → B, the strength and direction of an association between A and B).

In NetMiner 4.0, the matrix diagram shows the connections between factors. The cells in various shades (see Figure 4) at the intersections between factors with relative influence express the degree of influence between each factor in the matrix. Figure 4 below shows the analysis results of the techno-cultural factors' degrees of influence.

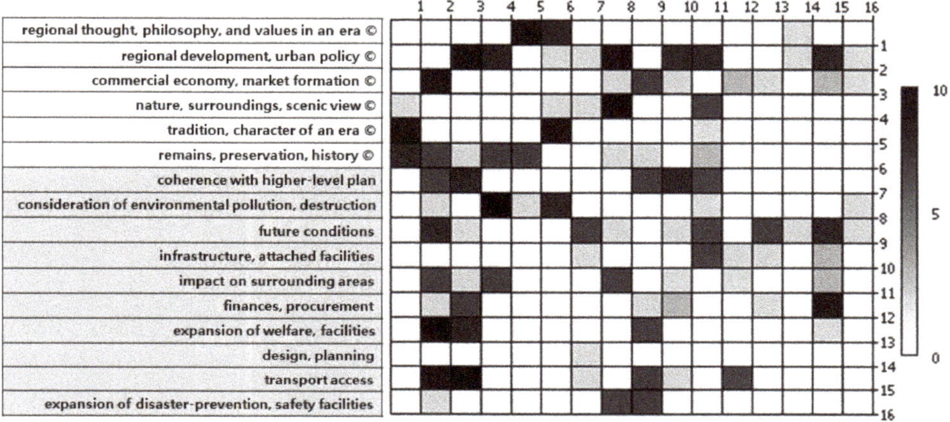

Figure 4. Matrix of technological and cultural factors.

3.3. Social Network Analysis

Combining the words 'social' and 'network', [21], a social network signifies a network of interconnected persons or objects and represents an organic network formed through the interaction of diverse individuals [22]. Social network analysis thus determines the correlational structure of individuals or groups, focusing on how entities are connected and on identifying the nature of their connective structure [23], as shown in Figure 5.

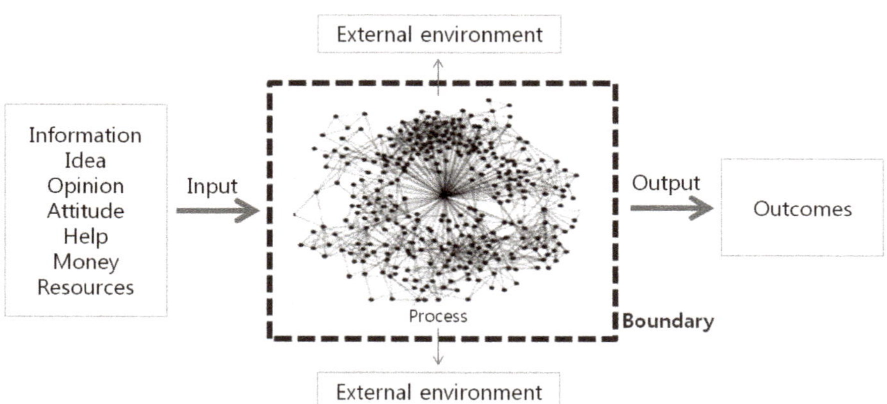

Figure 5. Network utilization in social network analysis.

As a method, social network analysis identifies and visualizes the centrality or importance of individual nodes by assigning weights to them [24]. These can subsequently be used to derive the various types of centrality—degree, closeness, betweenness, and eigenvector [25–27].

Degree centrality indicates the degree to which a node is connected to other nodes [28]. Closeness centrality expresses how close one node is to other nodes [29], and the higher the measure of this centrality, the easier it is for the node to be related to other nodes [30]. Betweenness centrality measures the degree to which one node acts as an intermediary in connecting other nodes into a network [31]. On the other hand, eigenvector centrality weighs the importance of nodes to obtain their centrality [32]. In a centrality analysis, the degree centrality is the basic measure, usually indicating the number of links connecting a given node to other nodes [33].

The equations below show how degree centrality and eigenvector centrality are used in social network analysis [34]:

$$\text{degree centrality} \quad = \quad \frac{\sum\limits_{i=1}^{n} g_i^{(1)}}{n(n-1)}$$
$$= \frac{\sum\limits_{i=1}^{n} g_i^{(2)}}{n(n-1)}, i = 1, \ldots, n \tag{1}$$

Here $g_i^{(1)}$ is the number of inbound edges to the ith entity, and $g_i^{(2)}$ is the number of outbound edges from the ith entity.

The eigenvector centrality of node x is defined in terms of the following formula:

$$\sigma_E(x) = v_x = \frac{1}{\lambda_{\max}(A)} \times \sum_{j=1}^{n} a_{jx} \times v_j \tag{2}$$

$v = (v1, v2, vn)$ is the eigenvector for the maximum eigenvalue $\lambda_{\max}(A)$ of the entire matrix, and a_{jx} represents the weight between node x and node j.

4. Results and Discussions: Correlation Analysis of Techno-Cultural Factors Affecting Policy Decision-Making through Social Network Analysis

Social network analysis of the decision-making factors in the LSPS was conducted using the constructed matrix. The 16 factors relevant to decision-making in the LSPS were analyzed in terms of degree centrality, eigenvector centrality, closeness centrality, betweenness centrality, and community structure. Visualizations of these analyses allowed for an intuitive grasp of the corresponding path models.

This study summarizes the results of social network analysis through the Appendices C–F (Tables A2–A5). Each Appendix's table shows the effects of the 16 factors derived by ranking. Each numerical value in the table represents the effect between the factors. The degree of influence with other nodes was expressed numerically around individual factors according to the analysis theme (degree centrality, eigenvector centrality, closeness centrality, betweenness centrality, and community structure). Each indicator can be interpreted that the higher the number, the easier it is to relate to other nodes.

4.1. Degree Centrality Analysis

Degree centrality involves two separate measures: in-degree centrality indicating the degree to which an entity is influenced by other factors; and out-degree centrality indicating the degree to which an entity influences other factors. The present study derived each factor's degree centrality by referring to the mean values of its in-degree and out-degree centrality.

Degree centrality is the most basic indicator of centrality in social network analysis that measures the importance of a given node in a network. A greater number of direct connections with other nodes determine a higher value of degree centrality. Previous studies used degree centrality to derive the importance of each node. However, this study cannot determine a node's influence throughout the entire network through exclusive consideration of its direct connections. To overcome this limitation, eigenvector centrality was used.

Analysis of degree centrality for the technological and cultural factors corresponding to the classification codes given in Table 7 showed that the following had the highest values in terms of in-degree centrality: 'regional development, urban policy', followed by 'future conditions', 'commercial economy, market formation', and 'impact on surrounding areas'.

Table 7. Classification of Cultural and Technological Factors: Round 1 of the Delphi Consultation.

Classification Codes		Associated Keywords
Cultural factors	ethos, thought, religion plan, system, policy society, economy geography, environment tradition, custom culture, inheritance	regional thought, philosophy, and values in an era regional development, urban policy commercial economy, market formation nature, surroundings, scenic view tradition, character of an era remains, preservation, history
Technological factors	connection with other plans environment growth potential amenities plan site scale, location development costs publicness scenic view transportation safety	coherence with higher-level plan consideration of environmental pollution, destruction future conditions infrastructure, attached facilities impact on surrounding areas finances, procurement expansion of welfare, facilities design, planning transport access expansion of disaster-prevention, safety facilities

Regarding out-degree centrality, which measures the degree to which a factor influences other factors, 'regional development, urban policy', 'future conditions', 'commercial economy, market formation', and 'impact on surrounding areas' ranked higher, following the case of in-degree centrality. 'Remains, preservation, history' and 'finances, procurement' have relatively higher out-degree centrality than in-degree centrality. The higher values of out-degree and in-degree centralities indicate a greater number of nodes that are directly connected with one another, or a greater number of factors that influence and are influenced by other factors. Moreover, network visualization analysis (Figure 6) showed that 'regional development, urban policy' has the highest weight, followed by 'future conditions', 'commercial economy, market formation', 'impact on surrounding area', and so on. Circle diagram analysis locates the core factors with a greater centrality of influence nearer the center, and it was shown that 'regional development, urban policy' has the greatest centrality.

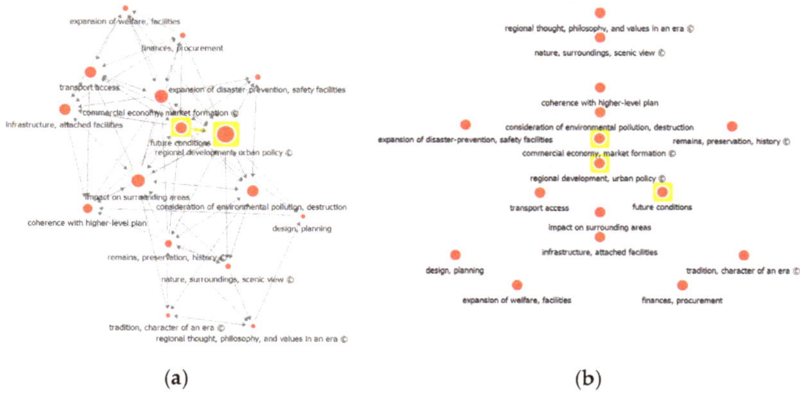

(a) (b)

Figure 6. Visualization of degree centrality analysis: (**a**) network visualization and (**b**) circle diagram.

4.2. Eigenvector Centrality Analysis

The Eigenvector centrality analysis supplemented the degree centrality analysis results. Eigenvector centrality measures the centrality of a given node by weighing the importance of other nodes connected to it. Most of the factors with greater in-degree and out-degree centrality values—'regional development, urban policy', 'future conditions', 'coherence with higher-level plan', 'impact on surrounding areas', and 'commercial economy, market formation'—also showed greater eigenvector centrality. Such results indicate that the said factors have more connections with other factors with great influence and, thus, relatively have a great influence on the entire network as well.

Analysis results for both degree and eigenvector centralities indicate high values for 'future conditions' and economic factors as items in the policy-related consensus. Thus, we can see that these factors exert great influence on matters involving regional development and social change, which can be interpreted as highlighting the importance of planned regional development. Network visualization of the eigenvalue centrality analysis (Figure 7) also indicated that 'regional development, urban policy' has the highest weight, followed by 'future conditions', 'coherence with higher-level plan', and 'impact on surrounding areas'. The circle diagram analysis locates the core factors with a greater centrality of influence nearer to the center; 'regional development, urban policy' showed the greatest centrality.

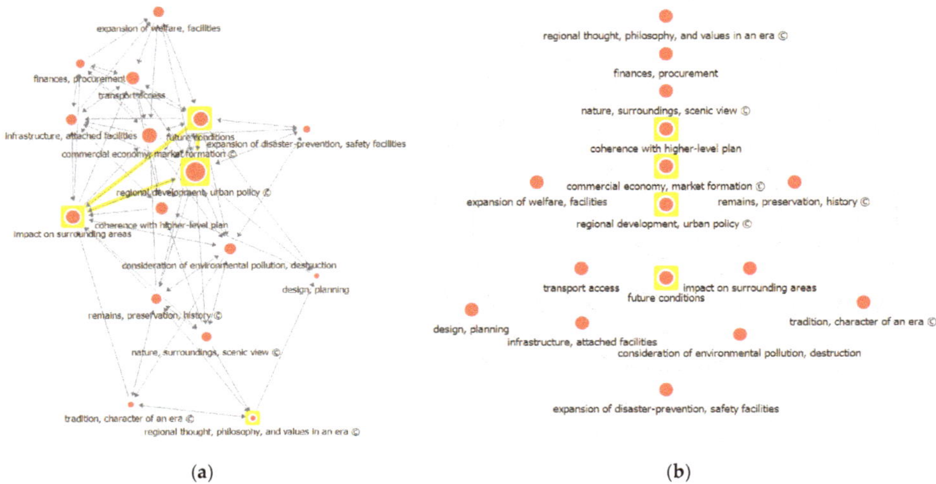

<table>
<tr><td>(a)</td><td>(b)</td></tr>
</table>

Figure 7. Visualization of eigenvector centrality analysis: (**a**) network visualization and (**b**) circle diagram.

4.3. Closeness Centrality Analysis

Closeness centrality shows the immediacy of connections between factors: the shorter the distance between a given node and others, the greater is the relationship between them. This is because a shorter distance implies greater accessibility, thus allowing relatively quick access to other nodes through that pathway than others.

Analysis results for in-closeness centrality indicated high values for 'regional development, urban policy', 'future conditions', 'commercial economy, market formation', and 'impact on surrounding areas'. These can be regarded as factors likely to be easily influenced by other factors. It is worth noting that 'remains, preservation, history' and 'commercial economy, market formation in an area' were shown to have relatively high centrality. As these are likely to be easily influenced by other factors, they may serve as important indicators for policy implications.

Analysis results for out-closeness centrality (Figure 8) also indicated high values for 'regional development, urban policy', 'future conditions', 'commercial economy, market formation', and 'impact on surrounding areas'. Note though that factors related to a region's cultural connections and historical influence—'remains, preservation, history' and 'regional thought, philosophy, and values in an era'—showed relatively high centrality. As these factors are also likely to influence other factors easily, they may serve as important indicators for social issues and implications related to decision-making in the LSPS.

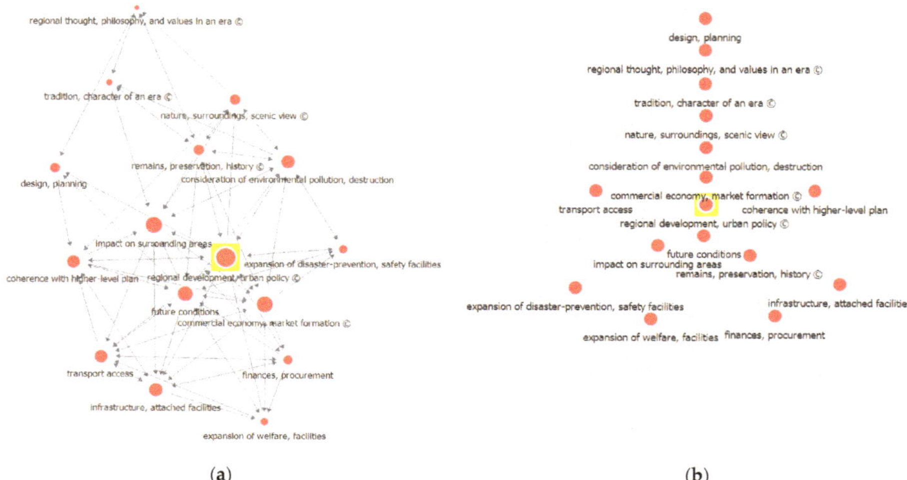

Figure 8. Visualization of closeness centrality analysis: (**a**) network visualization and (**b**) circle diagram.

Regarding the results of overall closeness centrality analysis, 'regional development, urban policy' showed the highest weight in decision-making during the LSPS, followed by 'future conditions' and 'commercial economy, market formation'. Circle diagram analysis locates the core factors with greater centrality of influence nearer to the center; 'regional development, urban policy' was at the center, along with 'coherence with higher-level plan' and 'commercial economy, market formation' possessing high scores in terms of in-closeness and out-closeness centralities.

4.4. Betweenness Centrality Analysis

Betweenness centrality indicates a given factor's intermediary role in connecting other factors. Factors frequently included in the shortest paths between other factors have high betweenness centrality, and they play a significant role in the network information flow. If such a factor is removed, it can have large repercussions on the overall flow and connectivity of the network. Betweenness centrality is thus an important measure of the intermediary role of factors.

Similar to the earlier results, analysis results for betweenness centrality indicated high values for 'regional development, urban policy', 'future conditions', and 'impact on surrounding areas'. As highly ranked factors in betweenness centrality, these occupy an important place in regulating the overall flow among the 16 factors.

Overall analysis results for betweenness centrality (Figure 9) showed that 'regional development, urban policy' has the highest weight, followed by 'future conditions' and 'impact on surrounding areas'. Circle diagram analysis locates the core factors with the greater centrality of influence nearer to the center. The yellow highlight in the diagram (see Figure 9) indicates that 'regional development, urban policy' has far greater centrality than the other factors.

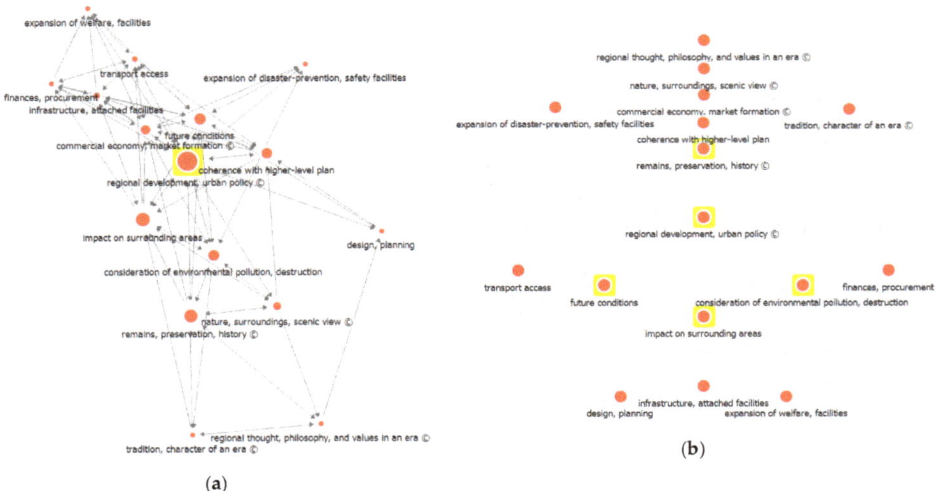

Figure 9. Visualization of betweenness centrality analysis: (**a**) network visualization and (**b**) circle diagram.

4.5. Community Structure Analysis

Community structure analysis facilitates the identification of a network structure that comprises different cohesive groups of closely connected nodes. This enables the understanding of the connective properties of a network through the characteristics of cohesive groups and the relationships obtained among them. Factors were sorted into different groups; the assigned numbers are for labeling purposes and does not indicate their ranking [35].

As shown in Table 8 and Figure 10, the results of the community structure analysis indicate that 'regional development, urban policy', was shown to have high values in centrality metrics, belonging to a separate group. 'Future conditions', 'expansion of welfare facilities', 'transport access', and 'commercial economy, market formation' adhere together in another group. In particular, three cultural factors, 'regional thought, philosophy, values in an era', 'tradition, character of an era', and 'remains, preservation, history' adhered together in a separate group. When overall relationships among the factors are considered, it was found that the important cultural factors had somewhat less centrality of influence on the main technological factors.

Note that groups of factors identified as important in the earlier centrality analysis, such as G1 (regional development, urban policy), G3 (coherence with higher-level plan), G4 (consideration of environmental pollution, destruction), and G6 (impact on surrounding areas), play a central role in the relationships among factors, albeit not as functional factors but as various convergent factors involved in the decision-making of the LSPS. They also constitute independent entities to which core meanings can be assigned.

The community structure analysis results show that various factors in the classification code have even influence on cultural influence factors. Understanding the decision-making in the LSPS is not only the visible 'technological factors' such as structure, form, and construction method but also the invisible 'cultural factors' such as the way of life projected in the process of space creation, the spirit of the times, religion and scholarship including intellectual insight into the arts.

Table 8. Community Structure Analysis Results.

Community	Included Factors ※(C): Cultural Factor
G1	regional development, urban policy (C)
G2	nature, surroundings, scenic view (C)
G3	coherence with higher-level plan
G4	consideration of environmental pollution, destruction
G5	infrastructure, attached facilities
G6	impact on surrounding areas
G7	finances, procurement
G8	design, planning
G9	expansion of disaster-prevention, safety facilities
G10	remains, preservation, history (C) + regional thought, philosophy, and values in an era (C) + tradition, character of an era (C)
G11	expansion of welfare, facilities + transport access + future conditions + commercial economy, market formation (C)

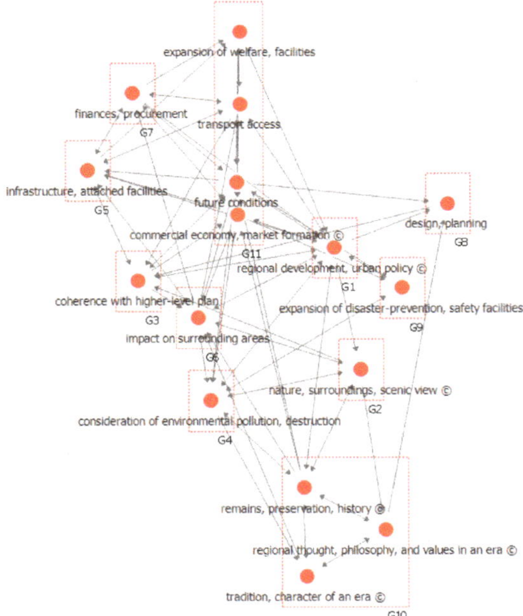

Figure 10. Visualization of community structure analysis.

In terms of social network analysis and utilization of big data conducted in this study, the use of various data-focused on local space is gradually increasing due to the development of information and communication technology for the change of local spaces and policy decision making scheduled for development. In other words, traditional urban spaces are being developed conceptually from intelligent city to smart city, with automation and the development of land communication technology. Additionally, with this trend, the local space is developing smarter by using various information.

In particular, in terms of the characteristics of convergence space data utilization in this study, various spatial information is used in planning and design coordination through social trend monitoring. Such information analysis can be used as data for policy decision-making, such as conflict management in the community and gathering opinions on public policies. In the future, big data analysis, including social network analysis and text mining analysis, is expected to be highly valuable as a methodology for regional development and spatial planning policies.

The convergent influencing factor of local construction policy is the element that comprehensively develops the elements of human lifestyle or phenomenal society, including industrial production and consumption activities from a technological and cultural point of view. This implies emotional factors that can lead to practical communication and empathy to realize cultural-heritage value creation, tourism resource development, and storytelling for the community.

Therefore, it is the product of the local space where the cultural value at the LSPS cannot be just explained as simple morphological and tangible information. Various studies that would deeply examine the existence of various intangible factors surrounding the space with the formation of technical space are thus necessary.

In another aspect, this study expects that it will be a policy material that can be used selectively in policy decision-making not only in Korea but also in areas similar to Korea. The information in this study is expected to be useful, especially as a result of cultural factors that cannot be ignored for regional development. In order to enhance this utilization, we will conduct future research by establishing an area that can be applied to the case based on the results obtained.

However, the applied methodology has limitations in comprehensively analyzing local policy decisions. There is also a limit to the application methodology to determine the local applicability of the analyzed cultural factors. In order to enhance this utilization, it is necessary to select areas that can be applied to the analyzed cases based on the results obtained and to conduct future research related to these verifications.

5. Conclusions

This study's approach derived more objective information from the data using text mining and social network analyses. To increase the objective reliability of the data, the Delphi method was used for a correlational analysis of the factors that can influence policy decision-making in the spatial planning stage.

Prior to the social network analysis, the Delphi technique and text mining data analytics were utilized to collect and investigate a wide range of convergent data. We ran a group of experts three times through the Delphi technique. We surveyed and interviewed a group of experts three times using the Delphi technique. A total of 31 experts in seven fields, including construction environment and spatial culture, were used to derive key indicators of techno-cultural factors. Social network analysis was then used to identify the connections between the factors relevant to decision-making in the LSPS from a convergent perspective, consequently determining the influence of these factors.

Through text mining analysis, 16 key factors that affect decision-making in the spatial planning stage were derived. The keyword found with the most influential factor was under the regional development and urban policy category, which is related to various local construction ideas based on: basic social issues concerning regional development such as policy agenda; mid- to long-term urban planning; and collection of various opinions for local development and policy-making. This factor is directly connected to various factors, exerting high influence, as well as being highly influenced by, other factors.

Looking at the cultural aspects of the 16 factors from a convergent perspective, 'commercial economy, market formation' and 'remains, preservation, history' were found to have a higher influence than expected. This showed that highly ranked factors are important in regulating the overall flow among the 16 factors. In particular, 'remains, preservation, history' has a higher incoming influence than outgoing influence; it is expected that this will have a relative impact on regional development.

Therefore, decision-making in regard to regional development should be re-examined to ensure that a socially conventional approach to cultural properties and historical sites in regional development is not confined to historical preservation. It was found that 'regional thought, philosophy, and values in an era' also has a relatively high level of incoming influence. Thus, it is a factor that can be likely influenced by other factors and can serve as a key indicator of policy decision-making.

Among the factors subjected to community structure analysis, 'regional development, urban policy', which exhibited high values in centrality metrics, was found to belong to a separate group, and 'future conditions', 'expansion of welfare facilities', 'transport access', and 'commercial economy, market formation' adhered together in another group. In particular, two cultural factors, namely, 'regional thought, philosophy, values in an era' and 'tradition, character of an era', adhered together in a separate group. When considering the overall relationships among the factors, it was found that the important cultural factors had somewhat less centrality of influence on the main technological factors.

It was found that the cultural factors 'regional thought, philosophy, values in an era' and 'tradition, character of an era' had little direct influence on 'regional development, urban policy', which was identified as the main influential factor. Such a phenomenon may be interpreted as indicating that cultural keywords or factors do not directly influence the core of construction policy. Such lack of cultural contents can inhibit regional growth; policy supplementation should thus be included. For a balanced expression of the diversity and uniqueness of each region, as the basis of regional development, convergent policies need to be utilized and strategic local policies that can actively promote industrial growth, tourism and local branding with an emphasis on bringing out the overall, traditional characteristics of each region should be implemented.

Author Contributions: Format analysis, E.S.P.; project administration, E.S.P.; writing—original draft preparation, E.S.P.; methodology, A.Y.L.; validation, A.Y.L.; writing—review and editing, A.Y.L. All authors have read and agreed to the published version of the manuscript.

Funding: This research was supported by Basic Science Research Program through the National Research Foundation of Korea (NRF) funded by the Ministry of Education (NRF-2016R1D1A1B03936379).

Conflicts of Interest: The authors declare no conflict of interest.

Appendix A

Table A1. Keyword frequency in the Target Research Data Set.

Rank	Keyword	Frequency	Rank	Keyword	Frequency
1	region	2827	26	identity	423
2	culture	2609	27	town	422
3	society	1516	28	education	421
4	space	1092	29	residents	420
5	development	1033	30	plan	405
6	economy	885	31	perception	403
7	growth	782	32	function	392
8	history	733	33	citizens	388
9	environment	697	34	life	380
10	change	665	35	policy measures	367
11	policy	663	36	region	367
12	design	654	37	tourism	354
13	characteristics	636	38	international	354
14	diversity	635	39	cultural property	354
15	state	615	40	nation	351
16	tradition	599	41	facilities	351
17	construction	587	42	place	349
18	China	553	43	politics	347
19	industry	527	44	technological support	345

Table A1. *Cont.*

Rank	Keyword	Frequency	Rank	Keyword	Frequency
20	world	490	45	technology	338
21	project	476	46	corporation	334
22	strategy	475	47	institutions	325
23	value	461	48	era	316
24	scenic view	457	49	promotion direction	315
25	government	450	50	system	315

Appendix B

```
   textmining.R                                                                          Source on Save                                                           Run       Source
 1  install.packages("rJava")
 2  install.packages("memoise")
 3  install.packages("multilinguer")
 4  install.packages(c('stringr', 'hash', 'tau', 'Sejong', 'RSQLite', 'devtools'), type = "binary")
 5  install.packages("KoNLP")
 6  install.packages("remotes")
 7  remotes::install_github('haven-jeon/KoNLP', upgrade = "never", INSTALL_opts=c("--no-multiarch"))
 8  library(KoNLP)
 9  install.packages("dplyr")
10  library(dplyr)
11
12  rm(sample_txt)
13  sample_txt1 <- readLines("sample1.txt")
14  sample_txt1
15  options(encoding = "UTF-8")
16
17  useNIADic()
18
19  nouns <- extractNoun(sample_txt1)
20
21  wordcount <- table(unlist(nouns))
22
23  df_word <- as.data.frame(wordcount, stringsAsFactors = F)
24
25  df_word <- rename(df_word, word = Var1, freq = Freq)
26
27  df_word <- filter(df_word, nchar(word) > 2)
```

Figure A1. Source Code for Text mining.

Appendix C

Table A2. In-Degree and out-Degree Centrality Results of Degree Centrality Analysis.

Factors ✳(C): Cultural Factor	In-Degree Centrality	Rank	Factors ✳(C): Cultural Factor	Out-Degree Centrality	Rank
regional development, urban policy (C)	0.667	1	regional development, urban policy (C)	0.733	1
future conditions	0.600	2	future conditions	0.667	2
commercial economy, market formation (C)	0.533	3	commercial economy, market formation (C)	0.600	3
impact on surrounding areas	0.533	3	remains, preservation, history (C)	0.533	4
coherence with higher-level plan	0.467	5	impact on surrounding areas	0.467	5
consideration of environmental pollution, destruction	0.467	5	finances, procurement	0.467	5
infrastructure, attached facilities	0.467	5	coherence with higher-level plan	0.400	7
transport access	0.467	5	consideration of environmental pollution, destruction	0.400	7
nature, surroundings, scenic view (C)	0.333	9	transport access	0.400	7
remains, preservation, history(C)	0.333	9	infrastructure, attached facilities	0.333	10
finances, procurement	0.333	9	nature, surroundings, scenic view (C)	0.333	10
expansion of welfare, facilities	0.267	12	expansion of welfare, facilities	0.267	12
design, planning	0.267	12	regional thought, philosophy, and values in an era (C)	0.267	12

Table A2. *Cont.*

Factors ❋(C): Cultural Factor	In-Degree Centrality	Rank	Factors ❋(C): Cultural Factor	Out-Degree Centrality	Rank
expansion of disaster-prevention, safety facilities	0.267	12	expansion of disaster-prevention, safety facilities	0.200	14
regional thought, philosophy, and values in an era (C)	0.200	15	tradition, character of an era (C)	0.200	14
tradition, character of an era (C)	0.200	15	design, planning	0.133	16

Appendix D

Table A3. Eigenvector Centrality Analysis Results.

Factors ❋(C): Cultural Factor	Eigenvector Centrality	Rank
regional development, urban policy (C)	0.48370	1
future conditions	0.32974	2
coherence with higher-level plan	0.32551	3
impact on surrounding areas	0.31794	4
commercial economy, market formation (C)	0.31218	5
transport access	0.30071	6
infrastructure, attached facilities	0.25420	7
consideration of environmental pollution, destruction	0.22437	8
nature, surroundings, scenic view (C)	0.21272	9
expansion of welfare, facilities	0.19055	10
finances, procurement	0.17505	11
remains, preservation, history	0.13545	12
expansion of disaster-prevention, safety facilities	0.08194	13
tradition, character of an era (C)	0.06341	14
regional thought, philosophy, and values in an era (C)	0.05853	15
design, planning	0.05202	16

Appendix E

Table A4. In-Closeness and out-Closeness Centrality Results of Closeness Centrality Analysis.

Factors ❋(C): Cultural Factor	In-Closeness Centrality	Rank	Factors ❋(C): Cultural Factor	Out-Closeness Centrality	Rank
regional development, urban policy (C)	0.750	1	regional development, urban policy (C)	0.789	1
future conditions	0.714	2	future conditions	0.714	2
commercial economy, market formation (C)	0.682	3	commercial economy, market formation (C)	0.682	3
impact on surrounding areas	0.682	3	remains, preservation, history (C)	0.682	4
consideration of environmental pollution, destruction	0.652	5	impact on surrounding areas	0.652	5
infrastructure, attached facilities	0.652	5	consideration of environmental pollution, destruction	0.600	6
coherence with higher-level plan	0.625	5	finances, procurement	0.600	6
transport access	0.625	5	coherence with higher-level plan	0.577	8
nature, surroundings, scenic view (C)	0.600	9	transport access	0.577	8
remains, preservation, history (C)	0.577	10	nature, surroundings, scenic view (C)	0.577	8
finances, procurement	0.556	11	regional thought, philosophy, and values in an era (C)	0.556	11
design, planning	0.556	11	expansion of welfare, facilities	0.536	12

Table A4. *Cont.*

Factors ※(C): Cultural Factor	In-Closeness Centrality	Rank	Factors ※(C): Cultural Factor	Out-Closeness Centrality	Rank
expansion of disaster-prevention, safety facilities	0.517	13	expansion of disaster-prevention, safety facilities	0.517	13
expansion of welfare, facilities	0.500	14	infrastructure, attached facilities	0.500	14
tradition, character of an era (C)	0.455	15	tradition, character of an era (C)	0.500	14
regional thought, philosophy, and values in an era (C)	0.429	16	design, planning	0.469	16

Appendix F

Table A5. Betweenness Centrality Analysis Results.

Factors ※(C): Cultural Factor	Betweenness Centrality	Rank
regional development, urban policy (C)	0.165	1
future conditions	0.112	2
impact on surrounding areas	0.101	3
consideration of environmental pollution, destruction	0.080	4
remains, preservation, history (C)	0.067	5
commercial economy, market formation (C)	0.064	6
nature, surroundings, scenic view (C)	0.062	7
coherence with higher-level plan	0.062	7
regional thought, philosophy, and values in an era (C)	0.021	9
infrastructure, attached facilities	0.020	10
finances, procurement	0.018	11
transport access	0.018	11
design, planning	0.011	13
tradition, character of an era (C)	0.006	14
expansion of welfare, facilities	0.006	14
expansion of disaster-prevention, safety facilities	0.002	16

Appendix G

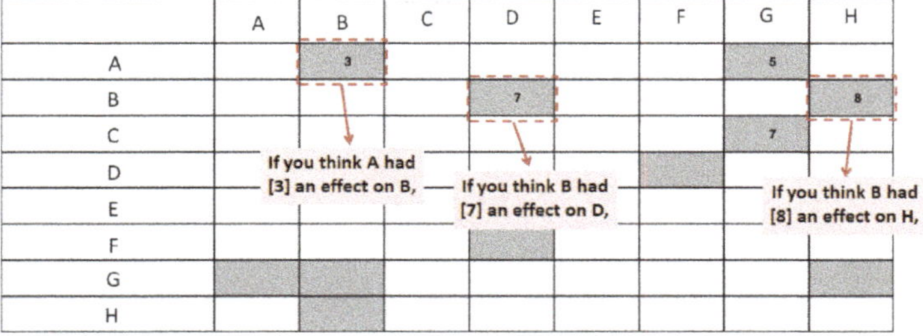

Figure A2. Questionnaires introduction page.

Appendix H

Directionality → (A → B, the directionality of influence by which A affects B)

A \ B	regional thought, philosophy, and values in an era ©	regional development, urban policy ©	commercial economy, market formation ©	nature, surroundings, scenic view ©	tradition, character of an era ©	remains, preservation, history ©	coherence with higher-level plan	consideration of environmental pollution, destruction	future conditions	infrastructure, attached facilities	impact on surrounding areas	finances, procurement	expansion of welfare, facilities	design, planning	transport access	expansion of disaster-prevention, safety facilities
regional thought, philosophy, and values in an era ©																
regional development, urban policy ©																
commercial economy, market formation ©																
nature, surroundings, scenic view ©																
tradition, character of an era ©																
remains, preservation, history ©																
coherence with higher-level plan																
consideration of environmental pollution, destruction																
future conditions																
infrastructure, attached facilities																
impact on surrounding areas																
finances, procurement																
expansion of welfare, facilities																
design, planning																
transport access																
expansion of disaster-prevention, safety																

Figure A3. Questionnaires sample for Delphi research.

References

1. Park, I.W.; Park, Y.K. A study on the establishing concept of pre-design in design process. *J. Archit. Inst. Korea Plan. Des.* **2001**, *17*, 67–74.
2. Julia, G.C.; Michael, V.A. *Field Project in Anthropology*; Waveland Press: Long Grove, IL, USA, 1997; ISBN 978-0881336856.
3. Rural Development Administration (RDA). *Recording System and Planning Model of Rural Amenity Resources*; Rural Development Administration: Drama, Greece, 2006.
4. Peter, J. The Concept of the Cultural Region and the Importance of Coincidence between Administrative and Cultural. *Rom. Rev. Reg. Stud.* **2005**, *1*, 13–18.
5. Sauer, C.O. *The Morphology of Landscape, Land and Life*; University of California Press: Davis, CA, USA, 1963.
6. Heo, Y.; Kim, S.; Yang, J.K. Establishment of analysis process and risk management model of planning phase in private development project. In Proceedings of the 5th Korean Institute of Construction Engineering Management (KICEM), Seoul, Korea, 6 November 2004.
7. Hong, Y.B. The cost planning process of pre-design phase. *Korean Archit.* **2000**, *373*, 60–65.
8. Lee, H.C.; Jang, S.J. A preliminary study on the concept of architectural planning. In Proceedings of the Autumn Annual Conference of Architectural Institute of Korea, Gyeonggido, Korea, 25 April 1998.
9. Shim, W.G. A study on architectural programming. *J. Archit. Inst. Korea* **1982**, *26*, 3–8.
10. Lee, J. A study on architectural programming phase during design process. *Rev. Archit. Build. Sci.* **1988**, *32*, 18–20.
11. Park, D.J.; Lee, S.B. Study on application BIM for performance support of construction pre-design. *J. Archit. Inst. Korea* **2010**, *12*, 191–198.
12. Choi, J.W.; Seo, H.; Kim, J. A study on support for the feasibility analysis of BIM-based construction pre-design. In Proceedings of the 2012 KiBIM Conference, Korean Institute of Building Information Modeling, Gangwondo, Korea, 1 February 2012. [CrossRef]
13. Jang, S.J. A comparative analysis on work report—Trends of architectural project plan services. *Res. Inst. Ind. Technol. Myongji Univ.* **1999**, *18*, 306–312.

14. Ministry of Land, Infrastructure and Transport (MOLIT). Presidential Decree No. 28211. Available online: https://elaw.klri.re.kr/eng_mobile/viewer.do?hseq=45354&type=sogan&key=4 (accessed on 26 July 2017).

15. Ministry of Land, Infrastructure and Transport (MOLIT). *A Study on the Preparation of Guidelines for Deliberation of the Urban Planning Committee*; Korea Planning Association: Seoul, Korea, 2014.

16. RISS. Research Information Sharing Service. Available online: http://www.riss.kr (accessed on 15 April 2020).

17. Goo, J.N. Text Mining for Korean: Characteristics and Application to Big Data. Master's Thesis, Sookmyung Women's University, Seoul, Korea, 2013.

18. Kim, S.Y.; Chung, Y.M. An experimental study on selecting association terms using text mining techniques. *J. Korea Soc. Inf. Manag.* **2006**, *23*, 147–165.

19. Lee, K.S.; Lee, T.H.; Shin, Y.K.; Kim, T.H.; Han, S.W. Quantified evaluation on the qualitative criteria for the selection of appropriate concrete slab form-works for residential buildings. *J. Korea Inst. Build. Constr.* **2011**, *11*, 136–144. [CrossRef]

20. Park, W.C. A Study on the Characteristics of Main-Contract Networks for Public Construction Projects Social Network Analysis. Master's Thesis, Hanyang University, Seoul, Korea, 2015.

21. Scott, N.; Baggio, R.; Cooper, C. *Network Analysis and Tourism: From Theory to Practice*; Channel View Publications: Clevedon, UK; Southport, QLD, Australia, 2008; ISBN 978-1-84541-088-9.

22. Jeon, Y.H.; Lee, J.S.; Kim, S.K.; Jeon, B.J. The effects of activities of daily living on the social networks of patients after a stroke. *J. Korean Soc. Occup. Ther.* **2013**, *21*, 49–60.

23. Chon, S.J. A Study on Social Network Affecting Job Effectiveness and Career Utility. Ph.D. Thesis, Sogang University, Seoul, Korea, 2005.

24. Newman, M.E.J. *Networks: An Introduction*; Oxford University Press: New York, NY, USA, 2010; ISBN 978-0-19-920665-0.

25. Freeman, L. (Ed.) *Social Networks Analysis*; Sage: London, UK, 2008; ISBN 978-1-59-457714-7.

26. Lee, S.Y. A study on port centrality of the east–west trunk service network—Based on social network analysis. *Ocean Policy Res.* **2015**, *30*, 73–103. [CrossRef]

27. Opsahla, T.; Agneessensb, B.; Skvoretzc, J. Node centrality in weighted. networks: Generalizing degree and shortest paths. *Soc. Netw.* **2010**, *32*, 245–251. [CrossRef]

28. Wasserman, S.; Faust, K. *Social Network Analysis: Methods & Applications*; Cambridge University Press: Cambridge, UK, 1994; ISBN 978-0521387071.

29. Borgatti, S.P. Centrality and network flow. *Soc. Netw.* **2005**, *27*, 55–71. [CrossRef]

30. Prell, C. *Social Network Analysis: History, Theory, and Methodology*; Sage: Groningen, The Netherlands, 2012; ISBN 978-1412947152.

31. Brandes, U. A faster algorithm for betweenness centrality. *J. Math. Sociol.* **2001**, *25*, 163–177. [CrossRef]

32. Landherr, A.; Friedl, B.; Heidemann, J. A critical review of centrality measures in social networks. *Bus. Inf. Syst. Eng.* **2010**, *2*, 371–385. [CrossRef]

33. Kang, B.U. Performance Analysis of Volleyball Games Using the Social Network and Text Mining Techniques. Master's Thesis, Dong-Eui University, Busan, Korea, 2013. [CrossRef]

34. Wang, W.; Wang, J.; Liu, K.; Wu., Y.J. Overcoming Barriers to Agriculture Green Technology Diffusion through Stakeholders in China: A Social Network Analysis. *Int. J. Environ. Res. Public Health* **2020**, *17*, 6976. [CrossRef]

35. Yun, H.J.; Seo, H.C.; Park, E.S.; Lee, Y.S.; Kim, J.J.; Lee, H.S.; Lim, S.B.; Lee, T.S. A study on developing the contents of historical education using social network analysis. *J. Korea Contents Assoc.* **2015**, *15*, 606–615. [CrossRef]

Publisher's Note: MDPI stays neutral with regard to jurisdictional claims in published maps and institutional affiliations.

International Journal of
***Environmental Research
and Public Health***

Article

Bridge Carbon Emissions and Driving Factors Based on a Life-Cycle Assessment Case Study: Cable-Stayed Bridge over Hun He River in Liaoning, China

ZhiWu Zhou, Julián Alcalá and Víctor Yepes *

Institute of Concrete Science and Technology (ICITECH), Universitat Politècnica de València, 46022 Valencia, Spain; zhizh2@doctor.upv.es (Z.Z.); jualgon@cst.upv.es (J.A.)
* Correspondence: vyepesp@cst.upv.es; Tel.: +34-96-387-9563

Received: 13 July 2020; Accepted: 12 August 2020; Published: 17 August 2020

Abstract: Due to the rapid growth of the construction industry's global environmental impact, especially the environmental impact contribution of bridge structures, it is necessary to study the detailed environmental impact of bridges at each stage of the full life cycle, which can provide optimal data support for sustainable development analysis. In this work, the environmental impact case of a three-tower cable-stayed bridge was analyzed through openLCA software, and more than 23,680 groups of data were analyzed using Markov chain and other research methods. It was concluded that the cable-stayed bridge contributed the most to the global warming potential value, which was mainly concentrated in the operation and maintenance phases. The conclusion shows that controlling the exhaust pollution of passing vehicles and improving the durability of building materials were the key to reducing carbon contribution and are also important directions for future research.

Keywords: greenhouse gas; environmental impact; cable-stayed bridge; life-cycle assessment; sustainable construction

---

1. Introduction

With the rapid development of the world economy, infrastructure construction has made a giant leap. The total greenhouse gas emissions associated with the multiple phases of an infrastructure's life cycle have accounted for 40% of global energy use [1]. According to the China Statistical Yearbook, it shows that in 2000, China consumed 56.929 million tons of oil in transportation, accounting for 24.9% of China's total oil consumption. The Development Research Centre of the State Council of the People's Republic of China forecasted that the country's transport oil consumption would reach 256 million tons by 2020 [2]. Huge energy consumption leads to serious pollution of the natural and living environment, and meanwhile, the amount of greenhouse gases increases. Scientists and institutions around the world have proposed a series of measures and policies to alleviate the problems caused by the greenhouse gas effect [3,4].

Larsson Ivanov et al. [5] have investigated air pollution and greenhouse gas emissions from the production of certain building materials and products. They demonstrated that road transport is also a major source of greenhouse gas emissions. The Swedish Transport Authority has planned that the investment of infrastructure projects (such as bridges and tunnels) would increase by at least 5 billion Euros from 2020 to 2029, and that carbon dioxide emissions must be cut by between 17% to 30%.

In 2006, the Elinkaareltaan Tarkoituksenmukainen Silta Project was launched in Finland, Sweden, and Norway [6]. In 2009, Denmark joined in. The project aimed to optimize a bridge's life cycle while covering economic, environmental, and aesthetic issues throughout the bridge's life cycle, and they developed a life-cycle assessment (LCA) tool for bridges [7].

Using the LCA, the project developed openLCA, Efootprint, Ebalance, and other software. The key aim of the software system is to establish a strong database, including the Center for Environmental Assessment of Product and Material Systems (SPINE@CPM) database of Sweden [8], Prozessorientierte Basisdaten (PROBAS) database of Germany [9], Environmental Management Association for Industry database of Japan (JEMAI) [10], The National Renewable Energy Laboratory of United States database (USNREL) [11], The Life Cycle Inventory of Universidad Real Instituto de Tecnologia de Melbourne(RMITLCI) database of Australia, the Swiss Ecoinvent database, and the European Reference Life Cycle Database (ELCD) have been established as complete databases [7].

The Ecoinvent database was created by several institutes using The Swiss Federal Institute of Technology Zurich domain name and the non-profit association Agroscope [12]. The database includes more than 2200 new data groups and 2500 updated data groups, which covers buildings, building materials, transportation, and so on. In addition to providing the summary data set, the database also includes the decomposed unit process data list, the data input and output of each production step, and the built data module. It provides a sufficient scientific research basis for LCA research in various fields.

In view of the increasing pollution of the environment by the construction industry, García-Segura et al. [13] and Itoh et al. [14] conducted carbon dioxide and cost assessments on box bridges. The study is a single example and lacks systematicity in the face of the construction of new types of bridges. Hong Wei [15] used life-cycle analysis and quantified the environmental impact of bridges. Heijungs et al. [16] presented research on framework modeling showing that there is insufficient practical guidance and proposed the establishment of a scientific framework for sustainable development life-cycle analysis in terms of products, materials, and technologies. Penadés-Plà et al. [17] used openLCA software to study the environmental impact of a box girder of two structural sizes, though the application reference value of actual engineering projects is insufficient. They studied the environmental impact contribution of box girder highway bridges under different maintenance schemes. In summary, the research results have laid the foundation for the research methods and ideas of the environmental impact of infrastructure. What is lacking is that the research is not comprehensive, systematic, and refined; the research and analysis are not comprehensive.

The comparison of case studies found that the combination of bridge structure design and aesthetics, human landscape and other concepts, the diversity of materials, the optimization of construction technology, rapid economic development, and the improvement of environmental requirements and other factors affect the bridge LCA, and a new assessment needs to be established. It is necessary to study the cause and effect process of "from the cradle to the end of life" at each stage of the entire life cycle. The comprehensive, meticulous, and rigorous research results that are in line with the bridge structure and bridge development form are more representative, important, and of higher quality data.

The above was the basis of the analysis of the thoughts and needs of this article. This study provides comprehensive research and analysis on the LCA of bridge structures and selected the comprehensive influencing factors of the four phases, from the cradle to the end of a completed three-tower cable-stayed bridge. In addition, the main causes and the mechanism of the environmental emission contribution in each stage were analyzed. Finally, the research results of the environmental emission contribution of cable-stayed bridges were obtained.

2. Methods

2.1. Research Framework and Method of LCA of a Cable-Stayed Bridge

This study used the openLCA 1.10.3 software [18], as well as the Ecoinvent database to study the contribution to the environmental impact of cable-stayed bridges. The LCA analysis of the cable-stayed bridge was divided into five phases: (1) cable-stayed bridge design, (2) cable-stayed bridge structural materials processing and construction, (3) cable-stayed bridge construction and installation, (4) cable-stayed bridge operation and maintenance, and (5) the decommissioning and dismantling of the cable-stayed bridge after the lifetime of the bridge. The reliability of the LCA results analysis mainly depended on the selection of a reasonable database and the accuracy of the parameters in each research stage. The study followed the ISO 14040:2006 framework [19] and the CML (Centrum voor Milieuweten schappen Leiden) 2001 standardized approach (Leiden University) [20]. According to the actual data of the whole process of cable-stayed bridges, the accuracy and effectiveness of the research and analysis were guaranteed. The verification information included the cable-stayed bridge design drawing, geological survey report, construction organization design, special plans, and the Ecoinvent database.

As shown in Figure 1, the LCA analysis of cable-stayed bridges was not carried out at the design phase. The bridge survey and design stage mainly consisted of the surveying and mapping of the engineering site by the design unit, as well as the interior design and production of the drawings. Large mechanical equipment and materials were not used at this stage, and only a small amount of prospecting and measuring equipment and the design work of building and structural engineers were required. As a whole, the environmental impact contribution of a cable-stayed bridge in this stage is not large; therefore, we did not analyze its effect.

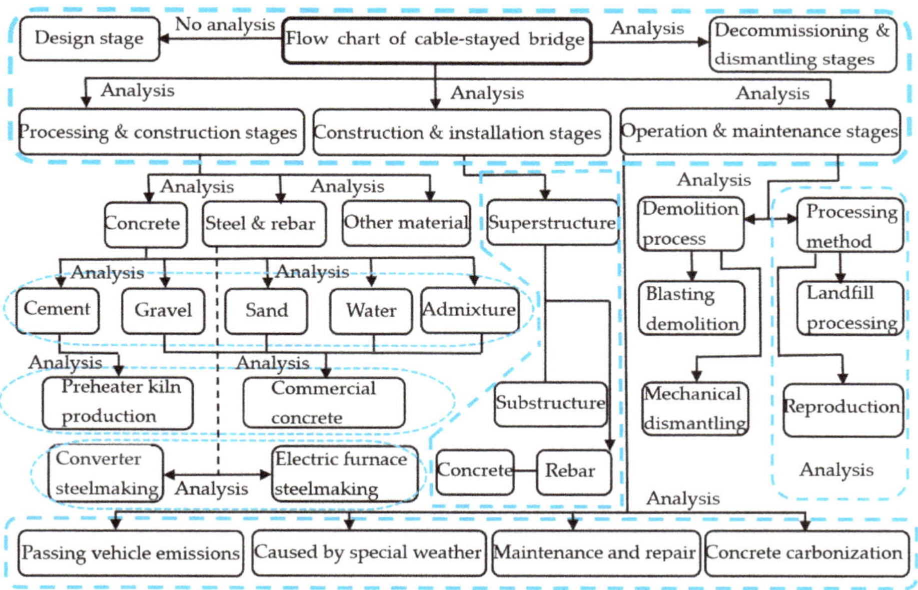

Figure 1. Life-cycle assessment (LCA) analysis flow chart of the three-tower cable-stayed bridge.

2.2. Definition of the Environmental Impact Scope of a Cable-Stayed Bridge

Throughout its life cycle, the contribution of cable-stayed bridges to the environment is the impact of the entire process from the cradle to the grave. It is necessary to input all the data of each stage of the entire life cycle as input analysis data into the software and use part of the data generated in the process as output analysis data; for example, use concrete production data in the construction phase as input data and use waste concrete and wastewater generated during the production process as output data analysis.

International Standard (ISO, 2006b) provides an explanation [19]. The defining principle of the time dimension of the analysis mainly considers phases (1), (2), and (3), which should be implemented in accordance with the provision times of the design drawings. Stage (4) should be implemented in accordance with the design life for 100 years. Stage (5) should start calculating short-time emissions in accordance with the specification for 100 years from the beginning of the demolition to the completion of the landfill.

In order to select the LCA impact assessment factor model framework of this article, the following three types of data were comprehensively studied: International Organization for Standardization (ISO), the International Society for Environmental Toxicology and Chemistry (SETAC), and the Danish Industrial Product Environmental Design Method (EDIP) to establish a framework (6 types), LCA software analysis factors (11 types), and midpoint modeling analysis (18 types), as shown in Figure 2 [17,19,21,22].

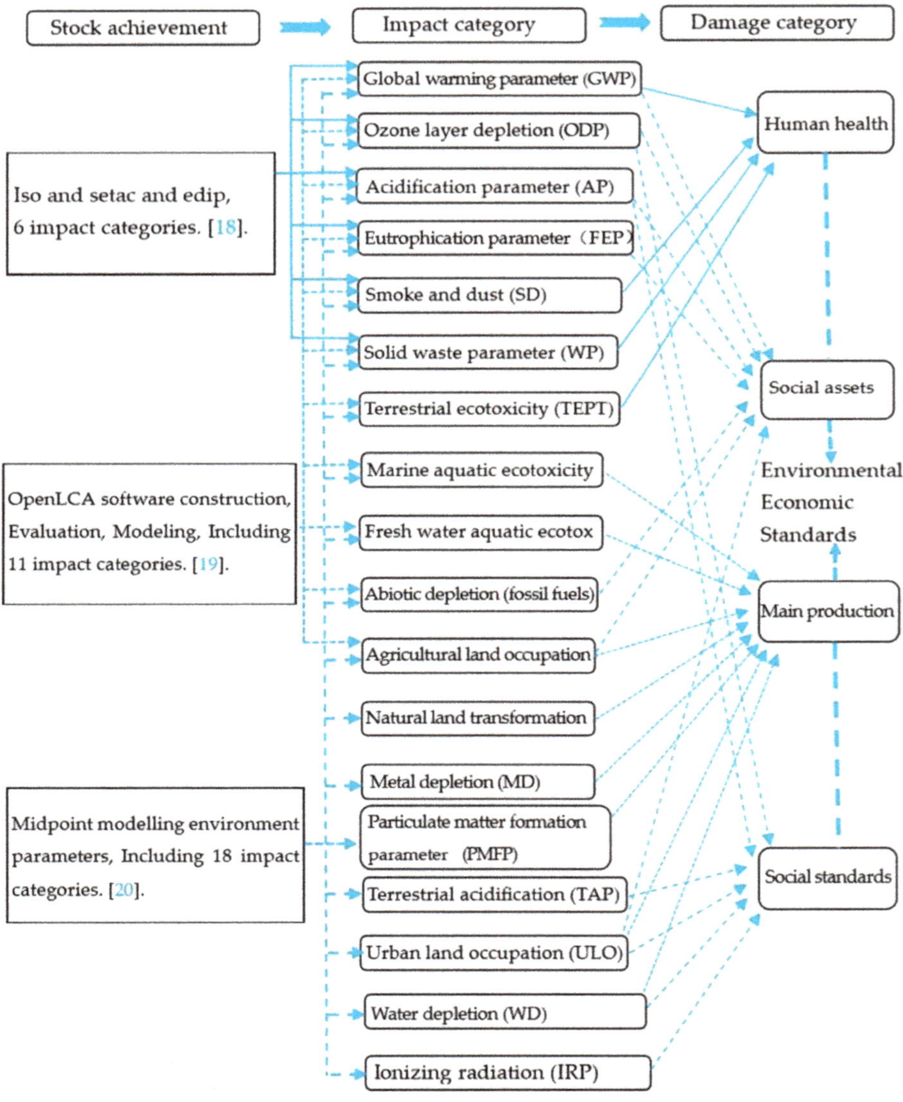

Figure 2. Schematic diagram of inventory and damage categories of construction materials' life cycle list. ISO: International Organization for Standardization, SETAC: International Society for Environmental Toxicology and Chemistry, EDIP: Danish Industrial Product Environmental Design Method

The main influencing factors and causes shown in Figure 2 show that the focus of building materials and whole-life research is on human health analysis and energy loss. Of the 18 types of influence parameters shown, this study focussed on the analysis of five major factors affecting the greenhouse effect, which according to Du et al. [23] and Kim et al. [24] are: global warming parameter (GWP), acidification parameter (AP), eutrophication parameter (FEP), particulate matter formation parameter (soot and dust PMFP), and solid waste parameter (WP).

2.3. Feature Modeling Method Selection and Weight Factor Analysis

In LCA modeling analysis, researchers mainly use two modeling methods: midpoint modeling and endpoint modeling [22,25]. In the full life-cycle analysis of the LCA process, the advantages and characteristics of the two methods should be comprehensively compared [26]. Each stage and each indicator adopts midpoint modeling, and the impact of bridge construction on human health and social assets adopts endpoint modeling. Penadés-Pla et al. [17] applied Kriging optimization and bridge modeling to find the midpoint and endpoint ranges.

Midpoint modeling typically involves selecting an indicator (the so-called midpoint) somewhere between the emissions and the endpoint in the environmental mechanism and modeling the impact of that indicator. The characteristic of midpoint modeling is that it does not pay attention to the overall environment mechanism, but the disadvantage is that there is uncertainty about the scope of the research, the duration of the research forecast, and the research model.

Endpoint modeling focusses on the representation of the contribution of LCA to the protected area. The representation model must include the entire environmental mechanism and attempt to model the process quantitatively. In the modeling process, the impact of modeling failure is not usually considered; thus, it is more uncertain and unknown. The potential benefit of this approach is that the effects at the endpoint level can be compared [12].

Considering the advantages and disadvantages of the two methods, the joint modeling and weighted analysis method of the midpoint and end-point were adopted in the modeling and analysis of a cable-stayed bridge [23] and parameter weighting was introduced into the LCA modeling process.

As shown in Figures 3 and 4, in the four phases of the cable-stayed bridge modeling and analysis process, the midpoint analysis modeling method was adopted, and the setting of weighted parameters was introduced. For the overall environmental impact assessment, the endpoint modeling research was adopted, the environment mechanism weighted parameters were introduced, and the feature modeling was introduced in the process of database selection and analysis process.

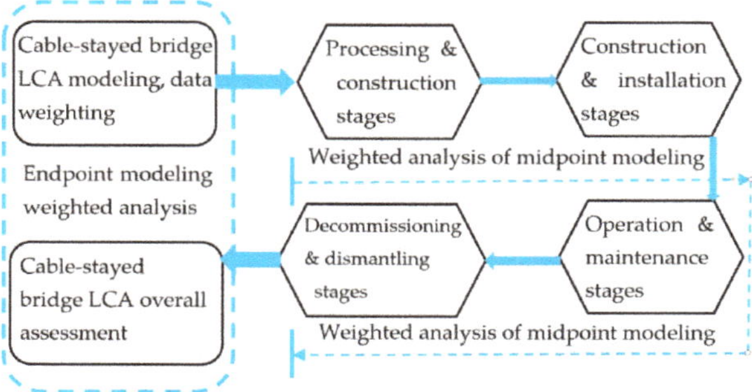

Figure 3. Schematic diagram of cable-stayed bridge modeling process.

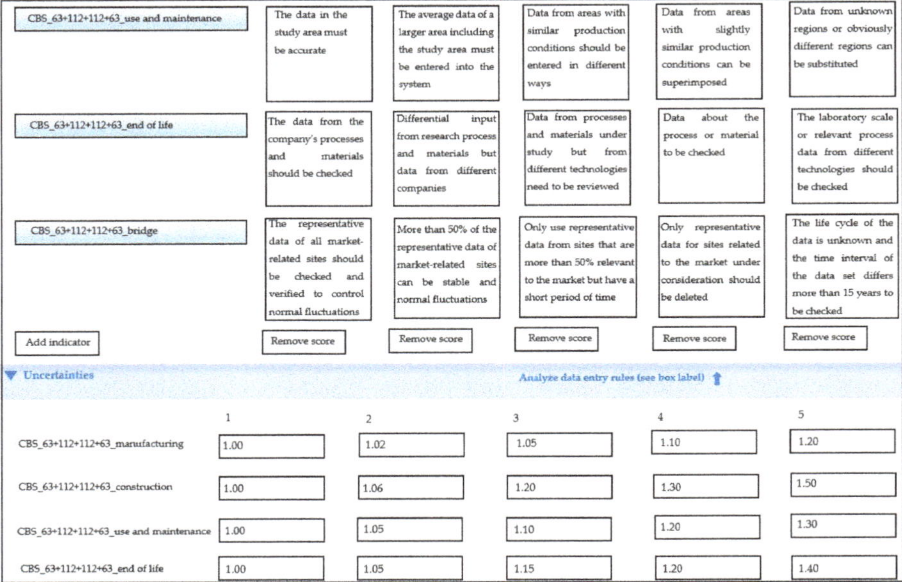

Figure 4. Schematic diagram of the LCA modeling process and weight coefficients.

3. LCA Assessment Process and Data Analysis

This study selected the municipal bridge across the Hun He River in the Liaoning province of China as the research object. The bridge is a three-tower concrete cable-stayed bridge with a single cable plane. The bridge is 360 m long (63 + 112 + 112 + 63 + 10 m) with a 38 m surface width and a 2 m cable anchorage zone. The central bridge tower is a beam–tower–pier consolidating system and the bridge towers on both sides are a beam–tower consolidation system. A single-box double-chamber structure is adopted in the main beam with a 2.4 m central height, a 1% cross slope, a 25 cm top-plate thickness, a 24 cm bottom-plate thickness, a 2.5 m central web thickness, and a 1 m side web thickness. At every 6 m interval on the main beam, a transverse separating beam is set up with a 70 cm thickness and transverse beams are set up at the ends. The cable anchor is fixed at the bottom of the central web of a box-type beam. The main towers have a 20 m height over the bridge surface. An I-shaped cross-section (with a 5.0 m × 3.8 m dimension) was adopted for the upper tower columns and a solid cross-section (with a 5.0 m × 2 m dimension) was adopted for the intermediate tower columns. A box-type thin-wall structure was adopted for the pier body, a transverse separating plate was set up inside the pier, and the base is an extensive one. For the stayed cable, high-tensile galvanized steel wire, a chill-cast anchor, and a hot-extruded polyethylene (PE) guard sleeve were adopted. The main beams used in the bridge construction were precast hollow reinforced concrete slabs with a 65 cm thickness and a 125 cm width.

As shown in Figure 5, the cable-stayed bridge is divided into three towers, four spans, and two cross-sections. The construction process was as follows: (1) First, adopting a cast-in-site caisson; second, hoisting to the pile position, digging, and discharging the soil in the well; third, sinking the pile to the designed bedrock position; and finally, pouring the slab concrete into the caisson. Continue to complete the pouring of the concrete of the dock. (2) The main beams were constructed with steel brackets with a "six + four" structure.

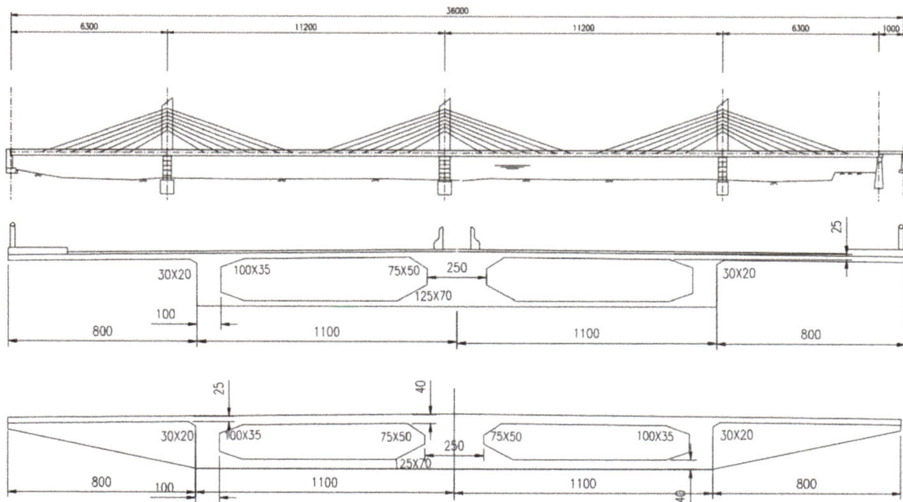

Figure 5. Three-tower cable-stayed bridge and the schematic diagram of the box girder structure.

The construction was divided into three sections: assembling the outer mold of the steel formwork, assembling the inner mold of the box girder using a wooden mold, and inverting the outer mold and bracket three times. The inner mold of the box girder (damaged during the dismantling) was processed three times to finish the construction of the main girder. (3) The stay cable tension of the previous process was completed, and at the same time, the concrete of the subsequent process was poured, and constructed in order. (4) The last section of main beam concrete construction, bridge deck pavement, railing installation work was completed, and finally, the bracket was removed.

3.1. Processing and Construction Stage

The main materials of the cable-stayed bridge construction included concrete (comprising cement, water, gravel, river sand, and admixture), asphalt, steel bar, plate, rubber bearings, steel strand, steel products (steel plates, steel template, six-four military support), anchorage, corrugated pipe, wooden templates, wood, and other subsidiary materials.

3.1.1. Concrete

The main ingredients of concrete are cement, water, gravel, river sand, and admixture. Cement is also the biggest contributor to environmental impacts. China's cement production mainly adopts the new suspended preheater kiln production process, and the cement output produced by the new suspension preheating kiln production process accounts for 95% of China's total annual cement output [27]. Each kind of building material has a physical and chemical environmental influence. According to the requirements of the construction drawings and the same-concrete data of the Ecoinvent database, ordinary Portland cement was selected. The database includes cement during the entire process, from the start to the finished product, including the source of upstream products (such as gypsum). The production of 1 kg of cement generates a waste heat emission of 0.135 MJ (standard deviation of 1.4918) [14]. There are four types of cable-stayed bridge concrete: C50, C40, C30, and asphalt concrete. The asphalt content is 6.5% and the density is 2.35 t/m^3 [28]. The commercial concrete used in the cable-stayed bridge was supplied by local manufacturers. According to the database, the loss in the production process was determined to be 24.5 kg of waste per 1 m^3 concrete. The sewage discharge value was 0.035 m^3/m^3. In the calculation of the environmental impact analysis, the concrete was divided into the production and construction phases, and the coefficient of the contribution of the emissions to the environmental impact is shown in Table 1.

Table 1. Environmental impact contribution coefficient of materials during the processing and construction [29–31].

Material Name	Unit	GWP	AP	FEP	PMFP	WP
P. I. 52. 5		1042.00	0.28	1.61	2.24	0.00
P. O. 42. 5	kg/t	920.00	0.25	1.43	2.02	0.00
P. S. 32. 5		678.00	0.20	1.09	1.57	0.00
River sand	kg/m^3	2.56	0.00	0.00	0.00	0.00
Gravel	kg/m^3	3.30	0.00	0.00	0.00	0.00
Flake	kg/m^3	3.37	0.00	0.00	0.00	0.00
C50		705.00	0.02	0.02	0.05	0.00
C40	kg/m^3	608.00	0.01	0.01	0.05	0.00
C30		565.00	0.01	0.01	0.04	0.00
Ordinary asphalt	kg/t	174.00	0.00	0.00	0.17	0.11
Modified asphalt		296.00	0.00	0.00	0.17	0.11
Grade I and II steel bars	kg/t	4524.00	46.10	28.60	158.40	258.00
Steel wire		3551.00	46.10	28.60	158.40	258.00
Large steel		4339.00	56.60	34.80	150.10	323.00
Medium steel	kg/t	3589.00	46.60	28.90	124.60	268.00
Small steel		3560.00	46.10	28.60	123.40	252.00
Diesel	kg/kg	4.62	0.00	0.00	0.00	0.00
Gasoline	kg/kg	4.36	0.00	0.00	0.00	0.00
Waterproof coating	kg/kg	0.41	0.16	0.02	0.00	0.02
Power consumption	kg/kwh	0.98	0.00	0.00	0.00	0.00
Consumption	kg/person/day	2.88	0.00	0.13	0.00	0.50

GWP: Global warming parameter; AP: Acidification parameter; FEP: Eutrophication parameter; PMFP: Particulate matter formation parameter; WP: Solid waste parameter.

3.1.2. Main Material

Rebar, steel, and pipe were the main materials. Steel smelting in China is divided into two types [32,33]: converter steel and electric steel. A total of 90% of the steel output is made using converter steel and about 10% is made using electric steel [14]. The environmental impact contribution of a 1 kg steel bar discharge, which was determined using the database, is specified in Table 1.

3.1.3. Material Transportation at the Manufacturing Stage

All the raw materials were ready to enter the site in the early phases of construction. According to the design plan, the transportation distance of concrete raw materials was 120 km, and the mixing water was tap water. The transportation distance of the commercial concrete was 30 km; the transportation distance of the steel bar, steel products, and steel strand was 160 km; the wood's transportation distance was 80 km; and other materials were provided from the non-ferrous metal market, which was 100 km away. The materials were transported using three types of trucks: 17.5 m (49 t), 6.8 m (18 t), and 4.2 m (4 t). A gantry crane, a 25 t crane, and six erection workers completed the loading and unloading. The machines and tools are shown in Table 2.

Table 2. The statistical table of used machinery and equipment at each stage [32,34].

Vehicle, Machinery Type	Fuel Consumption (kg/km)	Transport Distance (km)	Types of Shipping Materials
Heavy truck (49 t dead weight, length 13–17.5m)	0.67–0.84	120, 160	Cement-crushed stone-river sand; steel bar, steel strand, and other steel products
Medium-sized truck (18 t dead weight, length 5.8–6.8 m)	0.15–0.21	120, 80	Additives, wood
Light truck (4 t dead weight, length 2.6–4.2m)	0.10–0.14		
Concrete mixer truck (12 m^3)	0.13–0.16	30	C50, C40, C30
Gantry crane (50 t)	22–30 kW	0.5	Lifting steel bar, steel strand, and other steel products
Crane (25 t)	0.13–0.18	2	Lifting Rebar, steel strand, and other steel products

Remarks: 0 # Diesel = 0.835 kg/L, +10 # Diesel = 0.85 kg/L, −10 # Diesel = 0.84 kg/L. The truck used 0 # Diesel.

The energy consumption value of the environmental impact contribution during the construction stage is given as:

$$EC_m = \sum M_{qi} \times \lambda_i, \tag{1}$$

where EC_m is the value of environmental impact contribution of raw materials (kg), M_{qi} is the mass of material i (kg), and λ_i is the environmental emission coefficient of physicochemical material i (kg/kg).

The modeling calculation of the environmental impact of the transportation equipment is given as:

$$T_m = \sum \left[\left(G_{qi} + G_{si} \right) \times \left(\frac{D_{qi}}{100} \right) \times \left(M_{qi} / M_{ei} \right) \times \lambda_n \right] \tag{2}$$

where T_m is environmental impact contribution of the transport equipment (kg), G_{qi} is the fuel consumption of the truck load (L/100 km), D_{qi} is the freight car transport distance (km), G_{si} is the non-load fuel consumption of freight cars (L/100 km), M_{qi} is the total mas of the material, M_{ei} is the load capacity per vehicle (kg), and λ_n is the physicochemical environmental emission coefficient of of oil n (kg/kg).

The construction of the environmental impact model of loading and unloading machinery and personnel is given as:

$$M_m = \sum \left(T_{qi} \times K_{mi} \times \lambda_n \right) + \sum \left(P_{mi} \times \lambda_p \right), \tag{3}$$

where M_m is the environmental impact contribution of the loading and unloading machinery and personnel (kg), T_{qi} is the consumption per mechanical equipment (L/shift), K_{mi} is the total number of shifts (working days), P_{mi} is the number of stevedore shifts (working days), and λ_p is the numerical coefficient of environmental impact contribution of personnel per working day (kg/working day).

The total environmental impact contribution at the construction stage is given as:

$$C_m = EC_m + T_m + M_m. \tag{4}$$

3.2. Construction and Installation Phases

The construction and installation phases were the main phases of the cable-stayed bridge's environmental impact contribution to the research. The construction of various products required the joint work of a large quantity of mechanical equipment and construction personnel. The environmental impact mainly included the following parts.

3.2.1. Environmental Impact Contribution of Materials Processing

Due to the need for reinforcement, steel, steel wire, and other raw materials were sent to the site. The technicians performed the processing, lashing, and installation of the steel bars according to the construction design drawings. See Table 3 for the machinery and equipment used in the construction process.

Table 3. Summary table of the operation data of medium and small machines and equipment during construction and installation.

Device Name	Specification Model	Quantity (Table)	Power (KW)	Use Time (Month)	Oil (L/hour)
Excavator	PC400-1	1	228	6	35~45
Roller	YZ20Ton	1	249	2	25~35
Loader	ZL50	2	162	6	14~15
Dump Truck	CQ33 (L/100km)	2	380	2	50~60
Sprinkler	11.7m³ (L/100km)	1	22.5	10	15
Engineering rig	XR360 Rotary drilling rig	2	298	1	20~30
Engineering rig	Impact drill JK-6	1	200	1	10~15
Concrete pump truck	SY5125THB-9018III	1	176	1	0.5~0.6
Concrete transport truck	12m³ (L/100km)	6	240	6	8~10
Car crane	QT25	2	213	3	4~6
Mortar mixer	HZS180	2	120	2	120 kw
High frequency vibrator	ZG50	30	2	6	2 kw
Diesel generator sets	500KW	1	500	2	131
Rebar cutting machine	GT5-12	1	94	2	94 kw
Steel bending machine	GW 40	2	40	2	40 kw
Rebar cutting machine	GQ50	2	50	2	50 kw
Profile cutting machine	J3G-AL-400	1	400	1	400 kw

Remarks: The normal operation of the concrete pump truck is 40 cubic meters per hour.

3.2.2. Environmental Impact Contribution of the Machinery and Equipment in the Processing and Construction Stage

The calculation of the environmental impact model of mechanical equipment is given as:

$$C_m = \sum \left(E_{mj} \times T_j \times \lambda_j \right), \tag{5}$$

where C_m is the environmental impact contribution value of mechanical equipment (kg), E_{mj} is amount of fuel or power consumption of equipment j (L/hour, kW/hour), T_j is the total working hours of equipment j, and λ_j is the fuel or electricity physicochemical environmental emission coefficient of equipment j (kg/l, kg/kW).

3.2.3. Value of the Environmental Impact Contribution of Managers and Skilled Workers

The environmental impact modeling calculations for managers and skilled workers is done using:

$$P_m = W_m \times \lambda_p \times T_p, \tag{6}$$

where P_m is the value of the environmental impact contribution of skilled workers (kg), W_m is the total number of workers (people), λ_p is the environmental impact coefficient of workers (kg/day/worker), and T_p is the total time worked (days).

According to the construction organization design, there were 36 project management and technical staff, 180 technical workers on average, 24 logistics service staff, and the construction period was 14 months.

3.2.4. Contribution Value of the Electric Power Energy to the environment during the Construction Period

The calculation of the power energy environmental impact modeling during the construction period is given as:

$$E_m = \sum \left[T_i \times \lambda_i \times (1 + L_m) \right] + G_m \times \lambda_n \times T_m \times (1 + L_n), \tag{7}$$

where E_m is the environmental impact contribution value of the electricity and oil consumption during construction (kg), T_i is the power consumption (kWh/day) of personnel (managers, skilled workers, etc.),

λ_i is the physical-chemical environmental emission coefficient of the power consumption (kg/degree), L_m is the power loss value (degree/day), G_m is the amount of oil consumed by the generator in a power outage and during field operations (kg/hour), λ_n is the physicochemical environmental emission coefficient of oil class n, T_m is the total working time (hours) of the equipment, and L_n is the oil loss during the generator operations (kg/hour).

3.2.5. Values of the Contribution of Project Managers and Technical Workers to the Garbage and Sewage Environment

The calculation formula of the environmental impact model of waste and pollutants generated by managers and skilled workers is given as:

$$M_p = P_a \times T_m \times T_n \times \lambda_p + S_m \times P_a \times \lambda_x \times T_n, \tag{8}$$

where M_p is the contribution value of the garbage and sewage environmental impact from staff during the life of the project (kg), P_a is the total number of project personnel (people), T_m is the quantity of household garbage (kg/day), T_n is the working time of the staff on duty (days), λ_p is the environmental emission coefficient of the household garbage (kg/kg), S_m is the discharge quantity of personnel (kg/day), and λ_x is the environmental emission coefficient of the pollutant discharge (kg/kg).

The total environmental impact contribution during the construction stage is given as:

$$B_m = C_m + P_m + E_m + M_p. \tag{9}$$

The construction of the three-tower cable-stayed bridge was completed according to the flowchart shown in Figure 6, which saved materials, increased the working time of the mechanical equipment and the number of skilled workers, and improved the environmental impact contribution.

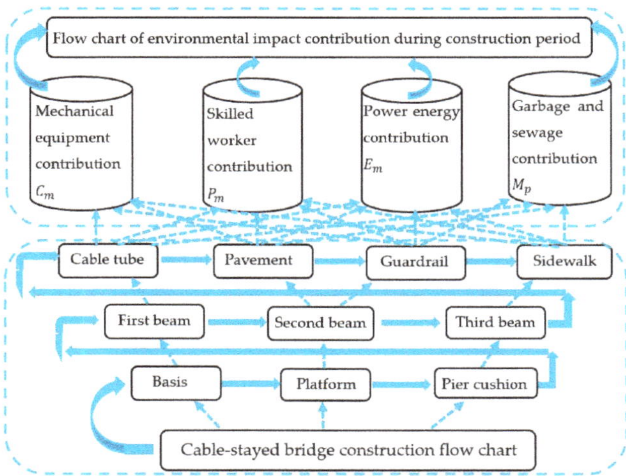

Figure 6. Schematic diagram of the process flow and environmental impact contribution during the construction.

3.3. Operation and Maintenance Stage

After the completion and acceptance of the cable-stayed bridge, it entered the operation and maintenance stage. The cable-stayed bridge is an integral part of the municipal road. After the cable-stayed bridge was completed and put into use, the local municipal road maintenance department was responsible for the daily maintenance and repairs of various types of damage. An analysis of the statistical data published by the maintenance department shows that the main content of the maintenance work is divided into daily maintenance for more than five years, monthly maintenance and inspection, annual maintenance, revision, and replacement [34]. After the beginning of the operation phase, a large number of motorized and non-motor vehicles pass every day; therefore, it was necessary to analyze the environmental pollution values of exhaust emissions [35].

3.3.1. The Amount of Environmental Impact Contribution for Maintenance of the Cable-Stayed Bridge

Table 4 summarises the period and content of the maintenance of the cable-stayed bridge. The data analysis was calculated according to the content of the table. The value of the contribution of the maintenance and maintenance environment caused by the impact of the natural environment in the table is uncertain. The analysis of attendance and the calculation of maintenance workers are also added.

Table 4. Summary table of the cable-stayed bridge maintenance, along with its maintenance cycle and causes [36–38].

Bridge Disease	Causes	Conservation Measures	Maintenance and Repair Cycle
Concrete carbonation, disease	Shock, vibration, overload, uneven settlement, chemical erosion, abrasion, blowing, freezing and thawing	Brush protective layer, repair cracks, recast pavement	The main beam is replaced every 50 years, the bridge deck pavement and the waterproof layer are replaced every 10 years, the main beam body is maintained every 5 years, and the common repair is every 2 years
Rebar disease	Chloride corrosion, concrete carbonation	Eliminate leakage points, crack closure, repair of protective layer, zinc coating protection	Consider carbonized corrosion, repair once every 70 years, the pre-stressed steel strands of the stay cables are replaced every 20 years
Component performance degradation and maintenance	Overload, aging	Local repair, component replacement	Piers and bearing caps are painted every 5 years, rubber bearings are replaced every 25 years, and expansion joints are replaced every 10 years
Probability of failure	External damage, exceeding design life	Repair, replacement	Deck drainage pipes are replaced every 50 years (the repair is every 2 years), anti-collision guardrails are replaced every 15 years (the repair is every 5 years), lighting devices are replaced every 50 years (the repair is every 5 years)
Unpredictable external environmental impact	Car accident, overload, collision, bad weather	Repair, replacement	Repair and replace at any time

The environmental impact contribution of the cable-stayed bridge maintenance:

$$C_m = \sum \left[B_m \times \left(\frac{T_m}{T_p} \right) \times \lambda_p \right] + \sum \left[P_m \times \left(\frac{T_m}{T_n} \right) \times \lambda_n \right] + C_v, \tag{10}$$

where C_m is the value of the environmental impact contribution to maintenance and repair (kg), B_m is the bridge deck pavement replacement area (m^2), T_m is the service life of the bridge design (years), T_p is each replacement time of the bridge deck pavement (years), λ_p is the environmental emission coefficient of the bridge deck pavement (kg/kg), P_m is the coated area of the pier column (m^2), T_n is the pier painting change time per time (years), λ_p is the environmental emission coefficient of the pier coating (kg/kg), and C_v is the environmental impact contribution value of the mechanical equipment during the maintenance stage (kg).

3.3.2. Environmental Impact Contribution of Vehicles during the Operation of the Cable-Stayed Bridge

Transportation accounts for 26% of the global energy consumption and 23% of greenhouse gas emissions are energy-related. Street traffic accounts for 74% of the world's transport sector traffic [36].

The cable-stayed bridge is part of a municipal road, which is used by a large number of vehicles every day and is a major contributor to global greenhouse gases. Colvile et al. [31] have shown that diesel, gasoline vehicle exhaust, liquid gasoline, and gasoline evaporation account for at least 50% of volatile organic compounds (VOC) in the environment. The chemical substance balance (CMB) and the proportion in the emissions inventory [39], which is determined using the chemical substance balance (CMB), are much greater than the proportion of paint and solvent contributions [40].

To obtain detailed data about the relevant traffic on the cable-stayed bridges, data can be searched for in the traffic database [41]. The total length of roads in Fushun in 2019 was 6911.4 kilometers, with an annual highway freight turnover of 1277.295 million tons and highway passenger turnover of 1052.48 million kilometers, with 262,000 civilian cars and 19,035 trucks [33]. According to statistical research results, carbon emissions from passenger cars in China are estimated to be 305.4 g/km, and carbon emissions from trucks are estimated to be 271.8 g/km [39].

The calculation of the environmental impact model during operation is given as:

$$T_m = \sum_1^{100} [C_m \times K_m \times \lambda_c \times V_m \times K_m \times \lambda_t](1 \pm \lambda_y), \tag{11}$$

where T_m is the environmental impact contribution during operation (kg); C_m and V_m are the annual toll of passenger cars and trucks (set), respectively; K_m is the passenger car journey distance on the cable-stayed bridge (km); λ_c and λ_t are the environmental emission coefficients of passenger cars and trucks (kg/kg), respectively; λ_y is the annual increase or decrease of passenger cars and freight cars (%); and 100 is design life (years).

3.3.3. External Environmental Impact Contribution Value of the Cable-Stayed Bridge

During the operational period, the environmental impact contribution of the cable-stayed bridge under the influence of special weather, such as snow, rain, and dust, was analyzed by referring to the monitoring data of the local environmental protection department. Monitoring data from 2010 to 2019 show that: GWP = 1.3~2.2 mg/m^3, AP = 12~39 mg/m^3, FEP = 22~39 mg/m^3, PMFP = 29~46 mg/m^3, and WP = 50~78 mg/m^3 [35].

The environmental impact contribution of the cable-stayed bridge should be determined according to the monitoring data. Considering the declining trend of the values of the five indices year by year, it was found from the statistical data over 10 years that the values kept changing by about 30%, and that the changes of the values decreased in the later period under a favorable environment. The influence of the changed values was not considered in this study.

3.3.4. The Value of the Contribution of Concrete Carbonation to the Environmental Impact of the Cable-Stayed Bridge

Concrete carbonation was mainly affected by its performance and external environmental factors. The temperature, carbon dioxide (CO_2) concentration, and relative humidity greatly influence carbonation, which also determines the carbonation depth and compressive strength of concrete [41]. Chen et al. [36] and some other studies have established a multi-field coupling numerical model and action quantization index for the carbonation analysis of Martínez-Muñoz et al. [42], and have quantitatively analyzed the influence of temperature, relative humidity, CO_2 concentration, and other factors on the concrete carbonation depth by introducing an environmental correction coefficient.

The model relation of the quantity of CO_2 absorbed by per unit volume of concrete is determined using:

$$m_0 = (1 - \alpha) \times 8.22B \tag{12}$$

where m_0 is amount of CO_2 absorbed by ordinary Portland cement concrete (mol/m^3), B is the amount of cement material per unit volume of concrete (kg/m^3), and α is the content of mixed materials in ordinary Portland cement (%).

The calculation of the numerical model of the total carbonation of concrete is done using:

$$K_m = N_{c50}m_{50} + N_{c40}m_{40} + N_{c30}m_{30}, \tag{13}$$

where K_m is total carbonation amount of concrete (kg); N_{c50}, N_{c40}, and N_{c30} are the volumes of C50, C40, C30 concrete (m³), respectively; and m_{50}, m_{40}, and m_{30} are the carbonation moduli of C50, C40, and C30 concrete (kg/m³), respectively.

The environmental impact contribution value during the operation and maintenance of a cable-stayed bridge (external environmental impact quantity in kg) is given as:

$$M_m = C_m + T_m + S_m + K_m. \tag{14}$$

3.4. Abandonment and Demolition Stage

The designed service life of highway bridges in China is 100 years. The service life of bridges is shortened under the condition of long-term exposure to the Cl^- or CO_2 harsh environments. After several bouts of maintenance, the designed service life of bridges is determined to enable the designed service life to be reached, before being abandoned and dismantled [38].

There are two commonly used demolition schemes: manual demolition with mechanical equipment and blasting demolition. The safety factor of demolition via blasting the cable-stayed bridge, which is located in the urban area, is low. Through a comprehensive evaluation, the plan of mechanical and manual demolition was adopted. The comprehensive plan of segmental cutting, segmental hoisting, site crushing, and freight car transportation to the pre-burial site and steel mill was determined from the aspects of technology, safety, economy, etc.

Referring to the demolition experience of similar bridges, the mechanical equipment requires 6 long-arm crushers, 4 loaders, 10 heavy-duty transport vehicles, 4 steel transport vehicles, and 30 management and technical workers. Demolition is scheduled to take three months. The generated environmental impact contribution coefficient was calculated by referring to Table 2. The crushed concrete waste is to be transported and buried in a landfill, which is 160 km away. The steel scrap is to be transported to a steel mill, 180 km away, for smelting. According to the study results of Kim et al. [43], and in combination with the bridge removal scheme, it was determined that the recovery rate of concrete is 95%, the recovery rate of steel is 72%, and the recovery rate of steel and steel strand is 85%.

4. Results and Discussion

4.1. Summary and Analysis of the Environmental Impact Contributions at Each Stage

The LCA analysis process of a cable-stayed bridge was completed, the data were summarized, and the impact of each stage on the environmental impact contribution was analyzed. Table 5 shows the statistics of the main engineering materials and auxiliary engineering materials project of the three-tower cable-stayed bridge.

Table 5. List of environmental impact criteria for a cable-stayed bridge at each stage.

Environmental Parameters	Unit	Processing and Construction Phases	Construction and Installation Phases	Operation and Maintenance Phases	Decommissioning and Dismanting Phases
GWP	kg	40,425,577.87	1,906,820.13	121,031,298.3	16,574,524.2
AP	kg	303,615.4	14.4	317,034.68	275.44
FEP	kg	193,738.31	40,772.73	192,615.69	8574.12
PMFP	kg	917,232.2	6.6169	979,739.53	187.68
WP	kg	8,191,263.94	156,810.04	1,740,820.94	32,862.67

Since the cross-sectional structure of the cable-stayed bridge was divided into two sections, for the analysis and calculations, 1 m^2 was selected as the LCA research unit, and the mix ratio of C50, C40, and C30 concrete was selected according to the ratio provided by the Ecoinvent database.

The statistical data of the main engineering materials and auxiliary engineering materials of the three-tower cable-stayed bridge are shown

As shown in Figure 7, the environmental impact contribution value of the cable-stayed bridge in each stage, and the environmental impact contribution value of the steel products and steel bar in the processing and construction phases, accounted for 36.64% and 36.35% of the total amount, respectively. The main reason for this result is that China's steelmaking process is mainly concentrated in the converter steelmaking process. Steel has a great impact on the environment during the production process, according to the study results of Zhu et al. [44]. Therefore, it is necessary to improve the steelmaking process and technical level and pay more attention to the development and application of low-carbon environmental protection technologies.

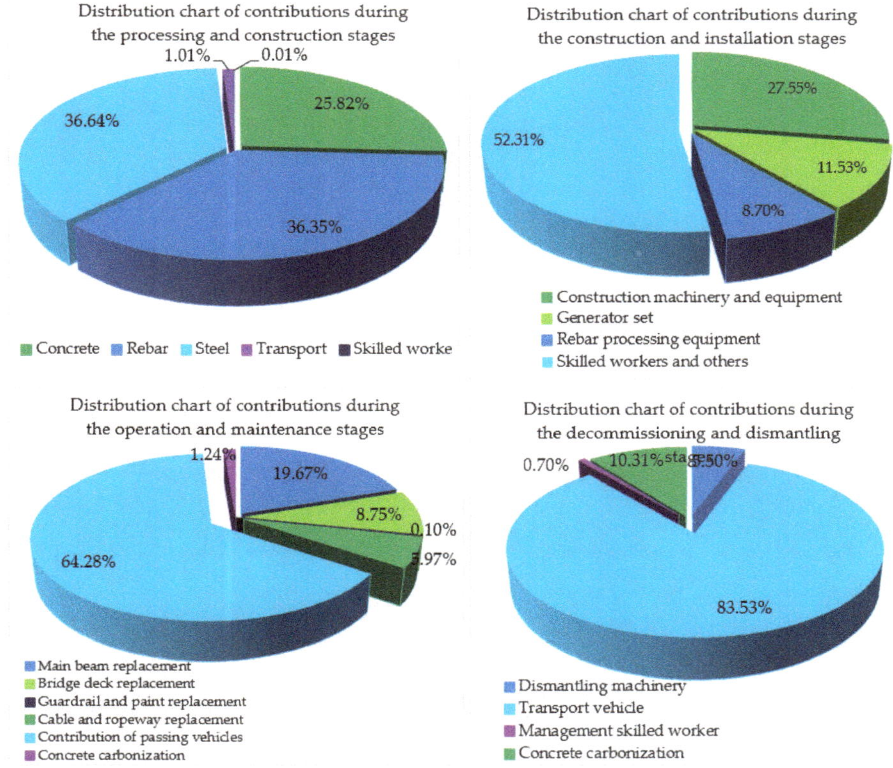

Figure 7. Schematic diagram of the environmental impact contribution of each stage of the cable-stayed bridge.

In the construction and installation phases, the environmental impact contribution was mainly due to skilled workers and the energy consumed by project participants (electricity, drinking water, accommodation materials), along with the wastewater dumped by the project participants for cooking and washing, which accounted for 52.31%. The cable-stayed bridge was a municipal project, which needed a large number of management technicians. All of these people who lived on the construction site for a long time were responsible for the large number of contributions to environmental impact.

The main reason for the increase in the value was that the environmental impact contribution of the materials was reduced but the environmental impact contribution of personnel and mechanical equipment was increased. The environmental impact contribution of the mechanical equipment reached 27.55% of the total amount.

In the operation and maintenance stage, the environmental impact contribution caused by vehicle traffic was dominant, accounting for 64.28% of the total, which is a number that needs to be taken seriously by the automotive and transportation sectors. The environmental impact contribution of vehicles can no longer be underestimated; although the designers and researchers are looking for ways to reduce the environmental impact contribution at other phases, there is still little they can do about it.

The value of the environmental impact contribution in the abandonment and demolition stage was mainly caused by concrete waste, steel, and steel waste removed by vehicle transportation, which accounted for 83.53% of the total amount.

The cable-stayed bridge will have a significant impact on the environment after 100 years of operation. This can be seen in the numerical results in Figure 6, which show that vehicle traffic was the main cause of environmental pollution and it also affected the environmental impact contribution of the entire cable-stayed bridge.

4.2. Summary and Analysis of the Environmental Impact Contributions of Five Indicators of the Cable-Stayed Bridge

As shown in Table 6, the total environmental impact contribution of GWP, AP, FEP, PMEP, and WP in the four phases of the three-tower cable-stayed bridge was 19,3013,584.7 t. As shown in Figure 8, the processing and construction phases accounted for 25.90%, the construction and installation phases accounted for 1.09%, the operation and maintenance phases accounted for 64.37%, and the abandonment and demolition phases accounted for 8.63%. The main reason for the huge impact on the environment during the operation and maintenance phases was that the replacement and maintenance of cable-stayed bridge structural components during its 100-year design life will affect the environment. According to the service life of its components (Table 4), the pre-stressed steel strands of the stay cables installed on the cable-stayed bridge need to be replaced three times (within 100 years of service life). The bridge deck pavement will be replaced 10 times. The drainpipe of the bridge will be replaced twice. The bridge anti-collision railing will be replaced twice. The exposed concrete waterproof layer of the bridge will be replaced 10 times.

Table 6. Statistical table of raw materials and accessory materials of the cable-stayed bridge.

Material Name	Unit	Quantity	Number	Material Name	Unit	C50	C40	C30
C50 concrete		9050	15	Cement		1337	213	560
C40 concrete	m^3	1761	16	Fly ash		315	54	124
C30 concrete		4714	17	Gravel	m^3	3693	770	1687
Asphalt concrete		447	18	Sand content		3102	629	2082
Rebar	Level I (Ton)	2037.5	19	Water		602	95	262
	Level II (Ton)	37	20	Steel	Ton		191.6	
Plate rubber support	m^3	2.33	21		ø127 (m)		3679	
Stranded wire	ø15.24 (Ton)	425.8	22	Bellows	ø90 (m)		7783	
Lasso	Ton	350.3	23		ø80 (m)		3670	
Cable anchor	Set	168	24		90 × 19 (m)		28,428	
Anchor	15——27 (Set)	240	25	Military beam, steel pipe bracket	Ton		2234	
	15——14 (Set)	126	26	Box beam steel formwork	Ton		384.1	
	15——9 (Set)	176	27	Box beam inner model	m^2		21,600	
	15——5 (Set)	1496	28	Fang Mu	m		9600	

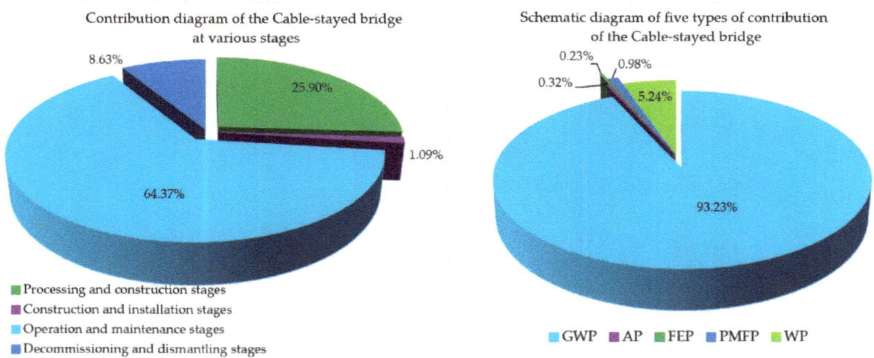

Figure 8. Schematic diagram of the environmental impact contribution of the four phases of the cable-stayed bridge.

Among the environmental impact values for the cable-stayed bridge, the carbon dioxide (CO_2) emissions accounted for 93.23% of the total emissions in Figure 7, which was one of the reasons for choosing the five types of research parameters. The other 13 types of environmental impact values were relatively small. It can be seen that, in the future, global warming due to gas emissions and its precise and detailed research and analysis should be the focus of researchers in the construction industry.

This is due to the following three reasons: the large amount of transport waste, the large number of used transport vehicles, and the long transport distance.

As shown in Figure 9, the environmental impact contribution of the cable-stayed bridge mainly focussed on the processing and construction phases and the operation and maintenance phases. The environmental impact contribution of these two phases was mainly concentrated on the production of steel bars and steel products and the contribution of the exhausts from the passing vehicles, accounting for 73% and 64.28% of the environmental impact contribution of each stage, respectively. How to better reduce the environmental impact contribution in the future is worthy of in-depth consideration by researchers, designers, and managers. The overall LCA environment contribution order of the cable-stayed bridge was: GWP (93.23%) > WP (5.24%) > PMFP (0.98%) > AP (0.32%) > FEP (0.23%). The highest proportions of GWP, PMFP, and AP in the operation and maintenance phases were 67.26%, 51.64%, and 51.05%. The highest proportions of WP and FEP in the processing and construction phases were 80.93% and 44.47%.

As shown in Figure 10, the environmental impact contribution of point 11 was the largest. The environmental impact contribution of point 1 was the second-largest, and the environmental contribution impact of point 16 was the third-largest; The environmental impact contributions of the other points were much lower.

For Figure 11, the percentages show that the GWP in the operation and maintenance stage accounted for 67.26% of the total GWP of the cable-stayed bridge, and accounted for 97.40% of the total environmental contribution during the operation and maintenance phases. For point 1, the percentages show that the GWP in the processing and construction phase accounted for 22.50% of the total GWP of the cable-stayed bridge and 80.80% of the total environmental contribution in the construction and installation phases. Finally, the environmental impact of point 16 was much lower, and percentages show that the demolition phase GWP accounts for 9.20% of the total GWP of the cable-stayed bridge, and the abandonment and demolition phase GWP accounted for 99.70% of the total environmental contribution in the abandon and demolition phase.

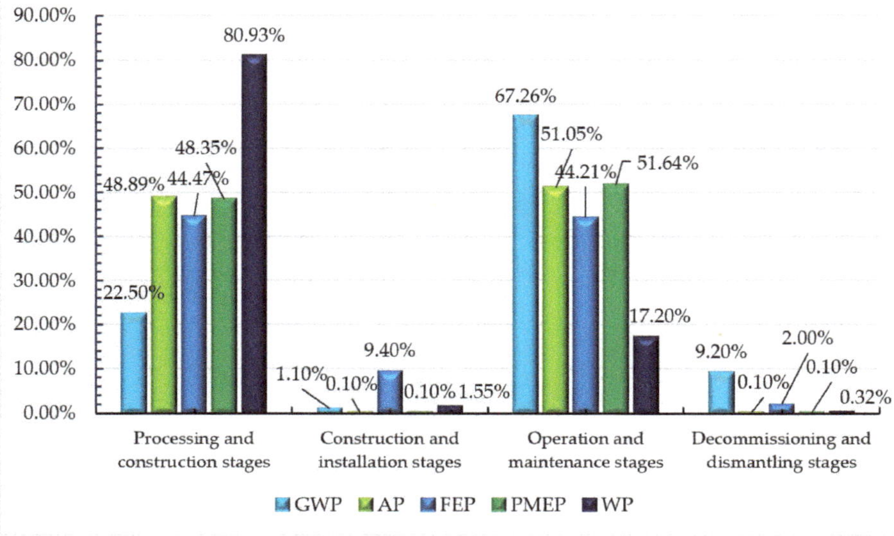

Figure 9. Schematic diagram of the environmental impact contribution of the four-stage cable-stayed bridge.

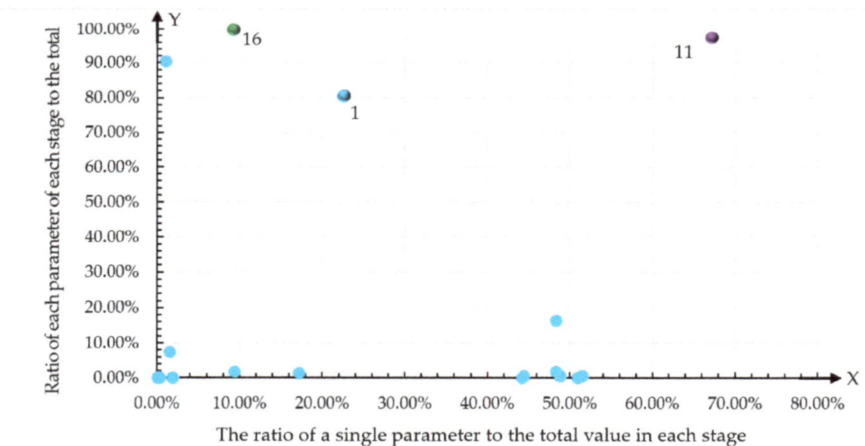

Figure 10. Distribution map of the points of five environmental impact contributions of the cable-stayed bridge.

Through the research and analysis in Sections 4.1 and 4.2, it was concluded that the environmental influence factors of cable-stayed bridges were divided into five index levels, and the influence factors of each index level had an impact on the environment.

The study proposed a Markov chain model capable of considering the maintenance factors used by Li et al. [45]. The Markov chain model can solve the problem of multiple factors. The given Markov chain probability diagram shows the ratio of the environmental impact factors at each stage of the cable-stayed bridge (Figure 11). The data comes from Table 5.

As shown in Figure 10, the environmental impact contributions of the cable-stayed bridge were mainly from the processing and construction phases, which manifested as FEP = 193,738.3 kg and WP

= 8191,263.9 kg, respectively. During the operation and maintenance phases, the contributions were GWP = 121,031,298 kg, AP = 317,034.6 kg, and PMEP = 979,739.5 kg, respectively.

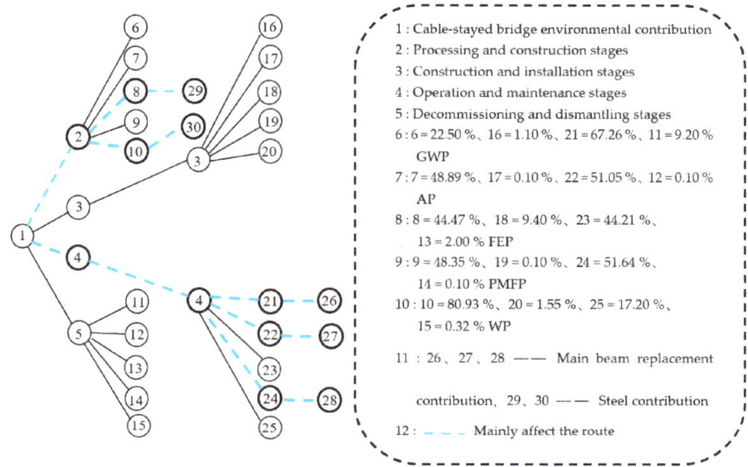

Figure 11. Schematic diagram of the Markov chain probabilities of the cable-stayed bridge.

5. Conclusions

This study analyzed a three-tower cable-stayed bridge in China. First, this research studied and analyzed the definition of environmental impact assessment parameters, omitted the set of 13 parameters with small impact, and focussed on the analysis of the five parameters by applying midpoint modeling. The final endpoint modeling analysis conclusion verified the accuracy and effectiveness of this method. The amount of CO_2 emission in the environmental impact value GWP of the cable-stayed bridge accounted for 93.23% of the total emissions.

Second, the analysis data of the cable-stayed bridge adopted the data analysis of the whole bridge. The bottom and topside parts of the bridge were all taken as the analysis object. The environmental impact contribution of the concrete production stage was classified into the production and construction phases for analysis because the cable-stayed bridge is a municipal project and it uses commercial concrete. Therefore, the environmental impact contribution of the construction and installation phases was mainly influenced by management technicians, accounting for 52.3% of the emissions during this stage. At the same time, it shows that in the process of the LCA analysis, data classification analysis should be set according to the actual situation of the project, which makes the results more scientific and practical.

Third, the operation and maintenance stage of the cable-stayed bridge was the main aspect that contributed to the environmental emissions, since most of the structural components of the cable-stayed bridge were replaced 2 to 10 times during their lifetime, and each change had a significant impact on the environmental impact contribution. Combined with vehicle exhaust emissions, this resulted in a 64.37% environmental impact contribution of the operations and maintenance phases, where the specific environmental impact contribution value was 124,307.2 t.

Finally, each stage contributed to the environmental emissions, and the numerical value reflected the degree of environmental impact. In the last two phases in particular, carbonation had a greater impact on the environmental impact contribution of the last stage. Especially in the operation and maintenance phases, concrete carbonation absorbed 1712.9 tons of CO_2, which made an important contribution to the environmental impact. At the same time, the carbonation of the concrete opened up corrosion channels for the steel bars of the structural components, resulting in maintenance and replacement during the operation stage, and finally, the cable-stayed bridge was demolished.

There are still some defects in this study. These lie in the insufficient analysis of the data of equipment loss and damage caused by daily accidents, and environmental impact assessment caused by the natural environment, such as earthquakes, tsunamis, and hurricanes. The data analysis, theories, and methods of modeling used in this study can be used as a reference for research in this field. Furthermore, the research results can provide ideas and references for researchers and managers to study the whole-life analysis of a basic bridge.

Author Contributions: Investigation, Z.Z. and J.A.; methodology, Z.Z, J.A., and V.Y.; supervision, J.A. and V.Y.; validation, Z.Z. and V.Y.; writing—original draft, Z.Z.; writing—review and editing, J.A. and V.Y. All authors have read and agreed to the published version of the manuscript.

Funding: This research was funded by the Spanish Ministry of Economy and Competitiveness, along with FEDER (Fondo Europeo de Desarrollo Regional), project grant number: BIA2017-85098-R.

References

1. U.S. Department of Energy Office; Murphy, C.; Cunliff, C. (Eds.) *Environment Baseline, Volume 1: Greenhouse Gas Emissions from the U.S. Power Sector*; Office of Energy Policy and Systems Analysis (EPSA) of the US Department of Energy: Washington, WA, USA, 2016.
2. The Intergovernmental Panel on Climate Change. Available online: https://www.ipcc.ch/2018/10/08/summary-for-policymakers-of-ipcc-special-report-on-global-warming-of-1-5c-approved-by-governments/ (accessed on 23 May 2020).
3. Sánchez-Garrido, A.J.; Yepes, V. Multi-criteria assessment of alternative sustainable structures for a self-promoted, single-family home. *J. Clean. Prod.* **2020**, *258*, 120556.
4. Kong, J.S.; Frangopol, D.M. Life-cycle reliability-based maintenance cost optimization of deteriorating structures with emphasis on bridges. *J. Struct. Eng.* **2003**, *129*, 818–828. [CrossRef]
5. Larsson Ivanov, O.; Honfi, D.; Santandrea, F.; Stripple, H. Consideration of uncertainties in LCA for infrastructure using probabilistic methods. *Struct. Infrastruct. Eng.* **2019**, *15*, 711–724. [CrossRef]
6. ETSI project-Stage III. Available online: http://etsi.aalto.fi/Etsi3/index.html (accessed on 29 July 2020).
7. Sonnemann, G.; Margni, M. (Eds.) *LCA Compendium The Complete World of Life Cycle Assessment*; Springer: Dordrecht, The Netherlands, 2015; Available online: http://www.springer.com/series/11776 (accessed on 23 May 2020).
8. Pålsson, A.C.; Carlson, R. Adaptations of the Swedish National LCA Database System. In Proceedings of the Sixth International Conference on EcoBalance, Tsukuba, Japan, 25–27 October 2004.
9. ProBas Prozessorientierte Basisdaten für Umweltmanagementsysteme. Available online: https://www.probas.umweltbundesamt.de/php/news.php?id=3 (accessed on 4 April 2020).
10. LCA Society of Japan. Japan Environmental Management Association for Industry. Available online: https://lca-forum.org/english/ (accessed on 24 April 2020).
11. USDA-National Agricultural Library. LCA Common Submission Guidelines. Available online: https://www.lcacommons.gov/sites/default/files/pdfs/DRAFT_lcaCommonsSubmissionGuidelines_Final_2018-07-25.pdf (accessed on 24 April 2020).
12. Ecoinvent database. Available online: https://www.ecoinvent.org/database/database.html (accessed on 2 April 2020).
13. García-Segura, T.; Yepes, V.; Frangopol, D.M.; Yang, D.Y. Lifetime reliability-based optimization of post-tensioned box-girder bridges. *Eng. Struct.* **2017**, *145*, 381–391. [CrossRef]
14. Itoh, Y.; Kitagawa, T. Using CO2 emission quantities in bridge lifecycle analysis. *Eng. Struct.* **2003**, *25*, 565–577. [CrossRef]
15. Hong Wei, L.M. Research on bridge environmental impact with life cycle assessment. *J. Wuhan Univ. Technol.* **2007**, *29*, 119–122.

16. Heijungs, R.; Huppes, G.; Guinée, J.B. Life cycle assessment and sustainability analysis of products, materials and technologies. Toward a scientific framework for sustainability life cycle analysis. *Polym. Degrad. Stab.* **2010**, *95*, 422–428. [CrossRef]
17. Penadés-Plà, V.; Martí, J.V.; García-Segura, T.; Yepes, V. Life-cycle assessment: A comparison between two optimal post-tensioned concrete box-girder road bridges. *Sustainability* **2017**, *9*, 1864. [CrossRef]
18. Jutta Hildenbrand OpenLCA 1.10. Available online: http://www.openlca.org/ (accessed on 30 July 2020).
19. ISO. *ISO 14044:2006 Environmental Management—Life Cycle Assessment—Requirements and Guidelines*; International Organization for Standardization: Geneva, Switzerland, 2006.
20. CML—Department of Industrial Ecology. CML-IA Characterisation Factors. Available online: https://www.universiteitleiden.nl/en/research/research-output/science/cml-ia-characterisation-factors (accessed on 4 May 2020).
21. Bare, J.; Hofstetter, P.; Pennington, D.W.; Udo de Haes, H.A. Life cycle impact assessment workshop summary. Midpoints versus endpoints: The sacrifices and benefits. *Int. J. Life Cycle Assess.* **2000**, *5*, 319–326. [CrossRef]
22. Wei, J.; Cen, K. A preliminary calculation of cement carbon dioxide in China from 1949 to 2050. *Mitig. Adapt. Strateg. Glob. Chang.* **2019**, *24*, 1343–1362. [CrossRef]
23. Du, G.; Safi, M.; Pettersson, L.; Karoumi, R. Life cycle assessment as a decision support tool for bridge procurement: Environmental impact comparison among five bridge designs. *Int. J. Life Cycle Assess.* **2014**, *19*, 1948–1964. [CrossRef]
24. Kim, T.H.; Tae, S.H. Proposal of environmental impact assessment method for concrete in South Korea: An application in LCA (life cycle assessment). *Int. J. Environ. Res. Public Health.* **2016**, *13*, 1074. [CrossRef]
25. Arbault, D.; Rivière, M.; Rugani, B.; Benetto, E.; Tiruta-Barna, L. Integrated earth system dynamic modeling for life cycle impact assessment of ecosystem services. *Sci. Total Environ.* **2014**, *472*, 262–272. [CrossRef] [PubMed]
26. Ogundipe, O.M. Marshall Stability and Flow of Lime-modified Asphalt Concrete. *Transp. Res. Proced.* **2016**, *14*, 685–693. [CrossRef]
27. Ming, N.H.; Kang, G.C.; Hua, G.Y. "Chinese-style" electric furnace steelmaking process carbon emission characteristics and source analysis. *J. Northeast. Univ. Nat. Sci.* **2010**, *40*, 212–217.
28. Liu, Y.; Wang, Y.; Li, D. Estimation and uncertainty analysis on carbon dioxide emissions from construction phase of real highway projects in China. *J. Clean. Prod.* **2017**, *144*, 337–346. [CrossRef]
29. Colvile, R.N.; Hutchinson, E.J.; Mindell, J.S.; Warren, R.F. The transport sector as a source of air pollution. *Atmos. Environ.* **2001**, *35*, 1537–1565. [CrossRef]
30. Fushun City Statistics Bureau. Fushun City 2019 National Economic and Social Development Statistical Bulletin. Available online: http://www.tjcn.org/tjgb/ (accessed on 9 April 2020).
31. The State Council of the People's Republic of China. Opinions on further strengthening the management of urban planning and construction. Available online: http://www.gov.cn/zhengce/2016-02/21/content_5044367.htm (accessed on 15 May 2020).
32. Wang, K.; Tian, H.; Hua, S.; Zhu, C.; Gao, J.; Xue, Y.; Wang, Y.; Zhou, J. A comprehensive emission inventory of multiple air pollutants from iron and steel industry in China: Temporal trends and spatial variation characteristics. *Sci. Total Environ.* **2016**, *559*, 7–14. [CrossRef]
33. Dai Wan Jiang, C.F. Discussion on Issues Related to Bridge Lifetime Design. *J. Wuhan Univ. Technol.* **2011**, *35*, 520–523.
34. Department of Ecology and Environment of Liaoning. Bulletin of the Ecological Environment Status of Liaoning Province from 2010 to 2019. Available online: http://sthj.ln.gov.cn/hjzl/ (accessed on 12 April 2020).
35. Hammervold, J.; Reenaas, M.; Brattebø, H. Environmental life cycle assessment of bridges. *J. Brid. Eng.* **2013**, *18*, 153–161. [CrossRef]
36. Chen, Y.; Liu, P.; Yu, Z. Effects of environmental factors on concrete carbonation depth and compressive strength. *Materials.* **2018**, *11*, 2167. [CrossRef] [PubMed]
37. Yu, B.; Cheng, D.; Yang, L. Quantification of environmental effect for carbonation and durability analysis of concrete structures. *China Civ. Eng. J.* **2015**, *48*, 51–59.
38. Liang, M.T.; Huang, R.; Feng, S.A.; Yeh, C.J. Service life prediction of pier for the existing reinforced concrete bridges in chloride-laden environment. *J. Mar. Sci. Technol.* **2009**, *17*, 312–319.

39. Ministry of Ecology and Environment of the People's Republic of China. List of Existing Chemical Substances in China. Available online: http://www.mee.gov.cn/gkml/hbb/bgg/201301/t20130131_245810.htm (accessed on 25 July 2020).
40. Watson, J.G.; Chow, J.C.; Fujita, E.M. Review of volatile organic compound source apportionment by chemical mass balance. *Atmos. Environ.* **2001**, *35*, 1567–1584. [CrossRef]
41. Development of Liaoning Provincial Highway and Waterway Transportation Industry during the "Twelfth Five-Year Plan" Liaoning Provincial Department of Transportation. Available online: http://jtt.ln.gov.cn/sj/tjxx/201608/t20160818_2504856.html (accessed on 31 July 2020).
42. Martínez-Muñoz, D.; Martí, J.V.; Yepes, V. Steel-concrete composite bridges: Design, life cycle assessment, maintenance and decision-making. *Adv. Civ. Eng.* **2020**, 8823370. [CrossRef]
43. Kim, K.J.; Yun, W.G.; Cho, N.; Ha, J. Life cycle assessment based environmental impact estimation model for pre-stressed concrete beam bridge in the early design phase. *Environ. Impact Assess. Rev.* **2017**, *64*, 47–56. [CrossRef]
44. Zhu, X.; Li, H.; Chen, J.; Jiang, F. Pollution control efficiency of China's iron and steel industry: Evidence from different manufacturing processes. *J. Clean. Prod.* **2019**, *240*, 118184. [CrossRef]
45. Li, L.; Sun, L.; Ning, G. Deterioration prediction of urban bridges on network level using Markov-chain model. *Math. Probl. Eng.* **2014**, *2014*, 728107. [CrossRef]

International Journal of
Environmental Research and Public Health

Article

Life Cycle Sustainability Performance Assessment Method for Comparison of Civil Engineering Works Design Concepts: Case Study of a Bridge

Kristine Ek [1,2,*], **Alexandre Mathern** [1,2], **Rasmus Rempling** [1,2], **Petra Brinkhoff** [1], **Mats Karlsson** [2,3] and **Malin Norin** [1]

[1] NCC AB, Gullbergs Strandgata 2, 405 14 Göteborg, Sweden; alexandre.mathern@chalmers.se (A.M.); rasmus.rempling@chalmers.se (R.R.); petra.brinkhoff@ncc.se (P.B.); malin.norin@ncc.se (M.N.)

[2] Department of Architecture and Civil Engineering, Chalmers University of Technology, 6, 412 96 Göteborg, Sweden; mats.d.karlsson@trafikverket.se

[3] Swedish Transport Administration, Bataljonsgatan 8, 553 05 Jönköping, Sweden

* Correspondence: kristine.ek@ncc.se

Received: 9 October 2020; Accepted: 23 October 2020; Published: 28 October 2020

Abstract: Standardized and transparent life cycle sustainability performance assessment methods are essential for improving the sustainability of civil engineering works. The purpose of this paper is to demonstrate the potential of using a life cycle sustainability assessment method in a road bridge case study. The method is in line with requirements of relevant standards, uses life cycle assessment, life cycle costs and incomes, and environmental externalities, and applies normalization and weighting of indicators. The case study involves a short-span bridge in a design-build infrastructure project, which was selected for its generality. Two bridge design concepts are assessed and compared: a concrete slab frame bridge and a soil-steel composite bridge. Data available in the contractor's tender phase are used. The two primary aims of this study are (1) to analyse the practical application potential of the method in carrying out transparent sustainability assessments of design concepts in the early planning and design stages, and (2) to examine the results obtained in the case study to identify indicators in different life cycle stages and elements of the civil engineering works project with the largest impacts on sustainability. The results show that the method facilitates comparisons of the life cycle sustainability performance of design concepts at the indicator and construction element levels, enabling better-informed and more impartial design decisions to be made.

Keywords: sustainability; life cycle assessment; life cycle costing; environmental externalities; indicator; multi-criteria decision analysis; civil engineering; bridge; design

---

1. Introduction

In civil engineering projects, life cycle environmental, social, and economic sustainability performance is becoming increasingly important, as reflected in the large number of standards published on the subject in recent years [1–4]. To make better-informed decisions regarding the impact of design choices on the sustainability of civil engineering works, sustainability performance assessment is recommended [5,6]. It is important that assessments are performed in a harmonized way and can be compared impartially. Current standards provide the general framework for the sustainability assessment of civil engineering works but do not give detailed guidance on the calculation of indicators and their aggregation [2,4]. In most studies on sustainability-based design and optimization of bridges, simplifications are used, and the assessment is based on one or two selected indicators and only covers certain life cycle stages [6], e.g., CO_2 emissions and the cost of construction materials [7] and of transport and installation [8] and embodied energy of construction materials [9].

The potential to influence the sustainability of a design is larger in the early stages of the design process than in later stages [10]. It is therefore important to define indicators that can support an iterative sustainability-driven design process from concept to final implementation. To enable the identification of sustainable designs, a formalized method that allows transparent, comparable, and automatable sustainability design and assessment is desired.

Ek et al. presented a harmonized method for life cycle sustainability assessment and comparison of civil engineering works design concepts [11]. The proposed method includes guidance on the calculation of environmental, social and economic indicators, based on life cycle assessment (LCA), life cycle costing (LCC) and external costs, and aggregation using normalization and weighting factors, in accordance with the principles and requirements of methods for sustainability performance assessment given in the standards [2] and [4].

This paper evaluates the previously proposed method by applying it in a road bridge case study. The study has two primary aims: (1) to analyse the practical application potential of the method in carrying out transparent sustainability assessments of design concepts in the early planning and design stages and (2) to examine the results obtained in the case study to identify critical indicators in different life cycle stages as well as critical elements in the civil engineering works project with the greatest impacts.

Life cycle stages are classified into so-called modules by the related standards [1–4,12–15], see Figure 1.

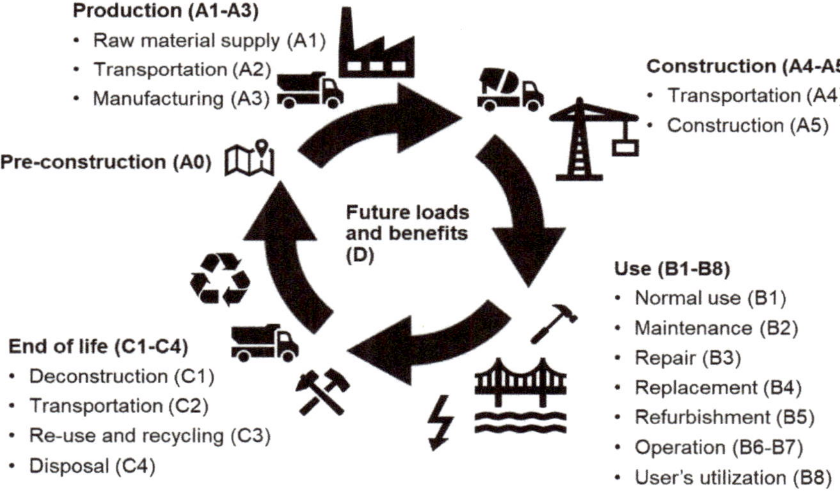

Figure 1. Schematic illustration of the life-cycle stages of a civil engineering works project and their classification in modules.

Module A0 is the pre-construction stage. Modules A1–A3 represent the production stage from raw material extraction to construction material manufacturing where A1 is material extraction, A2 is transport from extraction to manufacture, and A3 is material manufacture. A4–A5 represent the construction process stage where A4 is transport from material manufacture to the construction site and A5 is the construction site works. B1–B5 represent the use stage relating to maintenance where B1 is normal use of the bridge, B2 is maintenance, B3 is repair, B4 is replacement, and B5 is refurbishment. B6–B7 represent the use stage related to the operation where B6 is operational energy use and B7 is operational water use. B8 is the use stage related to the user's utilization of the civil engineering works. C1–C4 cover the end-of-life stage where C1 is deconstruction, C2 is transport from the deconstruction site to the waste management site, C3 is the waste processing of materials intended for reuse, recycling, and energy recovery, and C4 is waste disposal. Module D represents the benefits and loads beyond the system boundary of the civil engineering works.

2. Materials and Methods

The new method for life cycle sustainability assessment and comparison of civil engineering works design concepts presented by Ek et al. [11] was applied to a case study. The case study involved a bridge in a design-build infrastructure project and was selected for its generality. Two alternative bridge design concepts were assessed and compared: a concrete slab frame bridge (CSF bridge) and a soil-steel composite bridge (SSC bridge). The prerequisites for the assessment are presented in Table 1.

Table 1. Case study specific prerequisites for the assessment.

Characteristic	Case Study Prerequisite
Object of assessment	Bridge 6-1282-1 on Road 26, Sweden 6 m long, 9 m wide, 3 road lanes
Intended use of the assessment	Design concept comparison
Additional functions provided	-
Functional equivalent: (a) Type/use of the civil engineering works, (b) Capacity, (c) Reference study period and pattern of use, (d) Design life (required service life, RSL)	(a) Road bridge with fauna passage (b) 7200 AADT, 100 km/h (c) 80 years, see Appendices C and D (d) 80 years
Time of assessment in the life cycle	Detailed design/tender phase
Life cycle stages assessed	A1–A5, B1–B8, C1–C4, D
Justification of the exclusion of modules	A0 was excluded because of its insignificant impact on the sustainability performance and because it does not differ between the concepts.
Area of influence	Environmental, social, and economic dimensions (environmental externalities): The surroundings and people in the direct vicinity of the bridge, receiving emissions from fuel combustion and other activities during construction, use and deconstruction from passing vehicles across the length of the bridge. Economic dimension (Life cycle costs (LCC) and incomes): The users of passing vehicles on the bridge (module B8), the client of the constructed bridge (all other modules).
Energy and mass flows considered in the assessment	See Appendices B–D
General assumptions and scenarios used	See Appendices B–D
Sources of data for the indicators	See Appendices B–D
Statement about whether data are specific or generic	See Appendix A
Reference year for the cost data	2019

The functional unit is 1 m of bridge length and year of required service life (RSL). The reference study period is equal to the required service life: 80 years. A functional equivalent of 1 km of bridge and year of RSL is prescribed by the product category rules (PCR) of the International EPD System [13], but it was decided to use 1 m instead, because the short length of the bridge (6 m) would yield non-representative results if scaled to 1 km.

Life cycle assessment (LCA) was performed according to the standard EN 15804 [12] using the LCA software GaBi Professional, version 9.5 (Sphera Solutions GmbH, Leinfelden-Echterdingen, Germany) [16]. The GaBi datasets used are presented in Appendix A. Normalization and weighting factors presented in [11] (adapted from the factors used in the Product Environmental Footprint (PEF) method [17–19] were used (see Table 2). As proposed in [11], some indicators are categorized into the environmental dimension and some into the social dimension. In PEF, all indicators are in a single dimension. The PEF weighting factors of the indicators have thus been scaled to a total of 100 in the environmental and social dimension, respectively. The life cycle costing (LCC and incomes) was calculated according to the standard EN 15686-5 [14], and environmental externalities were calculated in accordance with ISO 14008 [15]. The economic indicators are presented separately in line with the standards' requirements. LCC and incomes as well as environmental externalities are presented as the net present value (NPV) using a discount rate of 3%. This discount rate was chosen as it is the rate prescribed by the currently available standard on calculation methods for economic performance (for buildings) [20]. Environmental externalities were calculated using the EPS 2015dx method [21]. Abbreviations and units of measurement used for the indicators included in the assessment are presented in Table 3.

Table 2. Normalization and weighting factors used for environmental and social indicators [11].

Dimension	Indicator	Normalization Factor (NF)	Weighting Factor (%)
	Acidification potential	55.6	8.43
	Eco-toxicity potential (freshwater)	42,683	2.61
	Potential soil quality index	819,498	10.80
	Global warming potential total (fossil + biogenic + luluc)	8096	28.63
	Abiotic depletion potential for non-fossil resources	0.0636	10.27
Environmental	Abiotic depletion potential for fossil resources	65,004	11.31
	Eutrophication potential (freshwater)	1.61	3.81
	Eutrophication potential (marine)	19.5	4.02
	Eutrophication potential (terrestrial)	177	5.04
	Ozone depletion potential	0.0536	8.58
	Photochemical ozone creation potential	40.6	6.50
	Potential ionizing radiation—human health	4220	18.94
	Human toxicity potential—cancer effects	0.0000169	8.05
Social	Human toxicity potential—non-cancer effects	0.000230	6.96
	Particulate matter emissions	0.000595	33.88
	Water user deprivation potential	11,469	32.17

The method requires the use of a life cycle inventory (LCI) to calculate the indicators. In the case study, the LCI was in the form of a bill of materials (BOM), which was calculated for each design concept for modules A1–A5 as well as for the use (B1–B8) and end-of-life (C1–C4 and D) stages. It was created based on data available in the tender phase. The BOM is presented in Appendix B. The values of modules B1–B8, C1–C4, and D were calculated based on the scenarios presented in Appendix C. Realistic and representative scenarios of the resource consumption and costs of modules B–D were developed for each design concept. Scenarios were developed based on project documentation and literature data, if available, as well as expert knowledge. Expert knowledge was based on both the project manager of the construction project used as case study, and the authors' previous experience from bridge design, construction and maintenance projects in Sweden.

For the LCC, the average market prices per unit for each of the resources included were used. Average prices were supplied by the project manager of the case study project.

Transport modes and distances are presented in Appendix D. Cut-offs were made for the transport of form oil, bitumen sheet, bitumen sealant, impregnation, geotextile, mortar, graffiti protection, polypropylene pipe, polyethylene foam, bituprimer, epoxy sealant, and plastic film.

Table 3. Sustainability dimensions, categories, indicator names, abbreviations, and units of measurement for the indicators.

Dimension	Category	Indicator Name	Abbreviation	Unit of Measurement
Environmental	Acidification	Acidification potential	AP	mol H + eq
	Biodiversity	Eco-toxicity potential (freshwater)	ETP-fw	CTUe
		Potential soil quality index	SQP	Dimensionless
	Climate change	Global warming potential total (fossil + biogenic + luluc)	GWP-total	kg CO_2 eq
	Depletion of abiotic resources—minerals and metals	Abiotic depletion potential for non-fossil resources	ADPE	kg Sb eq
	Depletion of abiotic resources—fossil fuels	Abiotic depletion potential for fossil resources	ADPF	MJ, net calorific value
	Eutrophication	Eutrophication potential (freshwater)	EP-freshwater	kg P eq
		Eutrophication potential (marine)	EP-marine	kg N eq
		Eutrophication potential (terrestrial)	EP-terrestrial	mol N eq
	Ozone depletion	Ozone depletion potential	ODP	kg CFC 11 eq
	Photochemical ozone creation	Photochemical ozone creation potential	POCP	kg NMVOC eq
Social	Health and comfort	Potential ionizing radiation—human health	PIR	kBq U235 eq
		Human toxicity potential—cancer effects	HTP c	CTUh
		Human toxicity potential—non-cancer effects	HTP nc	CTUh
		Particulate matter emissions	PM	Disease incidence
		Water user deprivation potential	WDP	m^3 world deprived eq
Economic	Life cycle economic balance	LCC and incomes	-	Euro
	External cost	Environmental externalities	-	Euro

3. Results

The results are first presented separately for the CSF bridge design concept and the SSC bridge design concept. The results for the two concepts are then compared. Positive economic indicator values indicate costs, while negative values indicate incomes. For the environmental and social indicators, positive values indicate a negative impact, and negative values indicate a positive impact.

The result for module B8 is presented in a separate figure, since it is relatively higher than that of the other modules. Furthermore, the result for module B8 was excluded from the comparison of the concepts, since it would have disguised the differences in concepts among the other modules. In addition, the result for module B8 was found to be equal for both concepts.

3.1. Concrete Slab Frame Bridge Design Concept

The results for each indicator of the sustainability assessment of the CSF bridge in units of measurement and per life cycle stage are presented in Table 4.

The results for the CSF bridge are presented per life cycle stage in Table 5. The results for the environmental and social dimensions are aggregated on the dimension level, using the normalization and weighting factors in Table 2, while the results for the economic dimension are summarized at the indicator level. The normalized and weighted results for the environmental and social dimensions are presented per life cycle stage and per indicator in Figure 2 (excluding module B8). Figure 3 shows the contribution of each resource to the total impact over the life cycle (modules A–C excluding B8) in the environmental dimension and the social dimension, respectively, for the resources with the greatest contributions. The normalized and weighted results for the environmental and social dimensions are presented for module B8 per indicator in Figure 4.

Table 4. Results for the concrete slab frame bridge (CSF) bridge design concept per indicator in units of measurement per life-cycle stage for the functional unit. Modules not assessed are abbreviated as "MNA".

Indicator	A0	A1–A3	A4–A5	B1–B5	B6–B7	B8	C1–C4	D
AP	MNA	0.78	0.57	0.049	0	4.2	0.29	−0.18
ETP-fw	MNA	1780	1642	133	0	4063	380	−333
SQP	MNA	1002	7588	22	0	22,083	167	−579
GWP-total	MNA	313	172	12	0	234	47	−70
ADPE	MNA	2.1×10^{-4}	1.6×10^{-5}	4.4×10^{-6}	0	5.4×10^{-5}	3.7×10^{-6}	-1.0×10^{-4}
ADPF	MNA	3740	2344	230	0	5229	584	−892
EP-freshwater	MNA	5.5×10^{-4}	5.6×10^{-4}	2.6×10^{-5}	0	3.110^{-2}	1.1×10^{-4}	-2.3×10^{-4}
EP-marine	MNA	0.24	0.25	0.015	0	1.68	0.10	−0.05
EP-terrestrial	MNA	2.61	2.82	0.17	0	20.4	1.1	−0.5
ODP	MNA	6.9×10^{-11}	2.6×10^{-11}	3.1×10^{-11}	0	4.9×10^{-10}	8.5×10^{-14}	-3.5×10^{-13}
POCP	MNA	0.64	0.71	0.046	0	2.7	0.28	−0.14
PIR	MNA	60	3.1	0.57	0	148.5	0.39	−13
HTP c	MNA	1.8×10^{-6}	5.6×10^{-8}	1.2×10^{-8}	0	7.1×10^{-7}	2.9×10^{-8}	-5.8×10^{-8}
HTP nc	MNA	4.8×10^{-6}	2.0×10^{-6}	4.3×10^{-7}	0	1.1×10^{-4}	2.9×10^{-6}	-6.9×10^{-7}
PM	MNA	1.5×10^{-5}	1.2×10^{-5}	5.8×10^{-7}	0	4.1×10^{-5}	2.4×10^{-6}	-3.2×10^{-6}
WDP	MNA	71	11	0.86	0	109	6.3	−16
LCC and incomes	MNA	210	287	18	0	86	12	−1.1
Environmental externalities	MNA	123	56	2.0	0	31	1.4	−7.8

Table 5. Results for the CSF bridge design concept per life cycle stage aggregated on the dimension level (and on the indicator level for the economic dimension) for the functional unit. The results for the environmental and social dimensions are normalized and weighted, and the results for the economic dimension are summarized. Modules not assessed are abbreviated as "MNA".

Dimension	Indicator (*Unit*)	A0	A1–A3	A4–A5	B1–B7	B8	C1–C4	D
Environmental	All (*dimensionless*)	MNA	2.3	1.6	0.12	4.4	0.43	−1.5
Social	All (*dimensionless*)	MNA	2.3	0.82	0.057	6.9	0.26	−1.2
Economic	LCC and incomes (*Euro*)	MNA	210	287	18	86	12	−1.1
Economic	Environmental externalities (*Euro*)	MNA	123	56	2.0	31	1.4	−7.8

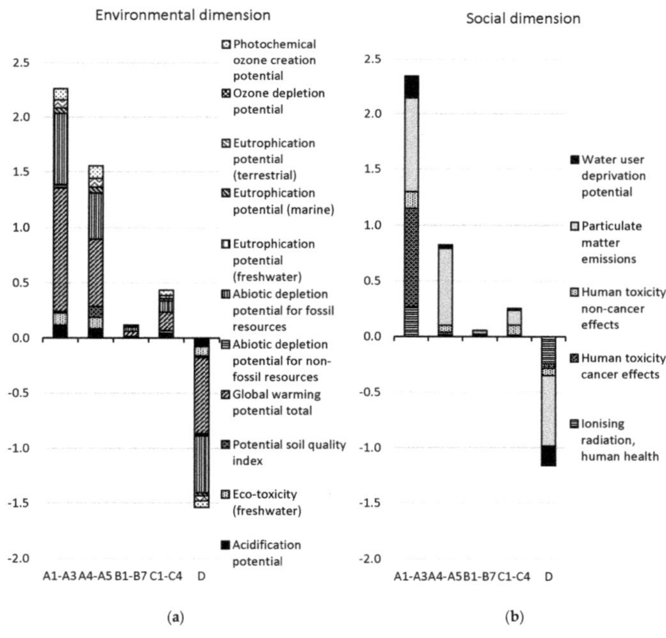

Figure 2. Normalized and weighted results for the CSF bridge design concept per life cycle stage in (**a**) the environmental dimension and (**b**) the social dimension for each indicator and for the functional unit. Module B8 is not included. See Figure 4 for module B8. Note that the ozone depletion potential indicator bar cannot be seen in the figure since it is very small.

In the environmental dimension, 52% of the total impact of life cycle stages A–C, excluding module B8, occurs in the production stage (modules A1–A3), followed by the construction stage (36%, modules A4–A5). If included, B8 would contribute to 50% of the total impact for life cycle stages A–C. Module D presents the potential to reduce the environmental impact of life cycle stages A–C (excluding module B8) by 35% through future re-use or recycling. The main contribution to the environmental impact over the life cycle (modules A–C, excluding B8) comes from the indicators "global warming potential" (44% of the total impact) and "abiotic depletion potential for fossil resources" (28% of the total impact, see Figure 2). Fifty-one percent of the global warming potential is caused by the production (29%) and transport to the construction site (22%) of 6833 tons of aggregates, while 23% is caused by the production of 472 tons of concrete, and 6% is caused by the production and combustion of diesel used on the site during construction. The production and transport of aggregates and the landfilling of aggregate waste contributes to 53% of the total environmental impact over the life cycle (see Figure 3).

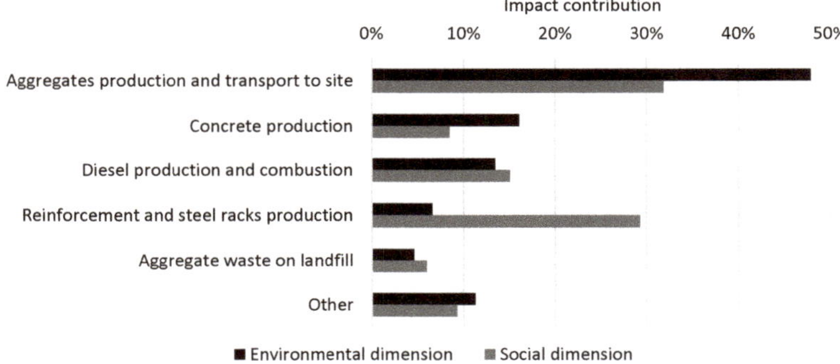

Figure 3. Contribution of resources to the total impact over the life cycle (modules A–C, excluding B8) for the CSF bridge design concept in the environmental and social dimensions.

In the social dimension, 67% of the total impact for life cycle stages A–C (excluding module B8) occurs in the production stage, followed by the construction stage (24%). If included, module B8 would contribute to 66% of the total impact for life cycle stages A–C. Module D presents the potential to reduce the social impact by 33% of the total impact for life cycle stages A–C (excluding module B8) by future re-use or recycling. The main contribution to social impact over the life cycle (modules A–C, excluding B8) comes from the indicators "particulate matter emissions" (49% of the total impact) and "human toxicity—cancer effects" (27% of the total impact). Thirty-five percent of the particulate matter emissions are caused by the production of aggregates and their transport to the construction site, and 28% is caused by the production and combustion of diesel at the construction site during construction. The production and transport of aggregates and the landfilling of aggregate waste contributes to 38% and reinforcement steel and steel rack production contributes to 29% of the total social impact over the life cycle (see Figure 3).

In the economic dimension, 54% of the total net cost for the indicator "LCC and incomes" in life cycle stages A–C (excluding module B8) occurs in the construction stage (mainly in A5), see Table 5. Forty percent occurs in the production stage. If module B8 was included, it would contribute to 14% of the total net cost for life cycle stages A–C. Module D presents the potential to reduce the future net cost (by future re-use or recycling) by 0.2% for life cycle stages A–C (excluding module B8). The main contributor (30%) to the cost of the LCC and incomes over the life cycle is the work costs during construction, followed by the costs for production and transport of aggregates (18%) and for the production and transport of concrete and reinforcement steel (17%). For the indicator "environmental externalities", 68% of the total cost in life cycle stages A–C (excluding module B8) occurs in the production stage, followed by the construction stage (31%). If module B8 was included, it would contribute to 12% of the total cost for life cycle stages A–C. Module D presents the potential to reduce future costs of the total impact for life cycle stages A–C (excluding module B8) by 4%. The main contributors (43%) to the environmental externalities over the life cycle are the production and transport of aggregates and transport to waste disposal and the landfilling of aggregate waste.

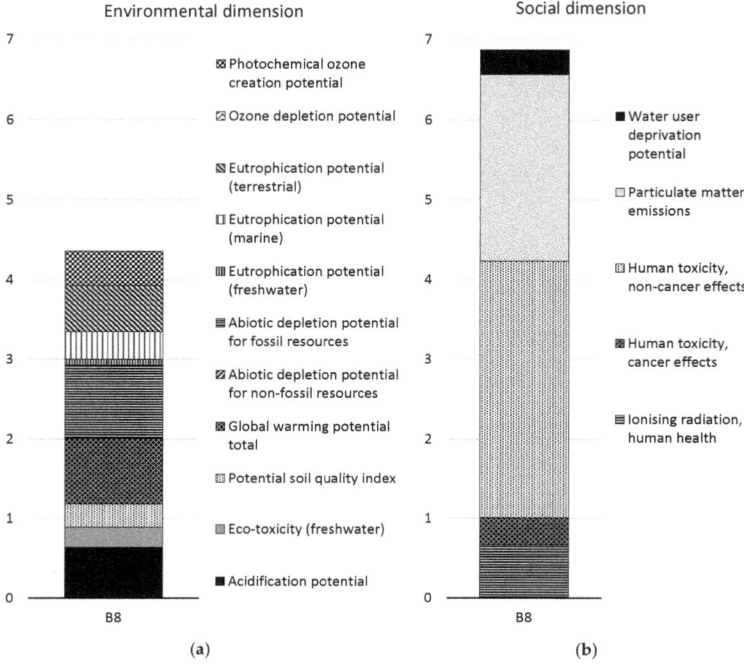

Figure 4. Normalized and weighted results for module B8 for both bridge design concepts in (**a**) the environmental dimension and (**b**) the social dimension for each indicator and for the functional unit. Note that the ozone depletion potential indicator bar cannot be seen in the figure since it is very small.

3.1.1. Modules A1–A3

In the production stage, greenhouse gas emissions have the largest impact in the environmental dimension (see Figure 2). The indicator "global warming potential" contributes to 49% of the total impact in this life cycle stage. Fifty percent of this originates from the production of 6833 tons of aggregates, 39% from the production of 472 tons of concrete, and 9% from the production of 23 tons of reinforcement steel (carbon and stainless). The production of asphalt, steel racks, and other built-in construction materials accounts for 2% of the global warming potential.

The second largest impact comes from the abiotic depletion potential for fossil resources; this constitutes 29% of the total environmental impact. Sixty-eight percent of this depletion is caused by the production of aggregates, 16% by the production of concrete, and 11% by the production of reinforcement steel.

The "acidification potential", the "eco-toxicity potential", and the "photochemical ozone creation potential" indicators also have significant impacts, each accounting for about 5% of the total environmental impact. The production of aggregates contributes to 46–69% of the total impact for each of these indicators.

The largest impact in the social dimension comes from the indicators "human toxicity—cancer effects" (38%) and "particulate matter emissions" (36%). For "human toxicity—cancer effects", 97% of the impact comes from the production of 553 kg of stainless-steel reinforcement. For "particulate matter emissions", 66% of the impact comes from the production of aggregates.

In the economic dimension, for the indicator "LCC and incomes", 39% of the total cost is for the aggregates, 23% is for the ready-mix concrete, and 19% is for carbon steel reinforcement. There is no income. For the indicator "environmental externalities", 36% of the total cost is for the production of aggregates, and 20% each is for the production of stainless-steel reinforcement and ready-mix concrete.

3.1.2. Modules A4–A5

In the construction process stage, greenhouse gas emissions have the largest impact in the environmental dimension (see Figure 2). The global warming potential accounts for 39% of the total impact in this life cycle stage. Seventy percent of this type of emission originates from the transport of aggregates from the production site to the construction site. Eighteen percent originates from the production and combustion of diesel in construction machines on the construction site.

The second largest impact comes from the abiotic depletion potential for fossil resources, having 26% of the total environmental impact. Sixty-eight percent of this is caused by the transport of aggregates to the site.

The third largest impact comes from the combination of the three indicators for eutrophication potential, together contributing to 9% of the total environmental impact. Approximately 72% of the eutrophication potential (marine) and eutrophication potential (terrestrial) comes from the production and combustion of diesel used on site, while 65% of the eutrophication potential (freshwater) comes from the transport of aggregates.

Significant impacts are also caused by the "eco-toxicity potential", the "potential soil quality index", and the "photochemical ozone creation potential" indicators, each accounting for 6–7% of the total environmental impact. The transport of aggregates contributes to 69% of the eco-toxicity potential. Ninety percent of the potential soil quality index value comes from the production of the ancillary materials wood, particleboard, and plywood for the formworks. The production and combustion of diesel used on the construction site contributes to 74% of the photochemical ozone formation potential.

The largest impact in the social dimension comes from particulate matter emissions (84%). Sixty-nine of this is caused by the production and combustion of diesel used on the construction site.

In the economic dimension, for the indicator "LCC and incomes", 65% of the total cost is for the construction workers and 11% is for the transport of aggregates. There is no income. For the indicator "environmental externalities", 62% of the total cost is for the transport of aggregates, and 25% is for the production and combustion of diesel used on the construction site.

3.1.3. Modules B1–B7

In the use stage, "global warming potential" (38%) has the largest impact in the environmental dimension, followed by "abiotic depletion for fossil resources" (35%), see Figure 2. Fifty-three percent of the global warming potential is caused by the production of 1.4 tons of steel racks, and 30% is caused by the production of 14 tons of ready-mix concrete. Thirty-two percent of the abiotic depletion potential for fossil resources is caused by the production of 10.5 tons of asphalt.

"Particulate matter emissions" (58% of the total impact) has the largest impact in the social dimension, followed the indicator "human toxicity—non-cancer effects" (23%). Thirty-nine percent of the particulate matter emissions is caused by the production of steel racks, and 21% is caused by the production of 35 kg of bituprimer. The largest contributor to "human toxicity—non-cancer effects" is the production of 77 kg of epoxy (45%).

In the economic dimension, for the indicator "LCC and incomes", 46% of the total cost is for steel racks. The cost for installation of the steel racks is 21% of the total cost. There is no income. For the indicator "environmental externalities", 61% of the total cost is for the production of steel racks, and 23% is for the production of asphalt.

3.1.4. Module B8

The impact of module B8, the stage relating to the user's utilization, is the same for both design concepts (see Figure 4, Tables 5 and 6). For the CSF bridge, the environmental impact of module B8 is equal to the environmental impact of all other modules together (excluding module D). The social impact of module B8 is 97% greater. The net cost for LCC and incomes is 84% lower for module B8 than for all other modules together (excluding module D). The "environmental externalities" indicator is 83% lower for module B8 than for all other modules together (excluding module D).

Table 6. Results for the SSC bridge design concept per life cycle stage and aggregated at the dimension level (or at the indicator level for the economic dimension) for the functional unit. The results for the environmental and social dimensions are normalized and weighted, and the results for the economic indicators are summarized. Modules not assessed are abbreviated as "MNA".

Dimension	Indicator (*Unit*)	A0	A1–A3	A4–A5	B1–B7	B8	C1–C4	D
Environmental	All *(dimensionless)*	MNA	5.0	1.4	0.078	4.4	0.44	−2.1
Social	All *(dimensionless)*	MNA	1.6	0.63	0.029	6.9	0.25	−1.4
Economic	LCC and incomes *(Euro)*	MNA	295	99	16	86	10	−0.8
	Environmental externalities *(Euro)*	MNA	1 087	52	1.5	31	1.3	−14

The indicators "abiotic depletion potential for fossil resources" and "global warming potential" account for 21% and 19% of the total impact, respectively, in the environmental dimension. Fifteen percent comes from the indicator "acidification potential", and 13% comes from the indicator "eutrophication potential" (terrestrial). However, if the results for all three eutrophication indicators are pooled, they account for 23% of the total environmental impact. Fifty-four percent of the global warming potential originates from the production and combustion of hydrogenated vegetable oil (HVO) (rapeseed methyl ester (RME) was used as a proxy for HVO). Fifty-eight percent of the abiotic depletion potential for fossil resources originates from the production of electricity. Seventy-six percent of the total eutrophication potential originates from the production and combustion of HVO.

Forty-seven percent of the total impact in the social dimension comes from the indicator "human toxicity potential—non-cancer effects", and 34% comes from "particulate matter emissions". Ninety-seven percent of the "human toxicity potential—non-cancer effects" originates from the production of HVO. For particulate matter emissions, 48% originates from the production of HVO, and 45% comes from the combustion of diesel.

In the economic dimension, for the indicator "LCC and incomes", 46% of the total cost is for steel racks. The cost for installation of the steel racks is 21% of the total cost. There is no income. For the indicator "environmental externalities", the largest contributor is the production of HVO.

3.1.5. Modules C1–C4

In the end-of-life stage, global warming potential has the largest impact in the environmental dimension (38%), followed by abiotic depletion potential for fossil resources (23%), the three eutrophication potential indicators together (12%), and photochemical ozone creation potential (11%), see Figure 2. Forty-six percent of the global warming potential and 49% of the abiotic depletion potential for fossil resources are caused by the landfilling of 683 tons of aggregate waste. Nineteen percent of the global warming potential and 21% of the abiotic depletion potential for fossil resources is caused by the production and combustion of diesel used in deconstruction.

Particulate matter emissions have the largest impact on the social dimension (56% of the total impact), followed by the indicator "human toxicity—non-cancer effects" (35%). Thirty-two percent of the particulate matter emissions is caused by the transport of 683 tons of aggregate waste, and 23% is caused by the production and combustion of diesel used during deconstruction of the bridge. Ninety percent of the impact of "human toxicity—non-cancer effects" is caused by the landfilling of aggregate waste.

In the economic dimension, for the indicator LCC and incomes, 93% of the total cost is for deconstruction workers. There is no income. For the indicator "environmental externalities", the largest part of the cost (52%) comes from the landfilling of aggregate waste and the second largest (20%) portion comes from the production and combustion of diesel used for deconstruction.

3.1.6. Module D

Regarding benefits and loads beyond the system boundary, the largest benefit in the environmental dimension is offered by the potential avoidance of contributing to global warming potential (45% of the total benefit) and the potential avoidance of contributing to the abiotic depletion potential for fossil resources (34%), see Figure 2. The main part of this benefit is due to the potential re-use of aggregates.

The largest benefit in the social dimension is offered by the potential avoidance of particulate matter emissions (55%) and the avoidance of ionizing radiation (21%). Here, as well, the main part of this benefit is due to the potential re-use of aggregates.

In the economic dimension, for the indicator "LCC and incomes", 41% of the total income comes from the potential recycling of concrete as a filling material, and 37% comes from the potential recycling of carbon steel reinforcement. For the indicator "environmental externalities", the largest benefit comes from the potential re-use of aggregates (48%).

3.2. Soil-Steel Composite Bridge Design Concept

The results for each indicator used in the sustainability assessment of the SSC bridge per metre of bridge in units of measurement and per life cycle stage are presented in Table 7.

Table 7. Results for the SSC bridge design concept per indicator in units of measurement per life-cycle stage for the functional unit. Modules not assessed are abbreviated as "MNA".

Indicator	A0	A1–A3	A4–A5	B1–B5	B6–B7	B8	C1–C4	D
AP	MNA	0.95	0.56	0.031	0	4.2	0.34	−0.8
ETP-fw	MNA	1773	1571	89	0	4063	360	−1348
SQP	MNA	977	781	13	0	22,083	156	−644
GWP-total	MNA	369	164	9.1	0	233	42	−309
ADPE	MNA	1.5×10^{-2}	1.4×10^{-5}	7.5×10^{-7}	0	5.4×10^{-5}	3.5×10^{-6}	-2.9×10^{-5}
ADPF	MNA	4667	2210	149	0	5222	554	−3646
EP-freshwater	MNA	6.3×10^{-4}	5.0×10^{-4}	1.2×10^{-5}	0	3.1×10^{-2}	9.8×10^{-5}	-4.1×10^{-4}
EP-marine	MNA	0.24	0.25	0.0089	0	1.7	0.12	−0.20
EP-terrestrial	MNA	2.6	2.8	0.10	0	20	1.4	−2.2
ODP	MNA	3.8×10^{-11}	2.6×10^{-11}	2.3×10^{-14}	0	4.9×10^{-10}	8.5×10^{-14}	-1.5×10^{-12}
POCP	MNA	0.69	0.70	0.028	0	2.7	0.36	−0.59
PIR	MNA	53	2.2	0.39	0	149	0.39	−43
HTP c	MNA	2.9×10^{-7}	3.3×10^{-8}	9.4×10^{-9}	0	7.1×10^{-7}	2.8×10^{-8}	-3.0×10^{-7}
HTP nc	MNA	4.4×10^{-6}	1.9×10^{-6}	1.4×10^{-7}	0	1.1×10^{-4}	2.9×10^{-6}	-2.6×10^{-6}
PM	MNA	1.7×10^{-5}	9.6×10^{-6}	2.9×10^{-7}	0	4.1×10^{-5}	2.5×10^{-6}	-1.5×10^{-5}
WDP	MNA	53	1.9	0.47	0	109	2.5	−48
LCC and incomes	MNA	295	99	16	0	86	10	−0.8
Environmental externalities	MNA	1087	52	1.5	0	31	1.3	−14

The results for the SSC bridge are presented per life cycle stage in Table 6. The results for the environmental and social dimensions are aggregated on the dimension level, while the results for the economic dimension are aggregated on the indicator level. The results for the environmental and social dimensions are normalized and weighted using the factors in Table 2, and the results for the economic indicators are summarized. Normalized and weighted results for the environmental and social dimensions per life cycle stage are presented per indicator in Figure 4 (only module B8) and Figure 5 (excluding module B8). Figure 6 shows the share of contribution of each resource to the total impact over the life cycle (modules A–C excluding B8) in the environmental dimension and the social dimension respectively, for the resources with the greatest contributions.

In the environmental dimension, 72% of the total impact for life cycle stages A–C (excluding module B8) occurs in the production stage (modules A1–A3), followed by the construction stage (20%, modules A4–A5). If module B8 was included, it would contribute to 38% of the total impact for life cycle stages A–C. Module D has the potential to reduce the future environmental impact (by future re-use or recycling) by 31% for life cycle stages A–C (excluding module B8). The main contribution to the environmental impact over the life cycle (modules A–C, excluding B8) is accounted for by the indicators "abiotic depletion potential for non-fossil resources" (35% of the total impact), "global warming potential" (30% of the total impact), and "abiotic depletion potential for fossil resources" (19% of the total impact). Practically all (99.6%) of the abiotic depletion potential for non-fossil resources and 36% of the global warming potential are caused by the production of structural steel for the bridge. Forty-seven percent of the global warming potential is caused by the production and transport of aggregates. The production of structural steel for the bridge contributes to 54% and the production and transport of aggregates contributes to 30% of the total environmental impact over the life cycle (see Figure 6).

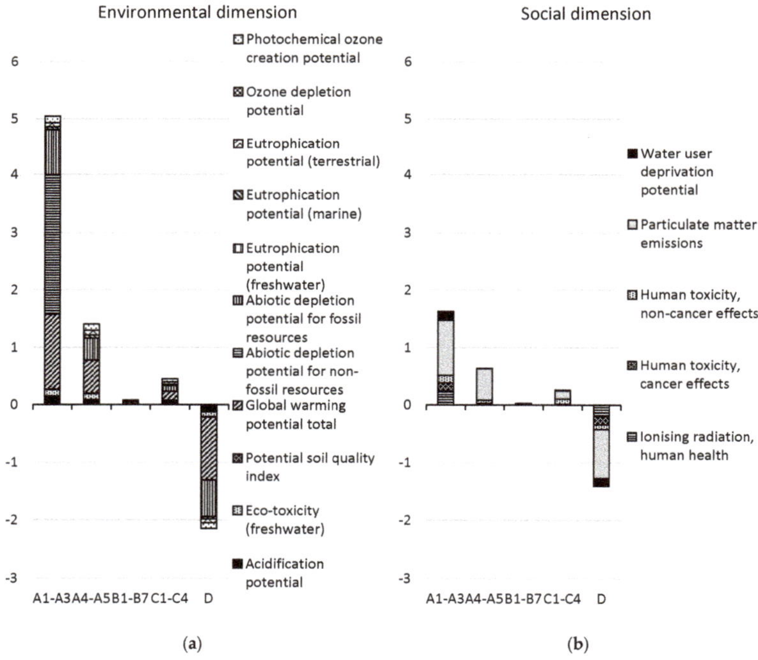

Figure 5. Normalized and weighted results for the SSC bridge design concept per life cycle stage in (**a**) the environmental dimension and (**b**) the social dimension per indicator and for the functional unit. Module B8 is not included. See Figure 4 for module B8. Note that the ozone depletion potential indicator bar cannot be seen in the figure since it is very small.

In the social dimension, 64% of the total impact of life cycle stages A–C (excluding module B8) occurs in the production stage, followed by the construction stage (25%). If module B8 was included, it would contribute to 73% of the total impact for life cycle stages A–C. Module D presents the potential to reduce the future social impact by 56% for life cycle stages A–C (excluding module B8). The main contributor to the social impact over the life cycle (modules A–C excluding B8) is accounted for by the indicators "particulate matter emissions" (66% of the total impact) and "human toxicity—non-cancer effects" (11% of the total impact). The largest portion of the particulate matter emissions is caused equally by the production of aggregates and the production and combustion of diesel used for the construction, maintenance, and deconstruction of the bridge over the life cycle (each 36%), followed by the production of structural steel (23%). Over the life cycle, the production and transport of aggregates contributes to 44%, the production of structural steel for the bridge contributes to 26% and the production and combustion of diesel used for the construction, maintenance, and deconstruction of the bridge over the life cycle contributes to 21% of the total social impact (see Figure 6).

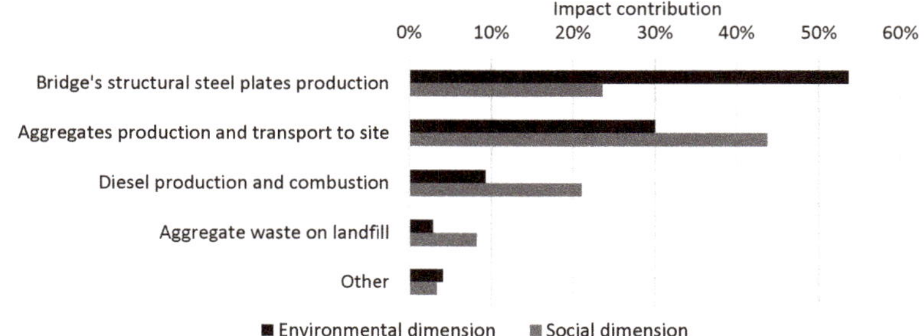

Figure 6. Contribution of resources to the total impact over the life cycle (modules A–C, excluding B8) for the SSC bridge design concept in the environmental and social dimensions.

In the economic dimension, 70% of the total net cost for the indicator "LCC and incomes" in life cycle in stages A–C (excluding module B8) occurs in the production stage, and 24% occurs in the construction stage (see Table 6). If module B8 was included, it would contribute to 17% of the total net cost for life cycle stages A–C. Module D presents the potential to reduce the future net cost (by future re-use or recycling) by 0.2% for life cycle stages A–C (excluding module B8). The greatest contributor (39%) to the cost of the LCC and incomes over the life cycle is the production of structural steel for the bridge. For the indicator environmental externalities, 68% of the total external cost in life cycle stages A–C (excluding module B8) occurs in the production stage. Ninety percent of the environmental externalities are caused by the production of structural steel for the bridge. If module B8 was included, it would contribute to 3% of the total external costs for life cycle stages A–C. Module D presents the potential to reduce the future external costs by 1% for life cycle stages A–C (excluding module B8).

3.2.1. Modules A1–A3

In the production stage, the abiotic depletion potential for non-fossil resources has the largest impact in the environmental dimension (see Figure 5). This indicator has 48% of the total impact in this life cycle stage. Almost all (99.7%) of the depletion is caused by the production of 40 tons of structural steel. The second largest contributor is the global warming potential; this constitutes 26% of the total environmental impact. Fifty-seven percent of this indicator is accounted for by the production of structural steel plates for the bridge and 42% by the production of aggregates. The abiotic depletion potential for fossil resources also has a significant impact (16% of the total environmental impact).

Particulate matter emissions have the largest impact in the social dimension (60%). Fifty-eight of this originates from the production of aggregates and 40% comes from the production of bridge steel.

In the economic dimension, for the indicator "LCC and incomes", 67% of the total cost is for structural steel and 28% is for aggregates. There is no income. For the indicator "environmental externalities", 96% of the cost is for the production of structural steel.

3.2.2. Modules A4–A5

In the construction stage, greenhouse gas emissions have the largest impact in the environmental dimension (see Figure 5). The global warming potential contributes to 41% of the total impact in this life cycle stage. Seventy-three percent of this originates from the transport of aggregates to the construction site. Nineteen percent originates from the production and combustion of diesel used in construction machines on the construction site. The second largest impact comes from the abiotic depletion potential for fossil resources, causing 27% of the total environmental impact. Seventy-two percent of this is caused by the transport of aggregates to the site. The three eutrophication potential indicators make a significant contribution, together contributing to 9% of the total environmental impact. The production and combustion of diesel used at the construction site contributes to 71% of the total eutrophication potential.

In the social dimension, particulate matter emissions have the largest impact by far (86%). Eighty-seven percent of these emissions are caused by the production and combustion of diesel used at the construction site.

In the economic dimension, for the indicator "LCC and incomes", 32% of the total cost is for the transport of aggregates, 28% is for the construction workers, and 20% is for transport of the bridge's structural steel plates. There is no income. For the indicator "environmental externalities", 66% of the total cost is for the transport of aggregates and 27% is for the production and combustion of diesel used at the construction site.

3.2.3. Modules B1–B7

In the use stage, greenhouse gas emissions have the largest impact in the environmental dimension (42%), followed by the depletion of fossil resources (33%) (see Figure 5). Seventy-three percent of the global warming potential is caused by the production of 1.4 tons of steel racks and 16% is caused by the production of 10.5 tons of asphalt. Fifty percent of the abiotic depletion potential of fossil resources is caused by the production of asphalt and 41% is caused by the production of steel racks.

In the social dimension, particulate matter emissions have the greatest impact (59% of the total impact), followed by 16% each from the indicators "human toxicity—cancer effects" and "human toxicity—non-cancer effects". Seventy-six percent of particulate matter emissions comes from the production of steel racks. The production of steel racks contributes to 87% of the factor "human toxicity—cancer effects" and 55% of "human toxicity—non-cancer effects". The production of asphalt contributes to 31% of "human toxicity—non-cancer effects".

In the economic dimension, for the indicator "LCC and incomes", 50% of the total cost is for steel racks. The installation of steel racks accounts for 23% of the total cost. There is no income. For the indicator "environmental externalities", 81% of the total cost is for the production of steel racks.

3.2.4. Module B8

The environmental impact of module B8, the stage relating to the user's utilization, is 38% lower than the environmental impact of all other modules together (excluding module D) for the SSC bridge, see Table 6. The social impact of module B8 is almost three times larger. The net cost for LCC and incomes is 80% lower for module B8 than for all other modules together (excluding module D). The "environmental externalities" indicator is 97% lower for module B8 than for all other modules together (excluding module D). For other aspects of module B8 that do not depend on the bridge type, see Section 3.1.4.

3.2.5. Modules C1–C4

In the end-of-life stage, global warming potential has the largest impact in the environmental dimension (34%), followed by the abiotic depletion potential for fossil resources (22%), the three eutrophication potential indicators together (14%), and the photochemical ozone creation potential

(12%), see Figure 5. Fifty-one percent of both the global warming potential and the abiotic depletion potential for fossil resources is caused by the landfilling of 683 tons of aggregate waste. Thirty-three percent of the global warming potential is caused by the production and combustion of diesel used for deconstruction. Thirty-four percent of the abiotic depletion potential for fossil resources is caused by the production of diesel used for deconstruction.

In the social dimension, particulate matter emissions have the greatest impact (56% of the total impact), followed by the indicator "human toxicity—non-cancer effects" (35%). Seventy-seven percent of particulate matter emissions and 90% of human toxicity—non-cancer effects are caused by the landfilling of aggregate waste.

In the economic dimension, for the indicator "LCC and incomes", 90% of the total cost is for deconstruction workers. There is no income. For the indicator "environmental externalities", the largest portion of the cost (55%) comes from the landfilling of aggregate waste and the second largest portion (33%) comes from the production and combustion of diesel used for deconstruction.

3.2.6. Module D

Regarding the benefits and loads beyond the system boundary, the largest potential future benefit in the environmental dimension is the avoidance of contributing to global warming potential (51% of the total benefit) and the avoidance of abiotic depletion for fossil resources (30%), see Figure 5. The main factors involved in the avoidance of contributing to global warming potential are the potential recycling of bridge steel (52%) and the potential re-use of aggregates (45%).

Table 8. Comparison of the design concepts for modules A–C and for module D for the functional unit. The results are aggregated at the dimension level for the environmental and social dimensions and at the indicator level for the economic dimension. The results for the environmental and social dimensions are normalized and weighted, while the results for the economic dimension are summarized. Module B8 is not included in the comparison. The best options are highlighted in grey.

Dimension, Indicator (*unit*)	CSF Bridge		SSC Bridge	
	A–C	D	A–C	D
Environmental, all (*dimensionless*)	4.4	−1.5	7.0	−2.1
Social, all (*dimensionless*)	3.5	−1.2	2.5	−1.4
Economic, LCC and incomes (*Euro*)	526	−1.1	421	−0.8
Economic, Environmental externalities (*Euro*)	182	−7.9	1 142	−14

The greatest potential future benefits in the social dimension are the avoidance of particulate matter emissions (61%) and the avoidance of ionising radiation—human health (14%). The main factors involved in the avoidance of particulate matter emissions are the potential re-use of aggregates (59%) and the potential recycling of bridge steel (39%).

In the economic dimension, for the indicator "LCC and incomes", 94% of the total income comes from the potential recycling of the bridge's structural steel plates. For the indicator "environmental externalities", the largest benefit comes from the potential recycling of the bridge's structural steel plates (70%).

3.3. Comparison of the Design Concepts

A comparison of the two design concepts over the life cycle (modules A–C excluding module B8) and in terms of the future re-use, recovery, and recycling potential (module D) is presented in Table 8 and Figures 7–9. The results are aggregated on the dimension level for the environmental and social dimensions and on the indicator level for the economic dimension. The results for the environmental and social dimensions are normalized and weighted, and the results for the economic dimension are summarized. The results for the environmental and social dimensions per life cycle stage for the two concepts are presented in Figure 10, and the summarized results for the economic indicators are presented in Figure 11 (excluding module B8).

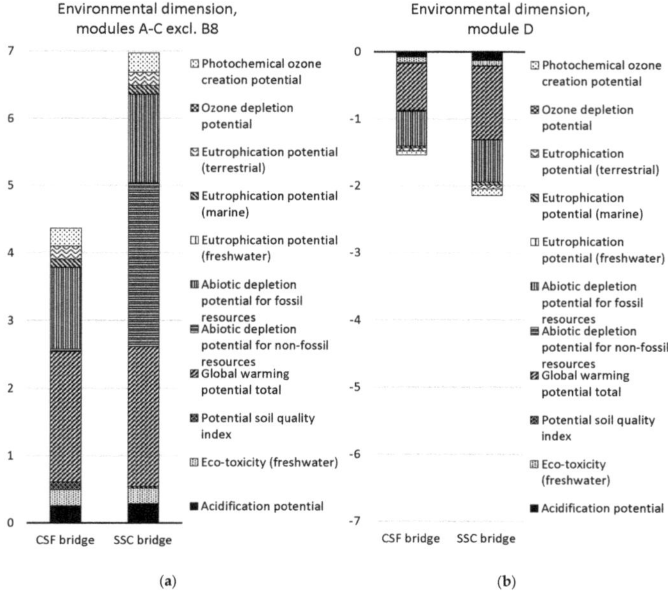

Figure 7. Comparison of the design concepts in the environmental dimension for (**a**) life cycle stages A–C excluding module B8 and (**b**) module D per indicator and for the functional unit. Note that the ozone depletion potential indicator bar cannot be seen in the figure since it is very small.

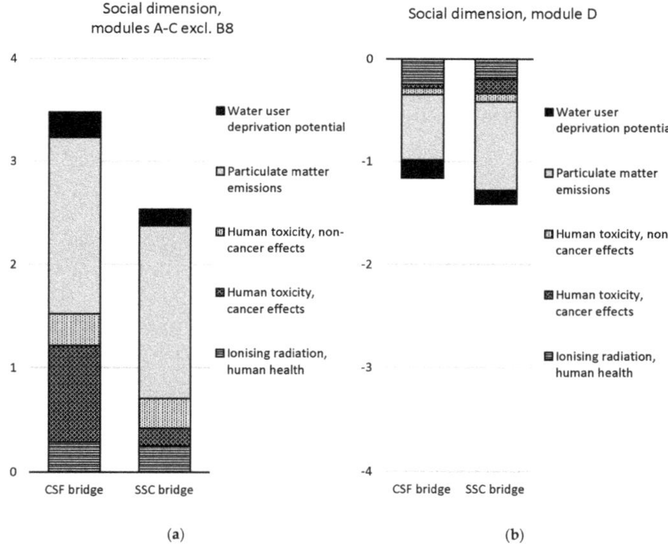

Figure 8. Comparison of the design concepts in the social dimension for (**a**) life cycle stages A–C excluding module B8 and (**b**) module D per indicator and for the functional unit.

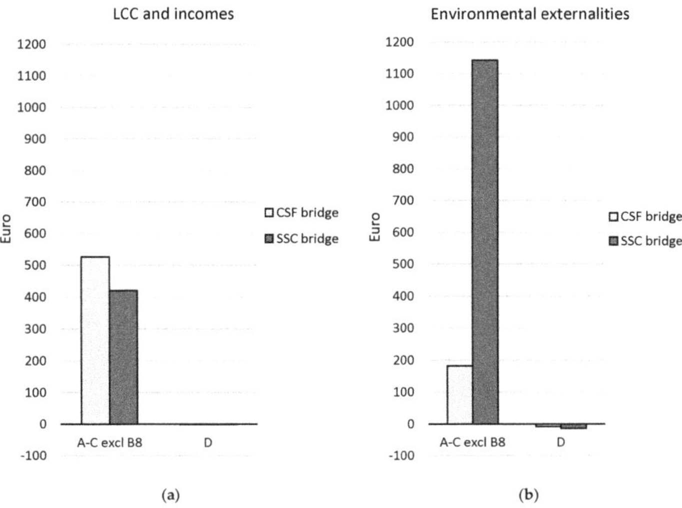

Figure 9. Comparison of the design concepts per life cycle stage for (**a**) the LCC and incomes and (**b**) the environmental externalities; presented in Euros for the functional unit. Module B8 is not included in the comparison.

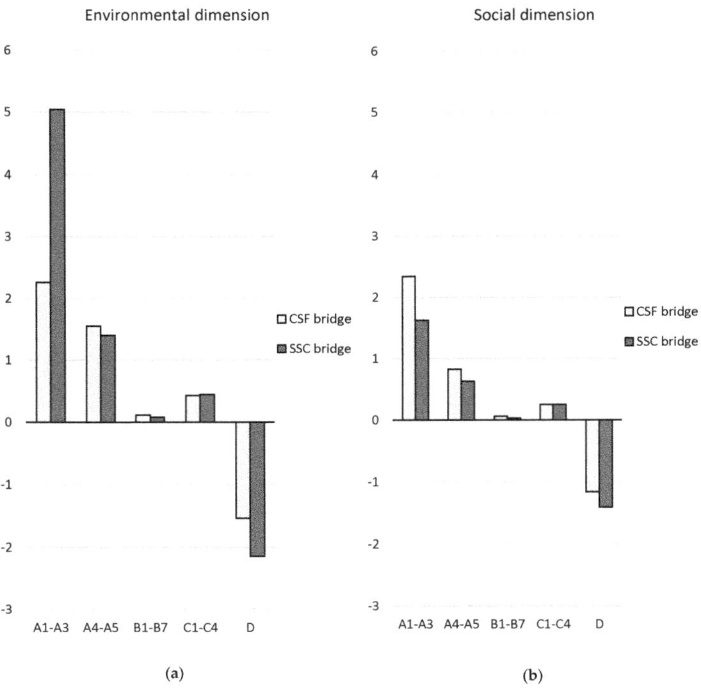

Figure 10. Comparison of the design concepts per life cycle stage in (**a**) the environmental dimension and (**b**) the social dimension for the functional unit. Module B8 is not included in the comparison.

In the environmental dimension, the CSF bridge performs better than the SSC bridge over the life cycle (see Table 8 and Figure 7). The environmental impact of the CSF bridge is approximately one-third lower than that of the SSC bridge. After the end-of-life stage (in module D), the potential avoidance of a negative environmental impact is 39% greater for the SSC bridge.

In the social dimension, the SSC bridge performs better than the CSF bridge over the life cycle (see Table 8 and Figure 8). The social impact of the SSC bridge is 27% lower than that of the CSF bridge. After the end-of-life stage (in module D), the potential avoidance of negative social impact is 22% greater for the SSC bridge.

In the economic dimension, the SSC bridge performs better than the CSF bridge over the life cycle for the indicator "LCC and incomes", but it has a significantly worse performance than the CSF bridge for the indicator "environmental externalities" (see Table 8 and Figure 9). The net cost of the SSC bridge is 20% lower than that of the CSF bridge when it comes to the indicator "LCC and incomes". In contrast, the impact of the factor "environmental externalities" is six times greater for the SSC bridge than for the CSF bridge. After the end-of-life stage (in module D), the potential income and the avoidance of environmental externalities have very small impacts for both concepts.

Considering the different life cycle stages, the SSC bridge has double the environmental impact in the material production phase (modules A1–A3) compared with the CSF bridge (see Figure 10). This is mainly due to the abiotic depletion potential of non-fossil resources caused by the manufacture of structural steel plates for the SSC bridge. This indicator contributes to almost half of the total environmental impact of the SSC bridge in the production stage (see Figure 7). The indicator global warming potential contributes to one-quarter of the environmental impact of the SSC bridge in the production stage.

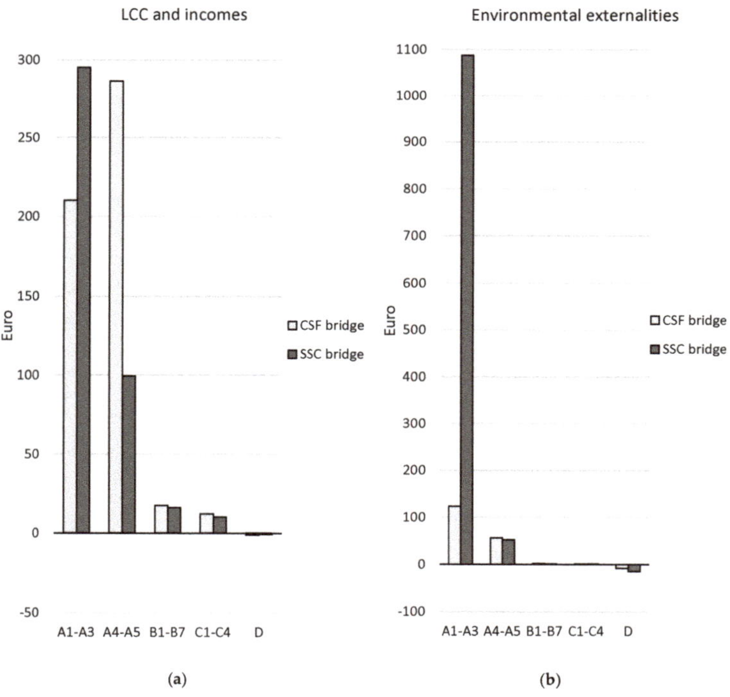

Figure 11. Comparison of the design concepts per life cycle stage for (**a**) the LCC and incomes and for (**b**) the environmental externalities presented in Euros for the functional unit. Module B8 is not included in the comparison. Note the different y-axis scales in the charts.

The CSF bridge has a 44% larger social impact in the material production phase than the SSC bridge (see Figure 10). This is primarily due to "human toxicity—cancer effects", which are mainly caused by production of stainless-steel reinforcement, and the particulate matter emissions, which are mainly caused by the production of aggregates.

In the production stage, the impact of the environmental externalities of the SSC bridge is almost nine times greater than that of the CSF bridge (see Figure 11). This is due to the use of non-renewable elements in the production of the bridge's structural steel plates. In the construction stage, the net cost for LCC and incomes is almost three times larger for the CSF bridge. This is because of the larger cost for construction workers for the CSF bridge compared with the SSC bridge.

4. Discussion

In the environmental dimension, the CSF bridge was found to perform better than the SSC bridge over the life cycle (see Table 8 and Figure 7). The environmental impact of the CSF bridge was 37% lower than that of the SSC bridge. Similar results were demonstrated in an LCA study comparing a steel box girder bridge and a concrete box girder bridge, where the concrete bridge alternative performed best environmentally overall [22]. However, another LCA study on four CSF bridges and four SSC bridges showed the opposite result: the SSC bridges performed better than the CSF bridges over the life cycle [23]. This is partly because only 54–74% of the structural steel mass was used in three of the SSC bridges in the study by [23], compared with the SSC bridge in this case study. It is also partly because 37% of the structural steel plates were assumed to be secondary steel produced in an electric arc furnace (EAF) route in [23], while in this case study, all of the structural steel plates are from primary steel produced through a blast furnace (BF) route.

In the social dimension, the SSC bridge was found to perform better than the CSF bridge over the life cycle (see Table 8 and Figure 8). The social impact of the SSC bridge was 27% lower than that of the CSF bridge. A similar result was demonstrated by [23], where particulate matter emissions were slightly lower for three of the SSC bridges compared with the CSF bridges.

In the economic dimension, the SSC bridge was found to perform better than the CSF bridge over the life cycle for the indicator LCC and incomes, but it performed significantly worse than the CSF bridge for the indicator environmental externalities (see Table 8 and Figure 9). The opposite result was shown for environmental externalities in [23]. Using the Ecotax02 and Ecovalue08 monetary weighting methods updated with the Ecovalue12 method indicators [24,25], the SSC bridges performed better than the CSF bridges. This might be explained by the fact that the depletion of abiotic resources indicator was not included in the calculation of environmental externalities in [23], even though it is part of both the Ecotax02 and the Ecovalue08 and Ecovalue12 methods. Non-renewable elements and non-renewable energy resources were found to be the major contributors contribute to the environmental externalities in this case study.

For both design concepts, the majority of the negative impact on sustainability was found to occur in the production stage (modules A1–A3). This was also shown in [23], where between 55% and 92% of the environmental impact occurred in the production stage, depending on the indicator. An LCA study of a steel box girder bridge and a concrete box girder bridge similarly showed that the production of materials for the bridge superstructure and the abutments accounted for the main share of the environmental impact, with a limited number of materials being important [22].

Furthermore, the case study demonstrated that 36% of the life-cycle environmental impact for the CSF bridge and 20% for the SSC bridge occurred in the construction stage (modules A4–A5). In [23], it was also shown that the environmental impact of the construction stage is significant; causing up to 34% of the life cycle impact for some indicators. However, in [22], the construction phase accounted for a relatively small part of the impact, and the use phase contributed more significantly, which is contrary to the results of this case study. A main difference between the two concepts in this case study is that a large part of the economic impact (LCC and incomes) was found to occur in the construction stage for the CSF bridge, and this was mainly due to the cost of construction workers.

Module B8 was found to contribute to 50% and 38% of the total environmental impact and 66% and 73% of the total social impact over the life cycle (modules A–C) for the CSF bridge and SSC bridge, respectively. This demonstrates that the environmental and social impacts of the bridge itself are, in fact, significant in comparison to the impact from traffic on the bridge. This was even more obvious for the economic impacts, as module B8 was shown to only contribute to between 3% and 17% for the two economic indicators considered for both bridge types.

The main contribution to the environmental impact over the life cycle (modules A–C excluding B8) was shown to come from the indicator abiotic depletion potential for non-fossil resources for the SSC bridge (35%) and the indicator global warming potential for the CSF bridge (44%). Similar results were demonstrated in an LCA study of a concrete box girder bridge and a steel box girder bridge where global warming, abiotic depletion, and acidification were found to be the indicators with the greatest contributions [22]. In [23], on the contrary, it was shown that the SSC bridges performed better than the CSF bridges regarding the indicator global warming potential. This was partly because less structural steel was used in the SSC bridges and because 37% of the structural steel plates were assumed to be secondary steel produced via an EAF route in [23] (see further explanation above). If only the indicator "global warming potential" had been considered in this case study, the SSC bridge would have performed only 7% worse than the CSF bridge in the environmental dimension. When considering all environmental indicators, the SSC bridge was found to perform 60% worse in the environmental dimension. This demonstrates the importance of including more indicators than only global warming potential, as shown in previous studies [26]. This is an important observation, as today, it is common practice to solely consider global warming potential (or one other indicator such as embodied energy) in assessments of environmental performance [9,26,27].

The main contributor to the social impact over the life cycle (modules A–C excluding B8) was shown to be the indicator particulate matter emissions for both concepts. This is, in part, because of the large weight given to this indicator, but also because construction activities are significant sources of particulate matter emissions, for example, when crushing aggregates [28,29]. For the CSF bridge, the indicator "human toxicity—cancer effects" was also shown to contribute to a large portion of the social impact. For the CSF bridge, it was shown that approximately one-third of the particulate matter emissions were caused by the production of aggregates and their transport to the construction site, and one-third were caused by the production and combustion of diesel used at the construction site. For the SSC bridge, it was shown that one-third of the particulate matter emissions were caused by the production of aggregates; one-third by the production and combustion of diesel used for the construction, maintenance, and deconstruction of the bridge over the life cycle; and one-quarter by the production of structural steel.

The net cost for the LCC and incomes indicator was found to be 25% higher for the CSF bridge than the SSC bridge over the life cycle. For the CSF bridge, the main costs came from cost for workers during construction, and the production and transport of aggregates, concrete, and reinforcement steel. For the SSC bridge, the main contributor to the cost of the LCC and incomes was the cost for the production of structural steel. The environmental externalities of the SSC bridge was six times greater than that of the CSF bridge over the life cycle because of the use of non-renewable elements in the production of the bridge's structural steel plates.

The production of aggregates and their transport to the construction site was shown to be the main factor in the environmental and social impacts of the CSF bridge and the social impact of the SSC bridge. It was also shown to be the second greatest factor in the environmental impact of the SSC bridge. Hence, there is great potential to reduce environmental and social impacts by re-using aggregates on site in the next life cycle to avoid the production of virgin aggregates and their transport to the construction site. The production of the bridge's structural steel plates plays the largest role in the environmental impact as well as the impact on environmental externalities for the SSC bridge due to the depletion of metals. Thus, there is great potential to reduce the impact by using steel produced from recycled steel. Regarding the LCC and incomes over the life cycle, the results show that costs can be reduced by lowering the work costs as well as the material costs for the materials purchased in large quantities, such as aggregates, steel and concrete.

It is important to keep in mind that the environmental and social impact results are highly dependent on the LCA datasets chosen for the calculations. It is possible to apply the method using generic licensed datasets or generic datasets from open online LCA databases or Environmental Product Declarations (EPDs), provided they follow the EN15804 + A2 standard. The use of supplier EPDs instead of generic datasets further increases the accuracy of the environmental and social assessment results, since EPDs contain supplier-specific declarations, while generic datasets may not be fully representative of the actual materials supplied in an assessed civil engineering works project. In this case study, only generic datasets were used. They are not completely representative of the resources purchased for the object of assessment. For example, the generic dataset used for aggregates differed from the aggregates purchased, especially regarding the distance between the mining site and crushing plant (10 km in the generic dataset, a few hundred metres for the actual supplier). Generic datasets were used because supplier specific EPDs that follow the EN15804 + A2 standard are not yet available.

As shown from the examination of results, this method allows the sustainability performance of design concepts to be compared at the life cycle stage and construction component level in the early design and planning stages. The data available in these early stages are sufficient for assessment. Through the examination possibilities made available by the transparency of the method, it is possible to identify the critical elements in a civil engineering works project with the greatest impacts on sustainability. This allows necessary adjustments to be made to achieve more sustainable design concepts. Due to its general character, the method can be applied to other types of civil engineering works, not only bridges.

For the social dimension, in particular, but also for the environmental dimension, further research is needed to define appropriate indicators for civil engineering projects. The sustainability dimensions could also be further aggregated using a multi-criteria decision analysis (MCDA) method to obtain an overall sustainability score [30–32]. Furthermore, scenarios for the construction, use, and end-of-life stages may be improved by collecting data from ongoing projects [5]. It is recommended that future studies carry out a sensitivity analysis to assess the influences of different scenarios and datasets whose uncertainty is considered important for the evaluated impacts.

5. Conclusions

The case study demonstrates that our method can be used to carry out comparable and transparent life cycle sustainability performance assessments in the early design and planning stages of a civil engineering works project. It allows the sustainability performance of design concepts to be compared at the life cycle stage and construction component level. The method enables the identification of critical indicators with the greatest impacts on sustainability at the different life cycle stages and for the critical elements. The method is transparent, because the underlying BOMs, scenarios, and datasets used for the assessment are clearly described, and the results can be evaluated down to the building material level. Since the method is based on quantitative indicators and fixed factors, the calculation process used in the assessment is automatable.

The use of supplier EPDs instead of generic datasets will further increase the accuracy of the environmental and social assessment results, since EPDs are supplier-specific declarations, while generic datasets may not be fully representative of the actual materials supplied in an assessed civil engineering works project.

The case study demonstrates the importance of including more indicators than global warming potential in environmental assessments. Environmental and sustainability performance is clearly dependent on several indicators, and care should be exercised when generalising results obtained in assessments that only take into account only one or a few indicators.

The method used in the case study includes state-of-the-art indicators according to current standard specifications and can be complemented with additional indicators. For the social dimension, in particular, but also for the environmental dimension, further research is needed to define appropriate indicators for civil engineering projects.

The results of the case study show the importance of the production stage (modules A1–A3) and the construction stage (modules A4–A5) on the sustainability performance over the life cycle. Production of structural steel for the SSC bridge has the greatest environmental impact and accounts for almost all of the environmental externalities, which explains the poorer performance of this bridge in the environmental dimension. However, the CSF bridge was shown to perform worse than the SSC bridge in the social dimension with a higher LCC. The former point can be mainly explained by the large global warming potential of the CSF bridge due to concrete production, and the latter point can be explained by higher costs during construction due to being more labour-intensive. The production and transport of aggregates have large negative environmental and social impacts for both bridge types.

In summary, the examination of the case study assessment results provides important knowledge on the indicators and life cycle stages associated with large sustainability impacts for each of the bridge concepts investigated. We conclude that to reduce the overall negative impact on sustainability, mitigation measures should primarily address the production and construction stages. Our findings contribute to the development of a better understanding of the sustainability impact of civil engineering works through the identification of elements with the greatest impacts. A special focus and adaptations of identified elements (e.g., origin and type of materials, equipment used, structural optimization) could significantly improve the sustainability performance of the design concepts. After necessary adaptations have been applied to a design concept, the assessment can be re-performed to assess the new conditions. If the assessment is automated, this step could be iterated until the most sustainable alternative is found.

Author Contributions: Conceptualization, K.E., A.M., M.K. and R.R.; methodology, K.E.; software, K.E.; validation, A.M. and M.K.; formal analysis, K.E. and A.M.; investigation, K.E. and A.M.; data curation, K.E.; writing—original draft preparation, K.E.; writing—review and editing, K.E., A.M. and P.B.; visualization, K.E.; supervision, R.R., M.K. and M.N.; project administration, R.R. and M.K.; funding acquisition, R.R. All authors have read and agreed to the published version of the manuscript.

Funding: This research was funded by VINNOVA and the Swedish Transport Administration, grant number 2017-03312 and 2017-037 respectively.

Acknowledgments: Case study data were provided by project manager Sonja Högvall, NCC.

Conflicts of Interest: The authors declare no conflict of interest.

Appendix A

The life cycle inventory (LCI) database provides the life cycle inventory data. The datasets used in the life cycle assessment (LCA) modelling are presented in Table A1.

Table A1. Assignments of inputs and outputs to LCIs from the GaBi database for the LCA.

Inventory	Description	Assumption	GaBi Dataset	Country	Source
Acetylene	Acetylene	-	Ethine (acetylene)	DE	Generic dataset from Sphera
Aggregate waste treatment	Aggregate and macadam waste	90% is re-used as aggregates on site, 10% is transported to landfill	Crushed stone grain 2–15 mm (undried) (EN15804 A1–A3)	EU-28	"
			Inert matter (Construction waste) on landfill	DE	"
Aggregates 0/16 and 0/90 mm	Crushed aggregates 0/16 and 0/90 from igneous rock in Sweden	European limestone 2/15 mm crushed stone	Crushed stone grain 2–15 mm (undried) (EN15804 A1–A3)	EU-28	"

Table A1. *Cont.*

Inventory	Description	Assumption	GaBi Dataset	Country	Source
Asphalt ABb	ABb asphalt from Sweden	European average supporting layer asphalt	Asphalt supporting layer (EN15804 A1–A3)	EU-28	"
Asphalt ABT	ABT asphalt from Sweden	European average asphalt pavement	Asphalt pavement (EN15804 A1–A3)	EU-28	"
Asphalt recycling	Recycling of ABb, ABT, and Viacogrip asphalt	0.7 kWh of Swedish grid mix electricity used per ton of recycled asphalt (crushing). 1000 kg of recycled asphalt replaces 740 kg of virgin aggregates and 60 kg of virgin bitumen.	Crushed stone grain 2–15 mm (undried) (EN15804 A1–A3)	DE	"
			Bitumen (Eurobitume LCI report 2019) w infrastructure	EU-28	Generic dataset based on Eurobitume report 2019 [33]
Asphalt ViacoGrip	ViacoGrip asphalt from Sweden	European average SMA asphalt	Stone mastic asphalt SMA (EN15804 A1–A3)	EU-28	Generic dataset from Sphera
Average electricity/diesel driven train	-	-	Rail transport cargo—average, average train, gross tonne weight 1000 t/726 t payload capacity	GLO	"
Bitumen sealant			Bitumen emulsion (EN 15804 A1–A3)	DE	"
Bitumen sheet	Icopal Membrane 5BRO (YEP 6500)	Produced in Germany	Bitumen sheets PYE-PV 200 S5 ns (slated) (EN15804 A1–A3)	DE	"
Bitumen sheet waste	Incineration in Swedish district heating plant (Jönköping)	Incineration of average municipal solid waste (MSW) in Germany	Commercial waste in municipal waste incineration plant	DE	"
Bituprimer	Degadur®112	-	Methacrylate resin products, highly-filled, flow coatings—Deutsche Bauchemie e.V. (DBC) (A1–A3)	DE	"
Carbon steel reinforcement recycling	-	Average German production	Recycling potential steel profile (D)	DE	"
Concrete elements	Concrete kerbstone	Bricks of concrete C20/25	Concrete bricks (EN15804 A1–A3)	DE	"
Concrete waste treatment	-	0.7 kWh of Swedish grid mix electricity used per ton of recycled concrete (crushing). 1 kg of recycled concrete replaces 1 kg of virgin aggregates.	Crushed stone grain 2–15 mm (undried) (EN15804 A1–A3)	DE	"
Container ship	-	-	Container ship, 5000 to 200,000 dwt payload capacity, ocean going	GLO	"
Diesel	Diesel 7% bioblend	Diesel 6,4% bioblend	Diesel mix at filling station	EU-28	"
Diesel combustion	Diesel combustion	Combustion of diesel (modified for diesel 7% bioblend)	Diesel combustion in construction machine	GLO	"
Electricity	Swedish grid mix electricity	-	Electricity grid mix	SE	"
Electricity generation from waste incineration	Electricity generation from incineration wood, particle board, plywood, plastic, bitumen sheet, and hazardous waste	Swedish grid mix electricity	Electricity grid mix	SE	"
Epoxy sealant	NM Försegling 62F Tix	Primer for exterior applications	Powder coating based on epoxy resin (EN15804 A1–A3)	DE	"
Form oil	Form oil	From crude oil	Lubricants at refinery	EU-28	"

Table A1. *Cont.*

Inventory	Description	Assumption	GaBi Dataset	Country	Source
Geotextile	Drefon ST 550 (polypropylene fibre geotextile)	Approximated by woven cotton fibre fabric	Textile Manufacturing —Woven Fabric	GLO	Generic dataset from CottonInc
Graffiti protection	Graffiti Shield wax emulsion	From crude oil	Wax/Paraffins at refinery	DE	Generic dataset from Sphera
Hazardous waste treatment	-	Incineration	Hazardous waste in waste incineration plant	SE	"
HDG steel racks	Birsta W, single sided safety barrier (HDG)	Produced by blast furnace (BF) route, average European production	Steel forged component (EN15804 A1–A3)	EU-28	"
HDG steel recycling	-	Average German production	Recycling potential steel sheet galvanised (EN15804 D)	DE	"
HDG structural steel plates	SSAB Hot-rolled coils S355MC, produced in a blast furnace (BF) route in Sweden and galvanized in Poland	Produced through a BF route and galvanized, German average	Steel sheet HDG (EN 15804 A1–A3)	DE	"
HVO combustion	HVO combustion	Approximated by a combination of biomass/regular diesel combustion	HVO combustion in car	GLO	Dataset based on CO2e emission data from the Swedish EPA 2018, combustion of "Other biomass", other emissions (SO2, NOx, PM etc.) based on data for regular diesel
Hydrogenated Vegetable Oil (HVO)	HVO combustion	Approximated by RME	Rapeseed Methyl Ester (RME)	DE	"
Impregnation (direct emissions)	SILRES® BS 1701	Silicate emulsion primer	Primer silicate emulsion (building, exterior, white) (EN15804 A5)	DE	"
Impregnation (production)	SILRES® BS 1701	Silicate emulsion prime coat	Primer silicate emulsion (building, exterior, white) (EN15804 A1–A3)	DE	"
Liquified Petroleum Gas (LPG)	Liquified Petroleum Gas	average European production	Thermal energy from LPG	EU-28	"
Macadam 8/16 mm	Crushed macadam 8/16 from igneous rock in Sweden	European limestone 16/32 mm crushed rock	Crushed rock 16–32 mm (undried) (EN15804 A1–A3)	EU-28	"
Mortar	Fine concrete K40 and expander concrete EXM 702	Average European production	Normal mortar (A1–A3)	EU-28	"
Particle board	Form board from Sweden	Average European P2 (Standard FPY)	Particle board	EU-28	"
Plastic film	-	-	Plastic Film (PE, PP, PVC)	GLO	"
Plastic waste treatment	Incineration in Swedish district heating plant (Jönköping)	Incineration in average European waste incineration plant	Plastic packaging in municipal waste incineration plant	EU-28	"
Plywood	Formply from Sweden	Pine plywood produced in Germany	Plywood board (EN15804 A1–A3)	DE	"
Plywood and particleboard waste treatment	Incineration in Swedish district heating plant (Jönköping)	Incineration in German waste incineration plant	Particle board in municipal waste incineration plant	DE	"
Polyethylene foam	Concrete carpet	Consisting of polyethylene foam	Polyethylene foam (EN15804 A1–A3)	DE	"
Polypropylene pipe	PP road drum	PP pipe produced in Germany	Polypropylene pipe (PP) (EN15804 A1–A3)	DE	"

Table A1. *Cont.*

Inventory	Description	Assumption	GaBi Dataset	Country	Source
PVC tube	PVC drain hose	PVC drain pipe produced in Germany	Rain drain pipe (PVC pipe) (EN15804 A1–A3)	DE	"
PVC waste treatment	PVC waste incineration in Swedish district heating plant (Jönköping)	Incineration in German waste incineration plant	Polyvinyl chloride (PVC) in waste incineration plant	DE	"
Ready-mix concrete C35/45	Betongindustri concrete C35/45 vct = 0,40	Average European production	Concrete C35/45 (Ready-mix concrete) (EN15804 A1–A3)	EU-28	"
Recovery of heat for district heating	Heat recovery from wood, particle board, plywood, plastic, bitumen sheet, and hazardous waste	District heating produced by plant in Jönköping	District heating mix Jönköping 2019	SE	Specific dataset based on fuel use for Jönköping district heating plant
Reinforcement (carbon steel)	B500B	Produced by the electric arc furnace (EAF) route, average European production	Reinforced steel (wire) (EN15804 A1–A3)	EU-28	Generic dataset from Sphera
Reinforcement (stainless steel)	LDX2101	Produced from 100% alloyed stainless steel scrap	Fixing material screws stainless steel (EN15804 A1–A3)	DE	"
Road salt	-	From rock salt	Sodium chloride (rock salt)	DE	"
Stainless steel reinforcement recycling	-	Average German production	Recycling potential stainless steel sheet (EN15804 D)	DE	"
Tap water	Tap water	Swedish tap water produced from groundwater	Tap water from groundwater (for regionalization)	GLO (SE chosen in dummy)	"
Truck, Euro 6	-	-	Truck, Euro 6, 20–26 t gross weight/17.3 t payload capacity	GLO	"
Untreated wood	Spruce wood from Sweden	Coniferous wood produced in Germany	Solid construction timber (softwood) (EN15804 A1–A3)	DE	"
Wood waste treatment	Wood waste incineration in Swedish district heating plant (Jönköping)	Incineration in German waste incineration plant	Wood (natural) in waste incineration plant	DE	"

" denotes 'Same as above'.

Appendix B

Table A2. Bill of materials (BOM) for the design concepts for each module. Amounts are representative of the required service life (RSL) and for the bridges as a whole. The amounts for modules B1–B8, C1–C4 and D were calculated based on the scenarios presented in Tables A3–A5.

Concept	Module	Resource/Waste	Amount	Unit
		Aggregates 0/16 and 0/90 mm	6,820,600	kg
		Macadam 8/16 mm	12,000	kg
	A1–A3	Asphalt Abb	6983	kg
		Asphalt ABT	4364	kg
		Asphalt ViacoGrip	5237	kg
		Concrete elements	8100	kg

Table A2. *Cont.*

Concept	Module	Resource/Waste	Amount	Unit
Both		Hot-dip galvanized (HDG) steel racks	250	kg
		Polypropylene pipe	106	kg
		Geotextile	17	kg
	A5	Electricity (Swedish grid mix)	4560	kWh
		Tap water	1500	liters
		PVC tube	105	kg
		PVC waste	105	kg
	B1	Zinc to fresh water	0.55	kg
		PAH to fresh water	0.112	kg
	B2	Tap water	400	liters
		Road salt	65	kg
	B3	HDG steel racks	692	kg
		HDG steel racks waste for recycling	692	kg
	B4	HDG steel racks	692	kg
		Asphalt	10,500	kg
		Diesel 7% bioblend	17	liters
		Tap water	10,000	liters
		HDG steel racks waste for recycling	692	kg
	B5	N/A	-	
	B6	N/A	-	
	B7	N/A	-	
	B8	Diesel 7% bioblend	11,797	liters
		Hydrogenated Vegetable Oil (HVO)	41,611	liters
		Electricity (Swedish grid mix)	357,209	kWh
		Particles to fresh water	10,950	kg
		Micro plastics to soil	131	kg
	C1	Diesel 7% bioblend	39	liters
	C2	See Table A6	-	
	C3	Aggregates for re-use	6,149,340	Kg
		Asphalt waste for recycling	16,580	Kg
		Polypropylene plastic waste for incineration	106	Kg
	C4	Aggregates on inert landfill	683,260	Kg
	D	Electricity (Swedish grid mix)	19	kWh
CSF bridge	A1–A3	Ready-mix concrete C35/45	463,700	Kg
		Reinforcement (carbon steel)	22,680	Kg
		Reinforcement (stainless steel)	553	kg
		Bitumen sheet	380	kg
		Mortar	180	kg
		Epoxy sealant	77	kg
		Bituprimer	35	kg
		Impregnation	19	liters
		Graffiti protection	19	liters
		Bitumen sealant	10	kg
		Polyethylene foam	5	kg
	A5	Untreated wood	5400	kg
		Diesel 7% bioblend	5280	liters
		Particle board	1600	kg
		Plywood	1000	kg
		Form oil	130	kg
		Plastic film	22	kg
		Concrete waste for recycling	23,185	kg
		Wood waste for incineration	5400	kg
		Plywood and particleboard waste for incineration	2600	kg

Table A2. *Cont.*

Concept	Module	Resource/Waste	Amount	Unit
		Reinforcement waste (carbon steel) for recycling	1134	kg
		Plastic waste for incineration	22	kg
		Bitumen sheet waste for incineration	19	kg
	B1	CO_2 uptake	1139	kg
		Zinc oxide to air	0.55	kg
	B2	Diesel 7% bioblend	70	liters
		Tap water	29,400	liters
		Graffiti shield	15	liters
	B3	Concrete C35/45	1344	kg
		Tap water	1000	liters
		Concrete waste	1411	kg
	B4	Concrete C35/45	12,600	kg
		Diesel 7% bioblend	27	liters
		Bitumen sealant	10	kg
		Bitumen sheet	380	kg
		Bituprimer	35	kg
		Epoxy sealant	77	kg
		Bitumen sealant waste for incineration	10	kg
		Bitumen sheet waste for incineration	380	kg
		Bituprimer waste for incineration	35	kg
		Epoxy sealant waste for incineration	77	kg
	C1	Diesel 7% bioblend	1453	liters
		Tap water	386,000	liters
	C2	See Table A6	-	
	C3	Concrete for recycling	448,615	kg
		HDG steel for recycling	250	kg
		Carbon steel reinforcement for recycling	22,680	kg
		Stainless steel reinforcement for recycling	553	kg
		Bitumen sealant waste for incineration	10	kg
		Bitumen sheet waste for incineration	380	kg
		Bituprimer waste for incineration	35	kg
		Epoxy sealant waste for incineration	77	kg
	C4	N/A	-	
	D	Electricity (Swedish grid mix)	340	kWh
SSC bridge	A1–A3	HDG structural steel plates incl. bolts and nuts	40,653	kg
	A5	Diesel 7% bioblend	5230	liters
	B1	Zinc oxide to air	3.35	kg
	B2	Diesel 7% bioblend	52	liters
		Tap water	26,200	liters
	B3	N/A	-	
	B4	N/A	-	
	C1	Acetylene	8300	liters
		Diesel 7% bioblend	2339	liters
	C2	See Table A6	-	
	C3	HDG structural steel plates incl. bolt and nuts for recycling	40,653	kg
	C4	N/A	-	
	D	N/A	-	

Appendix C

Common scenarios for both design concepts are presented in Table A3. Specific scenarios for the CSF bridge and the SSC bridge design concepts are presented in Tables A4 and A5, respectively.

Table A3. Common scenarios for both design concepts for each module (B–D).

Module	Scenario
B1	(1) Zinc in steel racks oxidizes and is released to air and water. The amount of zinc released from galvanized steel racks on a highway was calculated to be 0.95 kg/year/km [34]. It is assumed that 50% of this is released as zinc oxide particles to air and 50% is released in soluble form to water, giving a total of 0.55 kg of zinc released from the bridge to water and air respectively over 80 years. (2) Bitumen in asphalt is degraded and PAH leaches into the local environment at a rate of 0.5 mg PAH/m^2 asphalt over 25 years [35], giving a total of 112 mg of PAH is released from the bridge to soil over 80 years.
B2	(1) Washing of steel racks with drinking water occurs twice a year. Approximately 25 L is used per 100 m rack according to expertise within the Swedish Transport Administration, giving 400 L of water consumed over 80 years. The work cost is negligible and therefore was not included. (2) A total of 0.8 kg of salt is administered per year across the whole road surface of the bridge [36], giving 65 kg over 80 years. The work cost is negligible and therefore was not included. (3) A total of 0.07 h is needed for snow removal per km of lane per year [36]. The bridge has 3 lanes, and we assumed the use of a diesel-driven vehicle of 100 kW, giving 1.3 L of diesel over 80 years, which is considered negligible and was therefore not included.
B3	10% of the steel rack mass is repaired every second year, giving 692 kg over 80 years. The cost for this amount of steel is 8650 Euros. The energy used for the repairs is estimated to be negligible and was therefore not included. The work cost was calculated assuming 1 h of work per occasion and a salary of 60 Euros/h, giving 2400 Euros.
B4	(1) Steel racks are replaced completely every 20 years, giving 692 kg over 80 years. The cost for this amount of steel is 8650 Euro of which the work cost is 2 400 Euro. (2) A depth of 30 mm of the top asphalt layer is replaced every 40 years, giving 10,500 kg of asphalt over 80 years. The amount of diesel used for milling is 1.6 L/ton asphalt milled, giving 17 L over 80 years. The work cost is negligible and was therefore not included.
B5	No refurbishment needed
B6	No energy consumption
B7	No water consumption
B8	(1) Energy consumption of passing vehicles: It is estimated that the mean daily traffic on the bridge per year over the RSL will increase, as illustrated in Figure A1. It is assumed that 50% of the vehicles will be diesel-driven and 50% will be electric up until 2030; 20% will be diesel-driven, 20% will be HVO-driven, and 60% will be electric in 2030–2045; and 30% will be HVO-driven and 70% electric from 2045 onwards. The diesel and HVO consumption is assumed to 0,06 L/km and the electricity consumption is assumed to be 0.2 kWh/km, giving 11,797 L of diesel, 41,611 L of HVO, and 357,209 kWh of electricity over 80 years. (2) A total of 760 kg of asphalt is abraded per 100 m lane/year (at AADT 15,000, 100 km/h according to [37], giving 10,950 kg particles (of diameter 50–1000 µm) over 80 years. (3) Tires are abraded by 0.05 g/km and vehicles [38], giving 131 kg of microplastic over 80 years (25% styrene and 75% butadiene).
C1	(1) Steel racks are lifted away using a diesel-driven crane. It is estimated that 1 h is needed for 14 m of racks using a crane that consumes 12 L/h [39], giving 12 L of diesel. The work cost is negligible and was therefore not included. (2) Asphalt is milled using a machine that consumes 1.6 L of diesel/ton asphalt milled, giving 27 L. The work cost was calculated assuming 16 h of work and a salary of 60 Euros/h, giving 1000 Euro.

Table A3. *Cont.*

Module	Scenario
C2	(1) Steel racks are transported by truck to a storage facility. (2) Asphalt is transported by truck to recycling. (3) A total of 90% of the aggregates are not moved, and 10% are transported 39 km by truck to landfill.
C3	N/A, end-of-waste is reached for asphalt and steel racks before waste treatment takes place.
C4	A total of 10% of the aggregates are disposed on an inert landfill = 683,260 kg. The cost for landfilling is 30 Euros/ton.
D	(1) 100% of the steel racks are recycled. The selling price is 0.1 Euro/kg steel. (2) 100% of the asphalt is recycled. Crushing of the recycled asphalt is done with an electric crusher using 0.7 kWh/ton, giving 19 kWh in total. The selling price is 5 Euros/ton asphalt. (3) A total of 90% of the aggregates are not moved and are re-used as filling material on site.

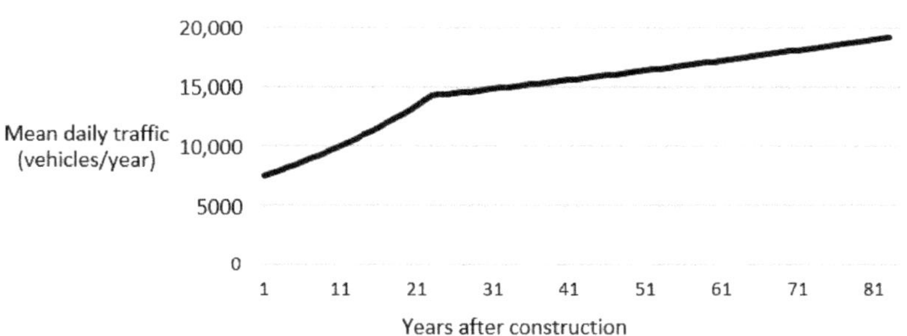

Figure A1. Estimated mean daily traffic on the bridge per year over the RSL (up to 80 years after construction).

Table A4. Specific scenarios for the CSF bridge design concept for each module (B–D).

Module	Scenario
B1	Carbonation of the concrete surfaces was calculated according to [40] with the following assumptions: • Area exposed to rain: 55 m^2 • Area protected from rain: 210 m^2 • Area in ground: 300 m^2 giving that 1139 kg CO_2 is taken up by the bridge over 80 years.
B2	Graffiti removal is done every 10 years by washing with hot water under high pressure. After washing, new graffiti protection is applied. It is assumed that 10% of the available surface area of 170 m^2 is covered by graffiti over 10 years, giving 17 m^2. It is estimated that it takes 15 min to wash 1 m^2, giving 4 h to wash in total per occasion. The equipment consumes 2.15 L of diesel/h, giving 70 L over 80 years. It consumes 15 L of drinking water/minute, giving 29.4 m^3 in total over 80 years. Graffiti protection of 10% of the surface, giving 15 L over 80 years. The work cost was calculated assuming 32 h of work and a salary of 60 Euros/h, giving 1 920 Euro.

Table A4. *Cont.*

Module	Scenario
B3	Minor repairs in the concrete parts are needed every 10 years. A 1 m^2 layer of concrete with a thickness of 70 mm is repaired per occasion, giving 1344 kg concrete repaired over 80 years. A total of 2000 L of water is used per m^3 of concrete repaired, giving 1000 L over 80 years. The amount of diesel consumed is negligible and was therefore not included. The work cost was calculated assuming 14 h of work and a salary of 60 Euros/h, giving 840 Euros.
B4	Replacement of edge beams is done every 40 years. The two edge beams are 400 mm wide, 500 mm high, and 6 m long, giving 5 m^3 of concrete is replaced over 80 years. The equipment has a diesel consumption of 5.3 L/h, and 1 m^3 concrete is removed/h, giving 27 L over 80 years. It also consumes 2000 L of drinking water/h [41], giving 10 m^3 in total over 80 years. The bridge insulation is totally replaced every 40 years, giving 77 kg of epoxy sealant, 10 kg of bitumen sealant, 35 kg of bituprimer, and 380 kg of bitumen sheet over 80 years. The work cost was calculated assuming 16 h of work and a salary of 60 Euros/h, giving 1000 Euro. The cost for 1 day of scaffolding rent is estimated to be 100 Euros.
C1	(1) Concrete is demolished using equipment that consumes 2000 L of drinking water and 5.3 L of diesel per m^3 of concrete, giving 386 m^3 water and 1023 L of diesel in total. The work cost was calculated assuming 193 h of work and a salary of 60 Euros/h, giving 11,600 Euros. (2) A total of 90% of the aggregates is not handled at all, and 10% is excavated using an excavator consuming 1.5 L/m^3 of excavated material, giving 430 L of diesel. The work cost was calculated assuming 290 h of work and a cost of 100 Euros/h, giving 29,000 Euros.
C2	(1) Reinforcement steel is transported by truck to a storage facility. (2) Concrete is transported by truck to a recycling facility.
C3	N/A, end-of-waste is reached for concrete and reinforcement steel before waste treatment takes place.
C4	N/A, no waste is disposed.
D	(1) All of the concrete is recycled into filling material. Crushing of the recycled concrete is done with an electric crusher using 0.7 kWh/ton, giving 340 kWh in total. The selling price is 5.3 Euro/ton concrete. (2) All of the reinforcement steel is recycled into reinforcement steel. The selling price is 0.1 Euro/kg of carbon steel and 0.9 Euro/kg of stainless steel.

Table A5. Specific scenarios for the SSC bridge design concept for each module (B–D).

Module	Scenario
B1	Zinc oxidizes on the bridge's structural steel plates into a powder which is assumed to disperse into the surrounding air (since the steel surface is protected from rain). Conservatively, it is assumed that 0.5 g of Zn is dispersed per m^2 plate and year [34,42]. The area is 70 m^2, giving 2.8 kg of Zn is released to air over 80 years.
B2	(1) Graffiti removal is done every 10 years by washing with hot water under high pressure. No graffiti protection is needed on steel. It was assumed that 10% of the available surface area of 126 m^2 is covered by graffiti over 10 years, giving 13 m^2. It was estimated that it takes 15 min to wash 1 m^2, giving 3 h to wash in total per occasion. The equipment consumes 2.15 L of diesel/h, giving 52 L over 80 years. It consumes 15 L of drinking water/minute, giving 22.4 m^3 in total over 80 years.

Table A5. *Cont.*

Module	Scenario
	The work cost was calculated assuming 32 h of work and a salary of 60 Euros/h, giving 1 920 Euro. Washing of the bridge's structural steel plates surface of 95 m^2 is also done by using hot water under high pressure. It is done once a year using 0.5 L of drinking water/m^2, giving 3800 L over 80 years.
B3	No repairs are needed.
B4	No replacements are needed.
C1	(1) Masses covering the bridge's steel construction are excavated. A total of 4600 m^3 is excavated using an excavator with a capacity of 40 m^3/h using 20 L of diesel/h, giving 2300 L of diesel. (2) Steel plates are cut with a cutting torch consuming 0.3 m^3 acethylene/h to cut 36 m of steel plate/h [43], giving 8.3 L/m. A length of 1000 m was assumed to be cut, giving 8300 L used over 80 years (corresponding to 9 kg of ethine). The work cost was calculated assuming 16 h of work and a salary of 80 Euros/h, giving 1280 Euros.
C2	The bridge's structural steel plates are transported by truck to a storage facility.
C3	N/A, end-of-waste is reached before waste treatment takes place.
C4	N/A, no waste is disposed.
D	All of the bridge's structural steel plates are recycled. The selling price is 0.1 Euro/kg steel.

Appendix D

Table A6. Transport modes and distances travelled for the resources (applicable for modules A4, A5, B1–B4 and C2).

Concept	Resource	Transport Mode	Distance (km)
Both	Aggregates (crushed rock), asphalt	Truck, Euro 6, 20–26 t gross weight/17.3 t payload capacity, 55% utilisation	39
	Concrete elements	"	66
	Diesel 7% bioblend	"	100
	Steel racks	"	646
	Aggregate waste, Asphalt waste, Concrete waste	"	39
	Steel waste (racks, reinforcement steel, stainless steel)	"	50
	Plastic waste	"	38
CSF bridge	Ready-mix concrete C35/45	"	40
	Reinforcement (carbon steel)	Average electricity/diesel driven train, gross tonne weight 1000 t/726 t payload capacity, 40% utilisation Truck Euro 6, 20–26 t gross weight/17.3 t payload capacity, 55% utilisation	1510 198
	Reinforcement (stainless steel)	Container ship, 5000 to 200,000 dwt payload capacity, ocean going, 70% utilisation Truck Euro 6, 20–26 t gross weight/17.3 t payload capacity, 55% utilisation	49 1045
	Untreated wood, particle board, plywood	Truck Euro 6, 20–26 t gross weight/17.3 t payload capacity, 55% utilisation	20

Table A6. *Cont.*

Concept	Resource	Transport Mode	Distance (km)
	Bitumen sheet waste, wood waste	"	38
	Hazardous waste	"	200
SSC bridge	Structural steel plates	Container ship, 5000 to 200,000 dwt payload capacity, ocean going, 70% utilisation Truck Euro 6, 20–26 t gross weight/17.3 t payload capacity, 55% utilisation	360 2027
	Steel waste (structural steel plates)	Truck Euro 6, 20–26 t gross weight/17.3 t payload capacity, 55% utilisation	50

" denotes 'Same as above'.

References

1. ISO. *ISO 15392:2019—Sustainability in Buildings and Civil Engineering Works—General Principles*, 2nd ed.; International Organization for Standardization (ISO): Geneva, Switzerland, 2019.
2. CEN. *EN 15643-5:2017—Sustainability of Construction Works—Sustainability Assessment of Buildings and Civil Engineering Works—Part 5: Framework on Specific Principles and Requirement for Civil Engineering Works*; European Committee for Standardization (CEN): Brussels, Belgium, 2017.
3. ISO. *ISO/TS 21929-2:2015—Sustainability in Building Construction—Sustainability Indicators—Part 2: Framework for the Development of Indicators for Civil Engineering Works*; International Organization for Standardization (ISO): Geneva, Switzerland, 2015; Volume 1.
4. ISO. *ISO 21931-2:2019—Sustainability in Buildings and Civil Engineering Works—Framework for Methods of Assessment of the Environmental, Social and Economic Performance of Construction Works as a Basis for Sustainability Assessment—Part 2: Civil Engineering*; International Organization for Standardization (ISO): Geneva, Switzerland, 2019.
5. Mathern, A.; Ek, K.; Rempling, R. Sustainability-driven structural design using artificial intelligence. In Proceedings of the IABSE Congress New York City—The Evolving Metropolis, New York, NY, USA, 4–6 September 2019; pp. 1–8.
6. Du, G.; Safi, M.; Pettersson, L.; Karoumi, R. Life cycle assessment as a decision support tool for bridge procurement: Environmental impact comparison among five bridge designs. *Int. J. Life Cycle Assess.* **2014**, *19*, 1948–1964. [CrossRef]
7. Rempling, R.; Mathern, A.; Ramos, D.T.; Fernández, S.L. Automatic structural design by a set-based parametric design method. *Autom. Constr.* **2019**, *108*, 102936. [CrossRef]
8. Yepes, V.; Martí, J.V.; García-Segura, T. Cost and CO_2 emission optimization of precast–prestressed concrete U-beam road bridges by a hybrid glowworm swarm algorithm. *Autom. Constr.* **2015**, *49*, 123–134. [CrossRef]
9. Penadés-Plà, V.; García-Segura, T.; Yepes, V. Accelerated optimization method for low-embodied energy concrete box-girder bridge design. *Eng. Struct.* **2019**, *179*, 556–565. [CrossRef]
10. Bragança, L.; Vieira, S.M.; Andrade, J.B. Early Stage Design Decisions: The Way to Achieve Sustainable Buildings at Lower Costs. *Sci. World J.* **2014**, *2014*, 1–8. [CrossRef] [PubMed]
11. Ek, K.; Mathern, A.; Rempling, R.; Karlsson, M.; Brinkhoff, P.; Norin, M.; Lindberg, J.; Rosén, L. A harmonized method for automatable life cycle sustainability performance assessment and comparison of civil engineering works design concepts. *IOP Conf. Ser. Earth Environ. Sci.* **2020**, in press.
12. CEN. *EN 15804:2012+A2:2019—Sustainability of Construction Works—Environmental Product Declarations—Core Rules for the Product Category of Construction Products*; European Committee for Standardization (CEN): Brussels, Belgium, 2019.
13. EPD International AB. Product Category Rules (PCR)—Bridges, Elevated Highways and Tunnels. Available online: https://www.environdec.com/PCR/Detail/?Pcr=12257 (accessed on 12 December 2019).
14. ISO. *ISO 15686-5:2017—Buildings and Constructed Assets—Service Life Planning—Part 5: Life-Cycle Costing*; International Organization for Standardization (ISO): Geneva, Switzerland, 2017.

15. ISO. *ISO 14008:2019—Monetary Valuation of Environmental Impacts and Related Environmental Aspects*; International Organization for Standardization (ISO): Geneva, Switzerland, 2019.

16. *GaBi Professional, Version 9.5.2.49, Software-System and Data-Base for Life Cycle Engineering*; Sphera: Leinfelden-Echterdingen, Germany, 2020.

17. Sala, S.; Crenna, E.; Secchi, M.; Pant, R. Global Normalisation Factors for the Environmental Footprint and Life Cycle Assessment. 2017. Available online: https://ec.europa.eu/jrc (accessed on 9 October 2020).

18. Sala, S.; Cerutti, A.K.; Pant, R. Development of a Weighting Approach for the Environmental Footprint. Luxembourg. 2018. Available online: https://ec.europa.eu/jrc (accessed on 9 October 2020).

19. European Commission. EF Reference Package 3.0. Available online: https://eplca.jrc.ec.europa.eu/LCDN/developerEF.xhtml (accessed on 22 September 2020).

20. CEN. *EN 16627:2015—Sustainability of Construction Works—Assessment of Economic Performance of Buildings—Calculation Methods*; European Committee for Standardization (CEN): Brussels, Belgium, 2015.

21. Steen, B. *The EPS 2015d Impact Assessment Method—An Overview*; Swedish Life Cycle Center: Gothenburg, Sweden, 2015.

22. Hammervold, J.; Reenaas, M.; Brattebø, H. Environmental Life Cycle Assessment of Bridges. *J. Bridg. Eng.* **2013**, *18*, 153–161. [CrossRef]

23. Du, G.; Pettersson, L.; Karoumi, R. Soil-steel composite bridge: An alternative design solution for short spans considering LCA. *J. Clean. Prod.* **2018**, *189*, 647–661. [CrossRef]

24. Ahlroth, S.; Finnveden, G. Ecovalue08–A new valuation set for environmental systems analysis tools. *J. Clean. Prod.* **2011**, *19*, 1994–2003. [CrossRef]

25. Finnveden, G.; Håkansson, C.; Noring, M. A new set of valuation factors for LCA and LCC based on damage costs—Ecovalue 2012. In Proceedings of the 6th International Conference on Life Cycle Management in Gothenburg, Gothenburg, Sweden, 25–28 August 2013.

26. Laurent, A.; Olsen, S.I.; Hauschild, M.Z. Limitations of Carbon Footprint as Indicator of Environmental Sustainability. *Environ. Sci. Technol.* **2012**, *46*, 4100–4108. [CrossRef] [PubMed]

27. Yepes, V.; García-Segura, T.; Moreno-Jiménez, J. A cognitive approach for the multi-objective optimization of RC structural problems. *Arch. Civ. Mech. Eng.* **2015**, *15*, 1024–1036. [CrossRef]

28. Giunta, M. Assessment of the environmental impact of road construction: Modelling and prediction of fine particulate matter emissions. *Build. Environ.* **2020**, *176*, 106865. [CrossRef]

29. Cheriyan, D.; Choi, J.-H. A review of research on particulate matter pollution in the construction industry. *J. Clean. Prod.* **2020**, *254*, 120077. [CrossRef]

30. Ek, K.; Mathern, A.; Rempling, R.; Rosén, L.; Claeson-Jonsson, C.; Brinkhoff, P.; Norin, M. Multi-criteria decision analysis methods to support sustainable infrastructure construction. In Proceedings of the IABSE Symposium 2019: Towards a Resilient Built Environment—Risk and Asset Management, Guimarães, Portugal, 27–29 March 2019; p. 8.

31. Belton, V.; Stewart, T.J. *Multiple Criteria Decision Analysis—An Integrated Approach*; Kluwer Academic Publishers: Dordrecht, The Netherlands, 2002.

32. Penadés-Plà, V.; García-Segura, T.; Martí, J.V.; Yepes, V. A Review of Multi-Criteria Decision-Making Methods Applied to the Sustainable Bridge Design. *Sustainability* **2016**, *8*, 1295. [CrossRef]

33. Eurobitume. *The Eurobitume Life-Cycle Inventory for Bitumen, Version 3.0*; European Bitumen Association: Brussels, Belgium, December 2019.

34. Legret, M.; Pagotto, C. Evaluation of pollutant loadings in the runoff waters from a major rural highway. *Sci. Total Environ.* **1999**, *235*, 143–150. [CrossRef] [PubMed]

35. Birgisdottir, H.; Gamst, J.; Christensen, T.H. Leaching of PAHs from Hot Mix Asphalt Pavements. *Environ. Eng. Sci.* **2007**, *24*, 1409–1422. [CrossRef]

36. Ihs, A.; Möller, S. *VTI Notat 53-2004 -Beräkningsmodell för Vinterväghållningskostnader*; VTI: Linköping, Sweden, 2004.

37. Swedish National Road and Transport Research Institute (VTI). VTI Slitagemodell. Available online: http://www.metodgruppen.nu/web/page.aspx?refid=127 (accessed on 8 October 2020).

38. Magnusson, K.; Eliasson, K.; Fråne, A.; Haikonen, K.; Hultén, J.; Olshammar, M.; Stadmark, J.; Voisin, A. *Swedish Sources and Pathways for Microplastics to the Marine Environment*; Swedish Environmental Protection Agency: Stockholm, Sweden, 2016; Volume C 183.

39. Erlandsson, M. *Miljödata för arbetsfordon*; Svenska Miljöinstitutet: Stockholm, Sweden, 2013; Volume BPI 13/1.
40. CEN. *EN 16757:2017—Sustainability of Construction Works—Environmental Product Declarations—Product Category Rules for Concrete and Concrete Elements*; European Committee for Standardization (CEN): Brussels, Belgium, 2017.
41. Aquajet Systems AB. Product sheet, Aqua Cutter 710H. Aquajet Systems AB. Available online: https://www.aquajet.se/wp-content/uploads/2014/06/leaflet_aqua-cutter-710H.pdf (accessed on 8 October 2020).
42. Lindstrom, D.; Wallinder, I.O. Long-term use of galvanized steel in external applications. Aspects of patina formation, zinc runoff, barrier properties of surface treatments, and coatings and environmental fate. *Environ. Monit. Assess.* **2010**, *173*, 139–153. [CrossRef] [PubMed]
43. AGA. Product Catalog. Gassvetsning, skärning, lödning och värmning. Available online: https://www.linde-gas.se/sv/images/Cutting_Welding_Catalogue_2018_SE_web_tcm586-137940.pdf (accessed on 8 October 2020).

Publisher's Note: MDPI stays neutral with regard to jurisdictional claims in published maps and institutional affiliations.

International Journal of
***Environmental Research
and Public Health***

Article

Proposal of Sustainability Indicators for the Design of Small-Span Bridges

Cleovir José Milani [1], Víctor Yepes [2,*] and Moacir Kripka [3]

[1] Department of Civil Engineering, Federal Technological University of Paraná, Via do Conhecimento, Km 1, Pato Branco, Paraná 80000, Brazil; milani@utfpr.edu.br
[2] Institute of Concrete Science and Technology (ICITECH), Universitat Politècnica de València, 46022 Valencia, Spain
[3] Civil and Environmental Engineering Graduate Program, University of Passo Fundo, BR 285, São José, Passo Fundo, Rio Grande do Sul 90000, Brazil; mkripka@upf.br
* Correspondence: vyepesp@cst.upv.es; Tel.: +34-963-879-563; Fax: +34-963-877-569

Received: 6 May 2020; Accepted: 17 June 2020; Published: 22 June 2020

Abstract: The application of techniques to analyze sustainability in the life cycle of small-span bridge superstructures is presented in this work. The objective was to obtain environmental and economic indicators for integration into the decision-making process to minimize the environmental impact, reduce resource consumption and minimize life cycle costs. Twenty-seven configurations of small-span bridges (6 to 20 m) of the following types were analyzed: steel–concrete composite bridges, cast in situ reinforced concrete bridges, precast bridges and prestressed concrete bridges, comprising a total of 405 structures. Environmental impacts and costs were quantified via life cycle environmental assessment and life cycle cost analysis following the boundaries of systems from the extraction of materials to the end of bridge life ("from cradle to grave"). In general, the results indicated that the environmental performance of the bridges was significantly linked to the material selection and bridge configuration. In addition, the study enabled the identification of the products and processes with the greatest impact in order to subsidize the design of more sustainable structures and government policies.

Keywords: bridges; sustainability; design; life cycle assessment

1. Introduction

Bridges have a major role in transportation infrastructure, supporting highway traffic loads, crossing various obstacles and enabling effective communication between two destinations [1]. Bridges generally represent a significant public resource; in Europe, for example, bridges account for approximately 2% of the road network length and 30% of its cost [2].

Bridges should be seen as a key part of the economic activity and well-being of a given community [3]. While a bridge is designed from economic, technical and safety perspectives, the environmental performance is often not considered in the decision-making process [4]. Several elements should be considered in the bridge life analysis process: factors concerning age, which are directly related to maintenance and various rehabilitation, repair or reinforcement procedures, and others related to the increasing weight of trucks and transportation facilities that pass over the bridge [5].

Given the importance of bridges and their relationship with the environment, the pillars of sustainability must be observed, meaning that the economic pillar of sustainable development must ensure the efficiency of natural resource consumption [3]. The potential environmental impacts must be measured by indicators because the sustainability of bridges includes the extension of maintenance and the end of the bridge's life [6]. The social pillar must be observed considering its positive and negative impacts [7].

The environmental performance of bridges is directly related to the choice of construction materials as well as to the type of bridge [8–10]. Currently, the use of sustainable materials in the construction of bridges is increasingly attempted considering economic, social and environmental factors [11]. Life cycle sustainability assessments applied to small-span bridge superstructures for side roads have revealed the good performance of composite steel and concrete structures (concrete deck and steel profile beams) [12].

The sustainable use of raw materials should be an objective of the construction industry [13], with emphasis on steel/concrete composite bridges (concrete deck and steel profile beams) [14]. Regardless of the criteria used to represent the sustainability of the structures, decision-making processes should include a complete analysis of the life cycle, from the cradle to the grave [15], considering the possibility of abiotic resource depletion, the acidification of the environment, the depletion of the stratospheric ozone layer, the eutrophication of water bodies and the photochemical creation of ozone [16].

With regard to life cycle cost analysis (LCCA), environmental, economic and social indicators [17] can be applied as a tool for calculating and comparing the life cycle cost of a product [18]. This allows LCCA to be interpreted as one of the three pillars of sustainability [19], and it is fundamental to the structure of sustainability due to its usefulness for determining the cost-effectiveness/cost-competitiveness ratio of various technological options that affect the environment [20].

The LCCA may include the total life cycle time, which takes into account the costs associated with the different phases through which the construction work proceeds over time [21], and the user's costs. It is characterized as total costs and evolves rapidly, meaning that the assessment of the structural performance and total accumulated cost throughout the life cycle determines the competent management of civil infrastructure [22].

The largest share of bridges is represented by small-span structures. For example, in a study conducted in Brazil, it was found that 74% of the bridges had spans smaller than 10 m, and 90% had lengths of up to 20 m [12]. Apparently, small-span bridges are generally underestimated considering the lack of executive design and technical monitoring in their construction, non-existent maintenance, evident signs of abandonment including poor conditions for use and the emergence of different pathologies or a lack of attention from the public authorities, thus generating sustainability issues [23]. In this regard, the rationalization of the process of building structures—including reducing costs [24–26], social impacts [27,28] and environmental impacts [29,30]—should be addressed in the design phase [31]. The environmental impact of bridges has been highlighted due to the consumption of resources [32]; it has been widely demonstrated that more sustainable construction benefits humanity [33].

The aim of this study was to determine the economic and environmental indicators related to bridge sustainability that should be incorporated into the decision-making process of design to alleviate environmental impacts, reduce resource consumption and minimize costs throughout the life cycle. Several small-span bridge configurations were analyzed, with their potential environmental impacts and costs quantified by life cycle environmental assessment and life cycle cost analysis, following the boundaries of systems from the extraction of materials to the end of bridge life (i.e., "from cradle to grave").

The study was performed using life cycle assessment (LCA) and life cycle cost analysis (LCCA), using the guidelines of ISO 14040 [34] and ISO 14044 [35]; the International Reference Life Cycle Data System (ILCD) [36]; the Ecoinvent 3 database. 5, 2018 [37]; the ReCiPe 2016 V1.1 global reach method [38]; the SimaPro software, version 9.0.0.32 [39,40]; the European methods IMPACT 2002 + V2.14 [41] and ILCD 2011 Midpoint + V1.10 [42]; and the American method Building for Environmental and Economic Sustainability (BEES + V4. 07 USA) [43].

Comparative analysis was carried out using the following the methods: the European method IMPACT 2002 + V2.14, which was formulated by the Swiss Federal Institute of Technology in Lausanne (EPFL) and proposes the combination of midpoint and endpoint approaches, characterizing the inventory results into 15 impact categories correlated to at least one of the four categories of damage [41]; the ILCD 2011 Midpoint + V1.10 method, which was launched by the European Commission, Joint Research Centre, in 2012 [42]; and the US method BEES + V4.07 USA, which measures the environmental performance of building products by using the life-cycle assessment approach specified in the ISO 14040 series of standards [43].

In this process of building sustainable bridges, the terms "sustainable development," "sustainability" and "sustainable" were cited in the Brundtland Report [44] in the context of economic, social and environmental issues [45], which attributed the genesis of terminology to the document registered by Hans Carl von Carlowitz in 1713 [46]. The life cycle assessment (LCA) method by Boulenger [47], in which some aspects of the assessment of sustainability [48] in bridge design were shown that aim at achieving a low impact on the life cycle [49], was also presented, observing the ecological, economic and social pillars of sustainability [50] with a management strategy that seeks to equalize the cost of annual maintenance for a stable budget [51].

The European Union formulated the three pillars of sustainability at the Copenhagen Summit and the Treaty of Amsterdam, 1997, forming a basis on which to build sustainable development [50], the performance of which is measured through environmental life cycle analysis, life cycle cost analysis and social life cycle analysis [52]. Regarding social analysis, there are a variety of criteria that indicate a lack of consensus and for which there is less certainty and discussions regarding the selection of more representative criteria are triggered [53]. Some of the important aspects of sustainability are associated with the extraction of construction raw materials, having direct positive and negative impacts on the environment, the economy and the social context as a whole [7].

The focus on sustainability in bridge design lies in the investigation of economic and environmental impacts with respect to the analysis of materials and the concept of sustainable development in the bridge industry [54]. The sustainability LCA of bridge structures is recommended to include the environmental, economic, social and functional qualities of bridges [55], although Hammervold, Reenaas and Brattebo [56] have proposed the inclusion of more categories, which impact the environment for comparative analyses.

Studies on the analyses of the life cycle of bridges have sought the explanation of the uncertainties associated with bridge maintenance [57–59], with reference to the application of holistic approaches in studies related to sustainable development in the bridge industry [60,61]. This view is different from that presented by Du et al. [62], who identified the variable influences of impact categories resulting from materials, structural elements and general design.

Tapia [63] evaluated the sustainability life cycle of deteriorated bridges to assess risk-based sustainability indicators of bridge performance. Pang et al. [64] had previously verified the issues related to old bridges in China, most of which were in need of maintenance due to the aging of road bridge materials over time, which has significant effects on structural performance [65].

Regarding the state of the art of sustainability and its pillars, the scarcity of studies on sustainable processes in bridge designs was considered as an opportunity to seek answers and to formulate alternatives for bridges that meet the needs of the population in their free transit and that can be built based on the parameters of sustainability.

Almeida, Teixeira and Delgado [66] developed a method including tests with combined approaches as a model for minimizing bridge life cycle costs, aiming at the optimization of maintenance intervention plans in a set of concrete bridges to calculate deterioration over time and to analyze costs. Zhang, Wu and Wang [9] assigned probability distributions for the parameters considered in the environmental impact criteria, evaluating the variability in inventory acquisition and detecting the key parameters with relevant environmental impacts. Penadés-Plá et al. [11] used several methods and sustainable criteria for decision-making in each phase of a bridge's life cycle, ranging from design to the recycling or demolition of the structure.

For existing bridges—specifically, steel/concrete composite bridges—the study by Bizjak et al. [67] indicated that the use of concrete reinforced with ultra-high-performance fiber in the construction and composition of the slab minimizes problems related to the adjustment of the sub-structure geometry, reducing the use of resources.

Huijbregts et al. [68] recommended the ReCiPe method as it provides a harmonized implementation of cause and effect paths for the calculation of the characterization factors of potential environmental impacts and damages (end point).

Although this study has a clear objective of proposing sustainability indicators for the economic and environmental pillars in the life cycle of small-span bridge superstructures, a brief statement is needed on the decisive issue of the decision-making process for the choice of small-span bridges with better sustainability performance. The use of a decision-making process enables solutions that meet normally conflicting objectives to be achieved, and this process can be conducted in various ways.

In the study undertaken by Kripka, Yepes and Milani [69] and the data collected at the beginning of this study, sustainable design alternatives for small-span bridges in Brazil were investigated. In this approach, the structures were evaluated while taking into account quantitative aspects (construction cost, assembly, materials transport, lifespan and environmental impact) and qualitative aspects (architecture and sense of safety for the user). Decision-making methods with several criteria were applied, minimizing the subjectivity implicit in the decision-making process. In this study, two decision-making methods with several criteria were adopted: the analytical hierarchy process (AHP) and the VIKOR method.

Many sustainability indicators have been applied in studies on bridges, in which the authors have chosen both the method and the indicators. Padgett and Tapia [70] applied a risk-based method for life-cycle environmental sustainability analysis. Du et al. [4] analyzed 20 environmental indicators presented in five bridge designs. Tapia [63] addressed the issue of deteriorated bridges with the objective of quantifying risk-based sustainability indicators of performance. Du highlighted global warming and energy consumption as two popular indicators in LCA [8]. Arya, Amiri and Vassie [71] used indicators such as climate change, resource use, waste, biodiversity and heritage. Hatami and Morcous [72] presented a proposal to undertake a life cycle cost analysis. Furuta used the Genetic Algorithm to minimize the total life cycle cost of a large number of bridges [51]. There is a perception that there is a lack of a standardized bridge life cycle assessment manual that can guide designers in their decision-making process for choosing the best design, with the goal of minimizing environmental and economic damage and a focus on sustainability.

In the construction of a bridge, a comparative study between various solutions must be carried out, with the final choice of a solution being made when all items of functionality, safety, esthetics, economy and durability are met [73]. For this research work, it was considered relevant to insert one more requirement; i.e., bridge sustainability indicators. The selection of the indicators must be in accordance with the objectives of the study, which require previous knowledge of the life cycle of the product or processes analyzed. Life cycle impact assessment (LCIA) provides the results of impact indicators related to human health, the natural environment and resource depletion [74].

There is an increase of global interest in environmental issues in general, and a growing number of studies aim to identify alternatives to improve production processes and the use of common resources. Although only few studies have been related to small-span bridges, significant findings were determined relative to bridges, as this topic concerns public equipment, which is relevant to society in daily use and is equally important for the purposes of sustainability when investigating the possibilities of the sustainable development of bridges in the design phase.

These possibilities can guide the development of bridge design based on the pillars of sustainability, enabling a study that analyzes bridge conditions by applying the available tools and resources to provide the opportunity to create alternatives to reduce environmental impacts.

The remainder of this paper is structured as follows. Section 2 introduces the proposed methodology and Section 3 presents results and discussions. Finally, in Section 4, the conclusions and final considerations are presented.

2. Materials and Methods

This section presents the methodology for the application of the LCA and LCCA techniques for small-span bridge superstructures, which are designed with different typologies and materials to identify their potential environmental impacts and characterize their effects on human health, ecosystems and resources and their costs in the life cycle of the bridges.

In recent years, many methods of impact characterization have been developed; examples are methods such as ILCD 2011 MIDPOINT + and IMPACT 2002 + in Europe and BEES and TRACI 2.1 in North America. However, the method that has become prominent for researchers is the global ReCiPe method (adopted in this work).

In this study, the application of the LCA on bridges was based on ISOs 14040 and 14044 and the ILCD guidelines, using the Ecoinvent database 3.5 from August 2018, the ReCiPe (H) 2016 V1.1 global reach method, which is a harmonized method for assessing the impact of the life cycle at the endpoint level, and the SimaPro software version 9.0.0.32. The results presented in this study reference damage to human health, ecosystems and resources.

ISO 14040 sets out the basic safety inspection guidelines for environmental management, life cycle assessment and their principles and framework. ISO 14044 deals with environmental management, life cycle assessment and requirements and guidelines, and ILCD provides a common basis for consistent, robust and quality-assured life cycle data and studies. Such data and studies support consistent sustainable consumption and production tools such as environmental labeling, eco-design, carbon footprint reduction and green public procurement.

Ecoinvent is a large database of LCIA in which data are transparently documented as inputs/outputs in the unitary process. ReCiPe is a method for life cycle impact assessment, and the primary objective of the ReCiPe method is to transform the long list of life cycle inventory results into a limited number of indicator scores. These indicator scores express the relative severity for an environmental impact category. In ReCiPe, indicators can be obtained at two levels: 18 midpoint indicators and three endpoint indicators. SimaPro, which is a type of software used for life cycle assessment (LCA), has the function of collecting data and analyzing the environmental performance of products and services and can model and analyze complex life cycles in a systematic and transparent way, following the recommendations of the ISO 14040 series.

At the damage assessment stage, impact category indicators with a common unit can be added. In the ReCiPe method, the 18 impact categories are listed under three endpoint damage categories:

1. Damage to human health: This is expressed as the number of years of life lost and the number of years of living with disability. These are combined as Disability Adjusted Life Years (DALYs)—an index that is also used by the World Bank and World Health Organization. Regarding the end point category of human health damage, the following items can be found at the mid-point that impact human health: global warming, stratospheric ozone depletion, ionizing radiation and ozone formation, fine particulate formation, human carcinogenic toxicity, non-human carcinogenic toxicity and water consumption.

2. Damage to ecosystems: This is expressed as the loss of species over a certain area for a certain time; the unit is species per year. Regarding the category of final ecosystem damage, the following can be found at an average level that impacts various ecosystems: global warming and terrestrial and freshwater ecosystems, and ozone formation and terrestrial ecosystems. In addition, terrestrial and aquatic ecosystems are affected by terrestrial acidification, eutrophication of freshwater and marine water, ecotoxicity of terrestrial, freshwater and marine ecosystems, land use and water consumption.

3. Resource scarcity: This is expressed as the excess costs of future resource production over an infinite period (assuming constant annual production), considering a discount ratio of 3%; the unit is USD2013. The endpoint damage in the resource category refers to the scarcity of mineral resources, as an average impact.

Furthermore, to validate the results obtained in the ReCiPe methodology, the sensitivity analysis was performed with the application of the following methods for characterization of environmental impacts: the European methods IMPACT 2002 + V2.14 and ILCD 2011 Midpoint + V1.10, and the American method BEES + V4.07 USA.

Initially, the objective of the LCA study was defined as follows: the intended application of the LCA results should be declared in a precise and unequivocal manner, along with the assumed data and methodological limitations of the study, reasons for conducting the LCA study, identification of the target audience of the study, whether the LCA study includes a comparative statement to be disseminated to the public, to whom the results of the study are intended to be communicated, the actors involved and the identification of those who commissioned the LCA/CAV study [74]. Next, the scope, inventory preparation, analysis and impact assessment were defined.

All stages were analyzed and interpreted considering steel/concrete composite bridges, reinforced concrete bridges cast in situ, precast reinforced concrete and prestressed bridges, all of which are in use in several Brazilian states and countries internationally.

The analysis methods include the identification of models of small-span bridge design used on rural and neighboring roads; furthermore, a standard model that fulfils the same function to compare environmental and economic performance is proposed.

Regarding the elaboration of bridge typology, research was conducted in several precast industries that produce bridges, and a survey was executed in databases of Brazilian and international transport infrastructure government agencies.

Twenty-seven bridge models were pre-selected, with spans ranging from 6 to 20 m, comprising a total of 405 bridges (Figure 1).

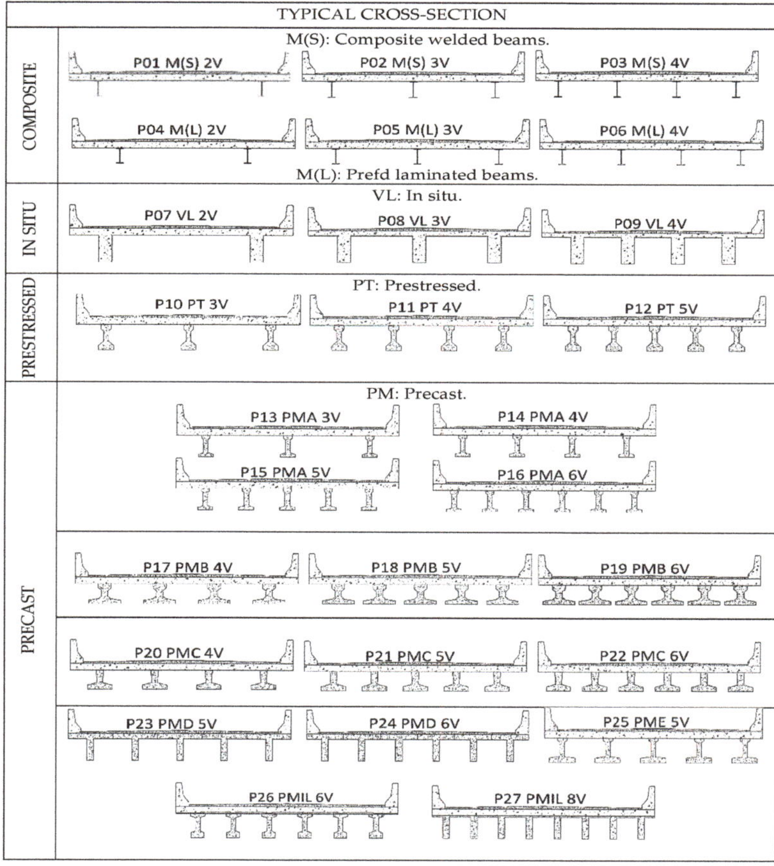

Figure 1. Number of bridges assessed: 27 typologies, 405 total bridges ranging from 6 to 20 m.

The nomenclature of the identification of the types of transversal bridges was adopted as follows: P01 to P27 represented the 27 types of bridges considered, 2V (two beams) to 6V (six beams) and 8V (eight beams) corresponded to quantities of beams, while M(S) corresponded to mixed welded beams, M(L) to prefabricated laminated beams, VL to in situ, PT to prestressed and PM corresponded to precast, with A, B, C, D and IL corresponding to the companies surveyed.

The following specifications were observed:

A load capacity of 450 kN: the selected roadway was bi-directional, with two traffic lanes with a width equal to 3 m each, with the typical vehicle traffic of the Brazilian standard NBR-7188, the TB-45 (450 kN); an Average Daily Volume (ADV) of 50 < ADV < 200 vehicles per day was considered for the calculation of the number of cycles that occurred in the bridge structure during its lifespan.

Beams: The selected beams were steel-rolled and electro-welded "I" profiles, in addition to pre-cast reinforced concrete and prestressed concrete beams and beams cast in situ.

Slabs: The adopted slabs were made of concrete, precast concrete and cast in situ, with a thickness of 20 cm and a reinforcement ratio that varied according to the quantities of beams in each bridge. Figure 2 presents a schematic model of the bridges analyzed; all bridges consisted of the same types of railings, rainwater drainage and asphalt paving.

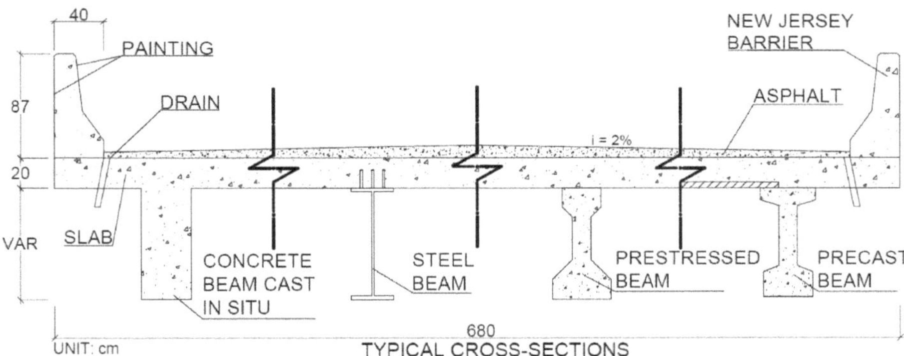

Figure 2. Typical cross-sections of bridge models with concrete beams cast in situ, steel beams, prestressed beams and precast beams.

To determine the LCCA, the amount of material and equipment hours and the number of personnel hours spent on the execution of each unit were compared and multiplied by the cost of materials, the hourly rent of equipment and the hourly wage of employees, respectively, which duly increased with social charges and budget difference income (BDI). Regarding the equipment, the hourly cost of transportation and operation involving the movement of materials and people within the site included trucks, cranes and breakers, among others. The cost of labor was represented by the wages of the workers who handled the materials, in addition to social charges and other expenses involving workers' participation.

The raw materials/products considered refer to the services necessary for the execution of the construction extraction stage. The services adopted are part of a list of procedures that begins with the support device, moving through the stages of precast element construction, concreting, asphalt paving, transportation, drains and painting. During the bridge use phase, the transportation services for the visual inspection, cleaning and demolition of asphalt paving as well as transportation to the sorting center (landfill) and pavement replacement, including machinery and equipment, materials and all transportation, were considered.

Finally, in the phase corresponding to the procedures at the end of the bridge life cycle, asphalt paving demolition services were considered. Reinforced concrete and steel beams, when applicable, were transported to the respective sorting centers (landfill). Budgets were prepared including all direct and indirect costs of the bridge construction phases, for all the bridges analyzed; subsequently, cost indicators were created for the small-span bridge superstructures.

A relationship was found between the services considered in reference to the bridge models of the following types: P1—steel/concrete composite bridges; P2—in situ cast reinforced concrete bridges; P3—prestressed concrete beam bridges; and P4—pre-cast reinforced concrete beams bridges. These are presented in Table 1.

Table 1. Products and processes evaluated by bridge type and related to the phases.

Service Description	Unit	P1	P2	P3	P4
Phase (1): Production and construction; Phase (2): use/maintenance; and Phase (3): end of life					
Grout 30 MPa (1)	m^3	x	x	x	x
Neoprene (1)	dm^3	x	x	x	x
Concrete production, 35 MPa (sand, basalt, cement, Portland, plasticizer, tap water) (1)	m^3	x	x	x	x
Concrete mixing factory—construction (1)	p	x	x	x	x
Precast concrete production parts (beams, slab) (1)	m^3	x	-	x	x
Wood forms (1)	m^3	x	x	x	x
Steel forms (usage 100 times) (1)	kg	x	x	x	x
Steel rebar (1)	kg	x	x	x	x
Producing I-beams (1)	kg	x	-	-	-
Welding, arc, steel—processing (1)	m	x	-	-	-
Metal working factory—construction (1)	p	x	-	-	-
Hot rolling, steel—processing (1)	kg	x	-	-	-
Building machine (1, 2 and 3)	p	x	x	x	x
Drainage pipes (1)	kg	x	x	x	x
Asphaltic pavement (1 and 2)	t	x	x	x	x
Painting (1 and 2)	m^2	x	x	x	x
Crane truck (1 and 3)	h	x	-	x	x
Transport (30 km) (1, 2 and 3)	t·km	x	x	x	x
Diesel (1, 2 and 3)	kg	x	x	x	x
Lubricating oil (1, 2 and 3)	kg	x	x	x	x
Electricity, medium voltage (1, 2 and 3)	kWh	x	x	x	x
Industrial machine, heavy, unspecified (1, 2 and 3)	kg	x	x	x	x
Tap water (1 and 2)	kg	x	x	x	x
Inspection (2)	h	x	x	x	x
Pavement demolition (2 and 3)	t	x	x	x	x
Asphalt pavement renovation (2)	t	x	x	x	x
Demolition Building machine (2 and 3)	t	x	x	x	x
Hydraulic digger (1, 2 and 3)	p	x	x	x	x
Waste reinforced concrete (3)	t	x	x	x	x
Treatment of waste asphalt—sanitary landfill (3)	t	x	x	x	x
Treatment of waste reinforcement steel—sorting plant (3)	t	x	-	-	-

Some scenarios were considered to support the decision-making process for the best performing bridge in terms of sustainability. Considering the scenarios, the environmental and economic scores were 50/50, 60/40, 40/60/70/30, 30/70 and 55/45, respectively.

Using life cycle cost analysis (LCCA), the cost of each bridge was calculated, generating a cost per m^2 of bridge. In the life cycle assessment (LCA), after modeling the ReCiPe method, indices per m^2 of the bridges in the categories of ecosystem damage, human health and resource depletion were obtained.

To calculate the economic and environmental performance of bridge superstructures, it was necessary to normalize the values of each parameter in order to compare values with the same unit. For that, the average of the values obtained was used as the conventional practice of cost and environmental indicators. The values of each bridge were divided by the mean, thus obtaining the indices for each m^2 of bridge. In determining the final score, scenarios were considered with the weights shown in Figure 3.

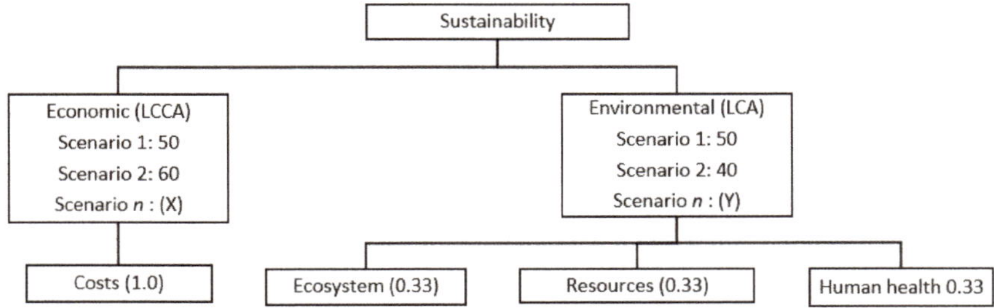

Figure 3. Criteria weights for the assessment of corporate sustainability.

The lowest economic coefficients (costs) and environmental damage (ecosystems, human health and resource depletion) represent the lowest impact in each category and thus represent the best performance of each bridge from the sustainability point of view.

3. Results and Discussion

Following the evaluation of the bridge inventories, environmental reports were generated to compare the performance between the analyzed models. In this section, comparisons of the 405 bridges analyzed for each of the three categories of damage at the endpoint are presented.

Figure 4 presents the results of the analyzed ecosystem damage category for all bridges in the study.

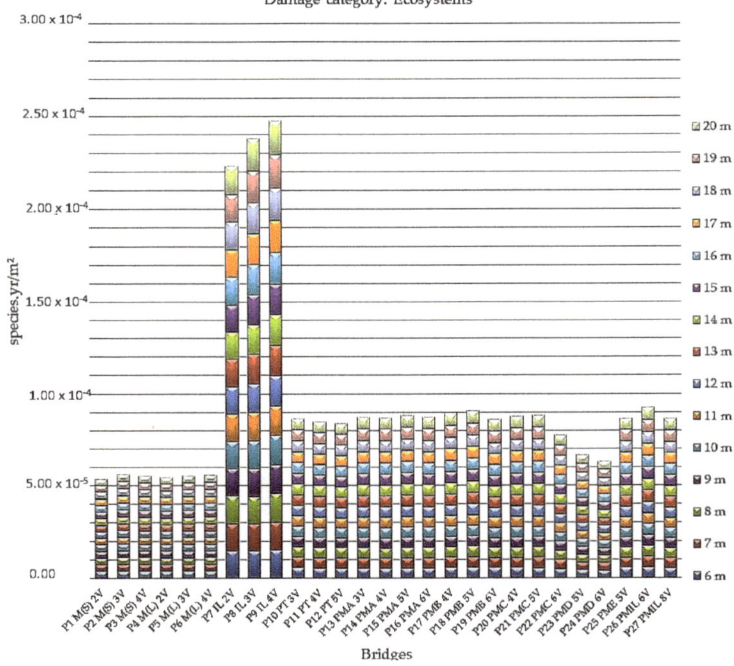

Figure 4. Indices of ecosystem damage in species per year per meter squared for all bridges analyzed.

The indices highlighted in this comparative assessment of damage to ecosystems confirm that the bridges cast in situ have the greatest impact on the loss of species over a given area, for a certain time, expressed in species per year per meter squared (species.yr/m^2). Notably, the bridges in reinforced concrete cast in situ of the P9 IL 4V model, which present the worst results in terms of performance in the ecosystems category, range from 1.49×10^{-5} species.yr/m^2 (6 m) to 1.79×10^{-5} species.yr/m^2 (20 m).

The analysis also indicates the best performance for steel/concrete composite bridges with two P1 M(S) 2V welded steel beams, with indexes ranging from 3.52×10^{-6} species.yr/m^2 (19 m and 20 m) to 3.67×10^{-6} species.yr/m^2 (17 m).

Figure 5 shows the results of the impacts on the resources category for all bridges in the study.

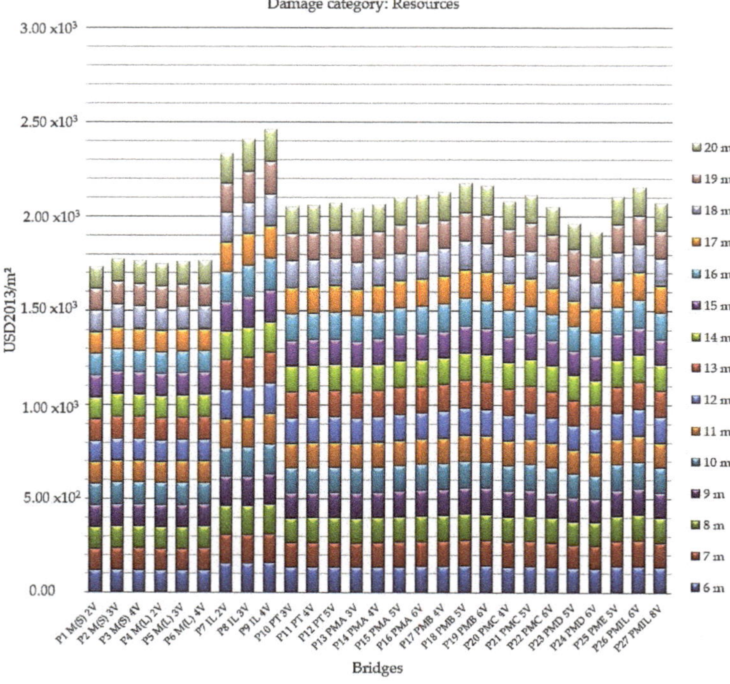

Figure 5. Resources impact indices in US dollars (in 2013) per meter squared for all bridges analyzed.

Considering the indices highlighted in this comparative assessment, the analysis of the graph indicates how the bridges that most impact the excess cost of future resources production over an infinite period, at US dollars (in 2013) per meter squared (USD 2013/m^2), are those cast in situ, emphasizing the reinforced concrete bridges cast in situ of the P9 IL 4V model. The worst results were in the category of impact son resources, with a variation from 1.55×10^2 USD 2013/m^2 (6 m) to 1.73×10^2 USD2013/m^2 (19 m and 20 m).

The bridges with better performance were highlighted as steel/concrete composite bridges with two welded steel beams, model P1 M(S) 2V. The index for these bridges varied from 1.15×10^2 USD2013/m^2 (9 m, 10 m, 11 m, 12 m, 13 m, 14 m, 15 m, 16 m, 18 m, 19 m, and 20 m) to 1.17×10^2 USD2013/m^2 (6 m and 17 m).

Figure 6 presents the results of the human health impact category, analyzed for all bridges in the study.

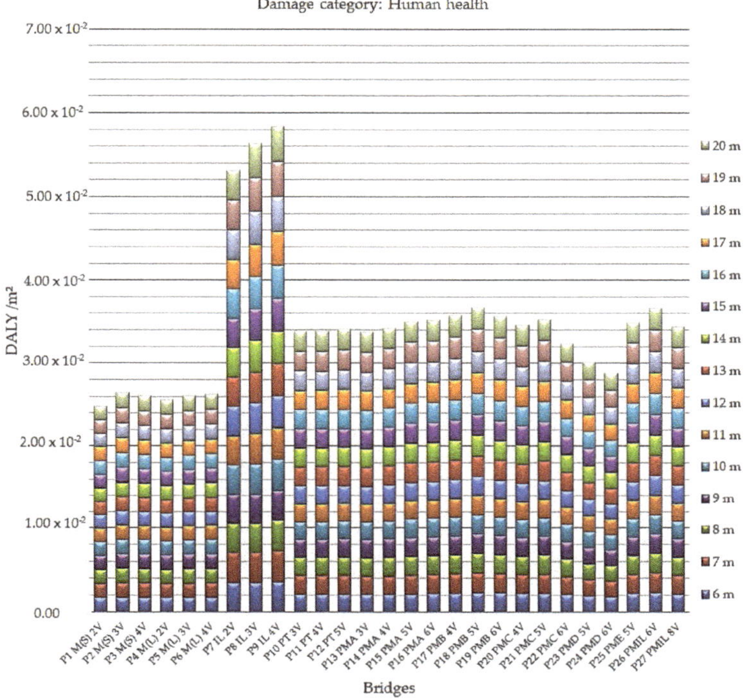

Figure 6. Human health impact indices in Disability Adjusted Life Years DALY per meter squared for all bridges analyzed.

The graph presents the indices in the comparative assessment. They confirm that the bridges that have the greatest impact on the number of years of life lost and the number of years lived with disability—expressed in DALY per meter squared (DALY/m²)—are the bridges cast in situ. Especially, the reinforced concrete bridges cast in situ of the P9 IL 4V model have the worst performance-related results in the category of human health impact, ranging from 3.56×10^{-3} DALY/m² (6 m) to 4.19×10^{-3} DALY/m² (20 m).

The bridges with the best performance indicated in the study are the steel/concrete composite bridges with two welded steel beams model P1 M(S) 2V, whose indexes varied from 1.63×10^{-3} DALY/m² (18 m, 19 m, and 20 m) to 1.73×10^{-3} DALY/m² (17 m).

After assessing the damage to ecosystems, human health and resources, we found that the concrete bridges cast "in situ" showed worse environmental performance. This occurs due to high cement consumption, because cement production—especially clinker—is the main factor among all analyzed factors that contribute to the cause of this damage. Composite bridges perform better, mainly due to low cement consumption. Another factor that contributes to this improvement is the type of process that the industry currently uses in steel production (the production of steel comprises about 28% of scrap steel).

The comparative analysis was performed with the application of other methods for the characterization of environmental impacts to validate the results obtained in the ReCiPe methodology. The data obtained with the application of the two methods that are also used in the characterization of environmental impacts are presented: the European methods IMPACT 2002 + V2.14 and ILCD 2011 Midpoint + V1.10, and the American method BEES + V4.07 USA.

Figure 7 presents the final single score results for the IMPACT 2002 method.

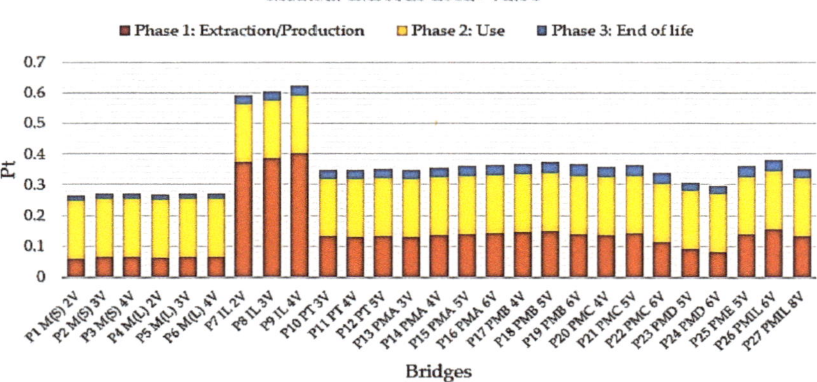

Figure 7. Single scored impact indices for the 2002 IMPACT method for the life cycle of bridges with a 10 m span.

Assessing the results generated by the IMPACT 2002 + 12.14 method, the lowest single score indicator for the entire life cycle of the bridge superstructure was shown to be the steel/concrete composite bridge model P1 M(S) 2V with two steel beams. The results presented by this method indicate the same bridge modeled in the ReCiPe methodology causes less environmental impact.

Figure 8 presents the final single score results for the European ILCD 2011 Midpoint + V1.10 method.

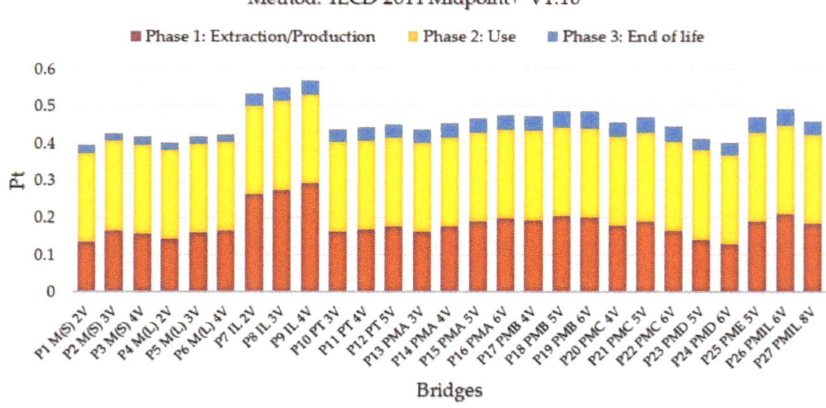

Figure 8. Single score impact indices for the International Reference Life Cycle Data System - ILCD 2011 method for the life cycle of bridges with a 10 m span.

Analyzing the data presented in the graph generated by the ILCD 2011 + V1.10 method, the two-beam steel/concrete composite bridge model P1 M(S) 2V has the lowest single score indicator. The results displayed by this method indicate the same P1 M(S) bridge modeled in the ReCiPe methodology causes the least environmental impact.

Figure 9 presents the final single score results for the BEES + V4.07 USA method.

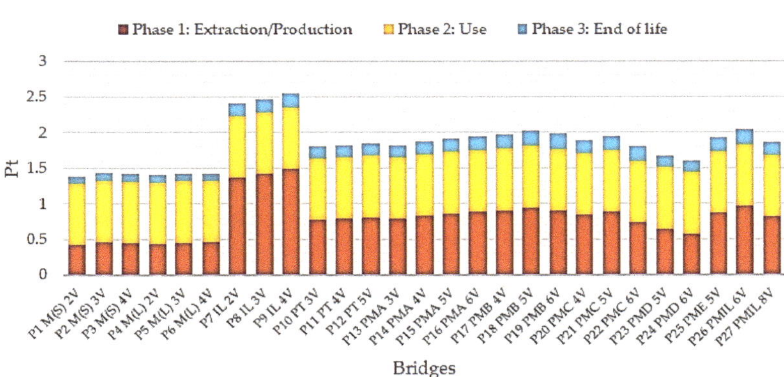

Figure 9. Single scored impact indices for the Building for Environmental and Economic Sustainability - BEES + V4.07 USA method for the life cycle of bridges with a 10 m span.

Interpreting the graph data generated by the BEES + V4.07 USA method, the steel/concrete composite bridge model with two P1 M(S) 2V welded steel beams exhibits the lowest single score indicator. The results shown by this method indicate that the P1 M(S) 2V bridge modeled by the ReCiPe methodology also causes less environmental impact.

After the comparison of the results obtained by the methods is applied to the life cycle inventory of the bridge superstructures, it is possible to confirm the results indicating that the steel/concrete composite bridge with two P1 M(S) 2V steel beams causes less impact on the environment. In Figure 10, the cost indices of each bridge model per m^2 in the entire life cycle are presented; i.e., the sum of the costs of Phases 1, 2, and 3.

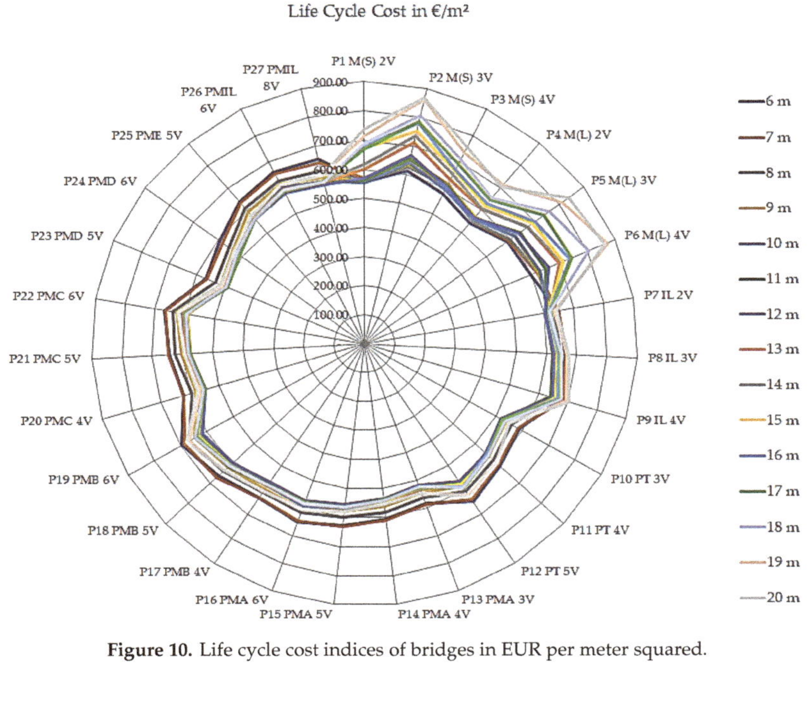

Figure 10. Life cycle cost indices of bridges in EUR per meter squared.

According to the indices presented in Figure 10 regarding the total cost, the bridge of the steel/concrete composite model with P4 M(L) 2V rolled beams is apparently the cheapest for the 6 m (€ 540.63/m^2) and 7 m spans (€ 552.87/m^2); however, the precast P23 PMD 5V model becomes the most accessible from the 8 m to 20 m span, with a variation in cost from € 534.49/m^2 (8 m) to € 490.60/m^2 (17 m). Regarding the highest cost, the P19 PMB 6V precast model is superior for the spans ranging from 6 m to 8 m, with its value varying from € 695.79/m^2 (6 m) to € 666.61/m^2 (8 m). For the 9 m (€ 651.76/m^2) and 10 m (€ 642.08/m^2) spans, the concrete bridge cast in situ P9 IL 4V model is the most expensive. For the other spans ranging from 11 m to 20 m, the P2 M(S) 3V steel/concrete composite bridge model with welded steel beams becomes the most expensive, varying from € 653.49/m^2 (11 m) to € 866.50/m^2 (20 m).

The comparisons are presented with the objective of subsidizing the choice of the most adequate typology according to the evaluated criteria, with a recommendation for sustainability indicators in the construction of bridges so that these models can meet the needs of small-span bridges related to rural and neighboring roads.

As shown, the cost of the bridges is compared to the damage categories (endpoint) in the 1 m^2 functional unit, allowing the visualization of the economic (LCCA) and environmental aspects (LCA). The span analyzed is 10 m, as this is the most used distance for small-span bridges. The comparison of the models based on cost parameters with the category of damage to ecosystems for the 10 m span led to the results presented in Figure 11.

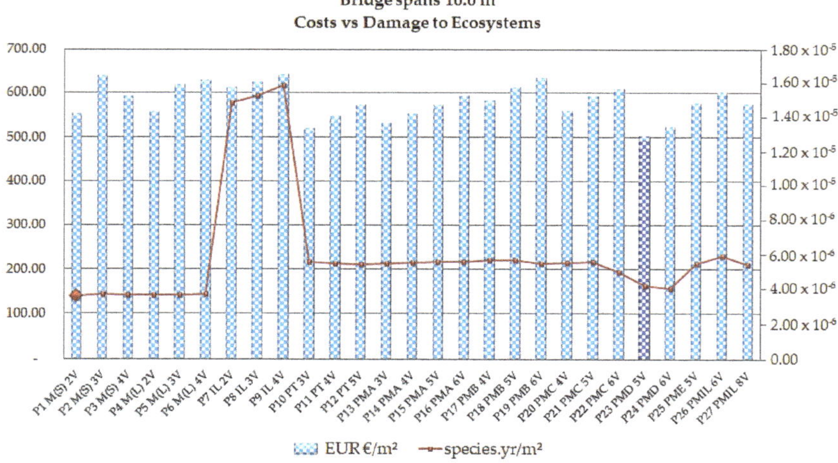

Figure 11. Cost x ecosystem damage indices in EUR/m^2 and species.yr/m^2 for the 10 m span.

As shown in Figure 11, the model with the lowest bridge cost for the 10 m span is the P23 PMD 5V five-beam model, with values in the order of € 502.12/m^2; its ecosystem damage indicator is 4.20×10^{-6} species.yr/m^2. Assessing the lowest indicator of ecosystem damage, the P1 M(S) 2V steel/concrete composite bridge model with two welded steel beams presents damage results in the order of 3.57×10^{-6} species.yr/m^2 with a cost of € 554.96/m^2.

The cost and environmental parameter differences between the two models are 9.52% and 15.07%, respectively. The P9 IL 4V cast in situ bridge model presents the worst scenario, due to both the higher cost and the fact that it presents the highest indicator of damage to ecosystems in the order of € 642.08/m^2 and 1.58×10^{-5} species.yr/m^2 for cost and damage, respectively.

Comparing the models in the cost parameters by the resource depletion category for the 10 m range, the results are presented in Figure 12.

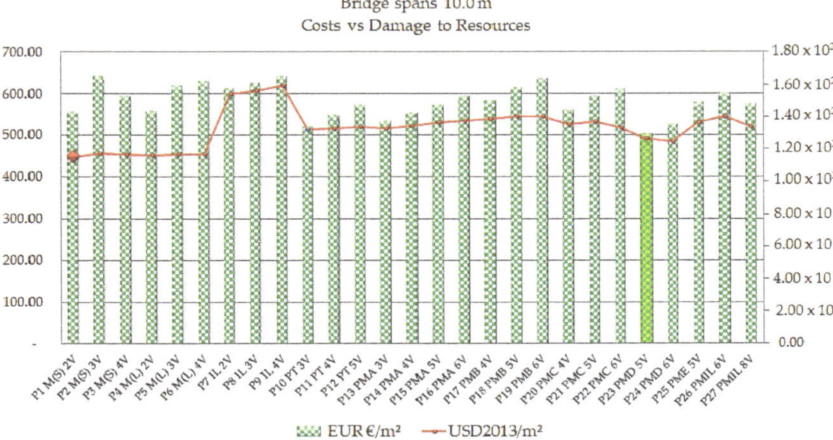

Figure 12. Cost times resources depletion ratios in EUR/m^2 and USD2013/m^2 for the 10 m span.

As shown in Figure 12, the model that presents the lowest cost for the bridges with a 10 m span is apparently the precast bridge with five beams, P23 PMD 5V, in the order of € 502.12/m^2, and its resource depletion indicator is 1.26×10^2 USD2013/m^2. Regarding the lowest resource depletion indicator, the steel/concrete composite bridge model with two P1 M(S) 2V welded steel beams presents depletion results in the order of 1.15×10^2 USD2013/m^2, at a cost of € 554.96/m^2. Comparing the two models, the cost difference is 9.52% and the difference in the environmental parameter is 8.92%. The P9 IL 4V cast in situ bridge model presents the worst situation, with the highest cost and resource depletion indicator, in the order of € 642.08/m^2 for cost and 1.60×10^2 USD2013/m^2 for depletion.

The results of the comparison of the models regarding the cost parameters combined with the category of damage to human health for the range of 10 m are presented in Figure 13.

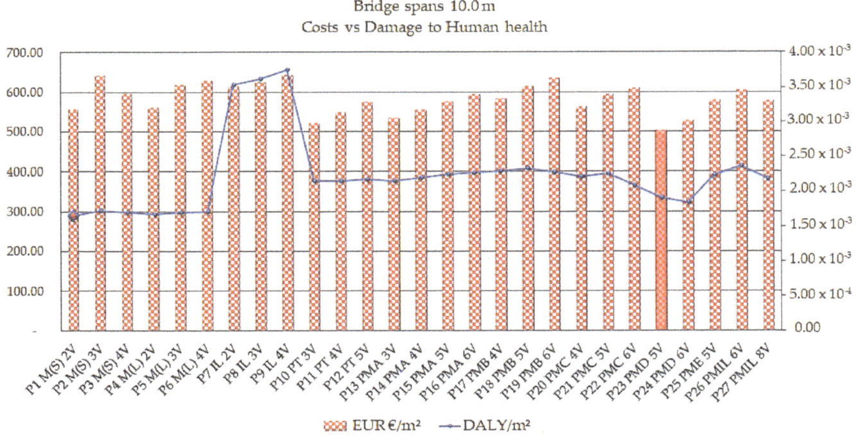

Figure 13. Cost times human health damage indices in EUR/m^2 and DALY/m^2 for the 10 m span.

The results shown in Figure 13 indicate the P23 PMD 5V precast concrete bridge as the model with the lowest cost between 10 m spans, in the order of € 502.12/m^2. In addition, the model has as an indicator of human health damage of 1.91×10^{-3} DALY/m^2. Assessing the lowest indicator of human health damage, the steel/concrete composite bridge model with 2 P1 M(S) 2V welded beams gives results in the order of 1.65×10^{-3} DALY/m^2 and a cost of € 554.96/m^2. Comparing the two models, the cost difference is 9.52% and the difference in the environmental parameter is 13.55%. The P9 IL 4V cast in situ concrete bridge model presents the worst situation, meaning that the cost is higher, presenting a cost of € 642.08/m^2 and a value of 3.74×10^{-3} DALY/m^2 for depletion.

The results show that the bridge with a span of 10 m that presented the lowest cost was the precast concrete bridge with five beams: smodel P23 PMD 5V. For the same span, the model that presented the best environmental performance in all categories of damage was the steel/concrete composite bridge with two beams: the type "I" welded model P1 M(S) 2V.

The influence of transport distances from the market (city) to work (bridge) was also verified with sensitivity analysis at different distances, calculated for 10 km, 30 km (standard adopted at work), 50 km, 80 km and 100 km. For the LCA, the results indicated an increase of 0.2% per km in terms of the damage to human health, ecosystems and resources; in relation to the LCCA, the costs increased by 0.3% per kilometer.

In the present study, several scenarios of downsizing were analyzed, with economic/environmental scores including 50/50, 60/40, 40/60, 70/30, 30/70 and 55/45.

The results indicate that the scenario presenting a score higher than 50% regarding the economic parameter—the precast P23 PMD 5V concrete bridge with five beams—exhibited the best performance. The scenario with a score higher than 50% regarding the environmental parameter—i.e., the composite bridge P1 M(S) 2V with two electro welded steel beams—showed the best environmental performance.

In this study, two decision-making scenarios were graphically presented; Scenario 1 was applied as 50% for each of the economic and environmental categories, and Scenario 2 was applied with a weight of 60% for the economic and 40% for the environmental categories.

The justification for presenting the 50/50 (sustainable) scenario was motivated by the theory of sustainability, which is evidenced when the damage indices show equal values for the environmental and cost issues. The 60/40 scenario (realistic) was presented taking into consideration the opinion of decision makers, who give more importance to the economic (costs) than environmental issues. This information was gathered from a questionnaire submitted to decision makers.

Figure 14 presents the classification with the best performance considering sustainability, with weights of 50/50 for bridges of 10 m in length.

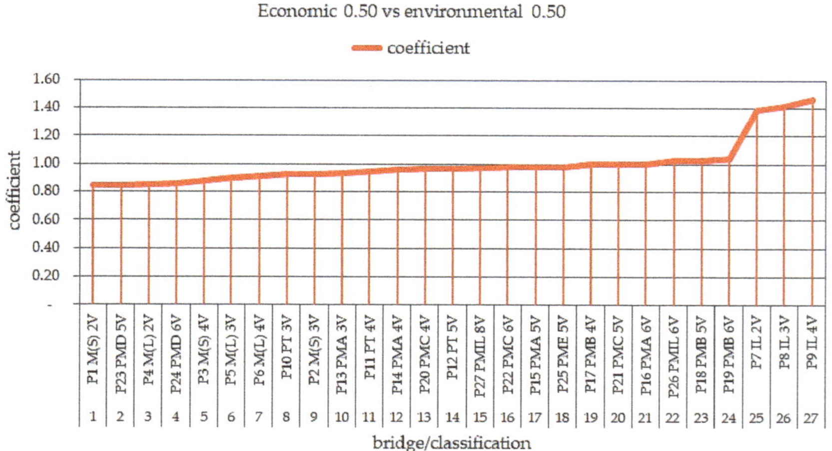

Figure 14. Classification of environmental performance with weights of 0.5 economic and 0.5 environmental per m^2 for bridges with lengths of 10 m.

Analyzing Figure 14, the model that presents the best sustainability performance according to the 50/50 scenario is apparently the steel/concrete composite bridge model with two P1 M(S) 2V welded steel beams, followed by the precast concrete bridge with five P23 PMD 5V type beams, which can be observed to be the two most efficient models.

Note that the precast concrete model with five beams, P23 PMD 5V, was derived from the civil defense kit program of the government of Santa Catarina, where the elements and dimensions were standardized to meet the purpose of the study. The P9 IL 4V in situ cast bridge model shows the worst situation, presenting the worst performance in terms of sustainability assessment in the life cycle of small-span bridge superstructures.

Figure 15 presents the classification with the best performance considering sustainability, with weights of 60/40 for bridges with a 10 m length.

Figure 15. Classification of environmental performance with weights of 0.6 for economic and 0.4 for environmental factors per m^2 for bridges with lengths of 10 m.

As can be seen Figure 15, the model with the best sustainability performance according to the 60/40 scenario is the precast concrete bridge, P23 PMD 5V; the second-best performing bridge is the steel/concrete composite bridge with two P1 M(S) 2V welded steel beams. Therefore, according to this methodology, these two models can be suggested to be the most efficient considering the sustainability assessment in the life cycle of small-span bridge superstructures.

After applying the proposed scenarios, one can suggest the steel/concrete composite bridge model with two P1 M(S) 2V beams, as shown in Figure 16, and the P23 PMD 5V type precast concrete bridge model, as shown in Figures 17 and 18.

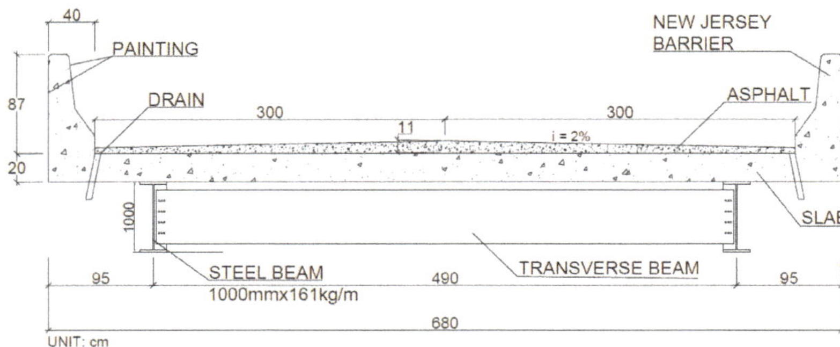

Figure 16. Cross-section of steel/concrete composite bridge P1 M(S) 2V, length 10 m.

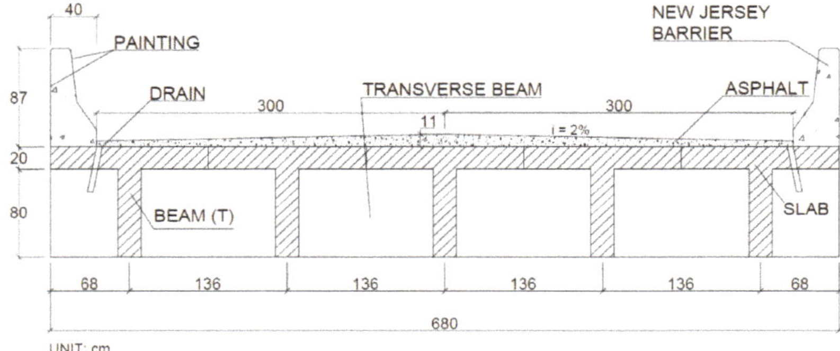

Figure 17. Cross-section of precast concrete bridge, P23 PMD 5V, length 10 m.

Figure 18. Obstacle transposition kit over Motucas River, at Estrada dos Portugueses, Vila Nova neighborhood, Joinville-SC - Type P23 PMD 5V. Source: Joinville City Hall [75].

4. Conclusions

The study's main objective was to apply sustainability assessment techniques to the life cycle of small-span bridge superstructures, followed by the proposal of environmental and economic indicators to be integrated in the decision-making process. To achieve this objective, 405 bridges were analyzed according to different categories of damage and sustainability scenarios.

The analysis conducted configured the object of the study with the purpose of finding answers that would make it possible to contribute to the mitigation of the environmental impact, the reduction of resource consumption and the minimization of costs in the life cycle of bridges. The obtained results indicate that, in terms of scenarios presenting a score higher than 50% regarding the economic parameter, the precast concrete bridge with five beams exhibited the best performance. On the other hand, scenarios with a score higher than 50% regarding the environmental parameter pointed to the composite bridge with two welded steel beams as the bridge with the best global performance.

Using the methodological structure of the LCA, the results of the study can contribute to the development of indicators to be applied in small-span bridge designs, assuming that considering the sustainability of bridge designs could ameliorate impacts on the environment. These impacts are currently seen as the response to damage verified in the ecological, economic and social spheres; the aim of sustainable design is a construction method that analyzes the adoption of existing protocols and the possibilities of a design based on the concept of sustainability.

It can be concluded that the evaluation of the sustainability of structures depends on many variables that can influence decision-making, such as the criteria considered, the weights and the decision-making methods with the various attributes used.

The results indicate that the environmental performance of bridges is highly correlated with the type of bridge material, taking into account the fact that different types of materials or bridges exhibit different levels of environmental performance, indicating the importance of authorities reformulating intervention policies in the decision-making process.

The study of alternatives that can be applied in bridge designs is an important contribution to the civil construction sector and thus also to the satisfaction, by the State, of the requirements presented by the population for the use of small-span bridges; thus, the results found in the research confirm that the use of sustainability indicators in the methodological basis of LCA in bridge design can have a positive impact on sustainable development.

In the authors' opinion, although a very important step towards the consideration of sustainability in bridge design can be made with the adoption of the indicators presented in this study, additional considerations should be included in future investigations, including, the eventual influence of social impacts on the obtained results according to different scenarios.

Author Contributions: This paper represents a result of teamwork. C.J.M., V.Y. and M.K. jointly designed the research. C.J.M. developed the methodology, carried out the investigation and drafted the original manuscript, and M.K. and V.Y. revised and improved the manuscript. All authors have read and agreed to the published version of the manuscript.

Funding: This research was funded by the Brazilian government in the form of CAPES and CNPq grants, as well as the Spanish Ministry of Economy and Competitiveness, along with FEDER funding (Project: BIA2017-85098-R).

Conflicts of Interest: The authors declare no conflict of interest.

References

1. Gonzalez, A.; Schorr, M.; Valdez, B.; Mungaray, A. Bridges: Structures and Materials, Ancient and Modern. *Intechopen.* 2020. Available online: https://www.intechopen.com/online-first/bridges-structures-and-materials-ancient-and-modern (accessed on 8 April 2020).
2. Almeida, J.C.C. Estudo do Ciclo de Vida de Pontes Rodoviárias Sistema de Apoio a Análise Comparativa dos Custos ao Longo do Ciclo de Vida de Diferentes Soluções para Pontes Rodoviárias de Betão Armado. Ph.D. Thesis, Universidade do Minho, Braga, Portugal, 2013.

3. Yadollahia, M.; Ansari, R.; Majid, M.A.A.; Yih, C.H. A multi-criteria analysis for bridge sustainability assessment: A case study of Penang Second Bridge, Malaysia. *Life-Cycle Des. Perform.* **2014**, *11*, 638–654.

4. Du, G.; Safi, M.; Pettersson, L.; Karoumi, R. Life cycle assessment as a decision support tool for bridge procurement: Environmental impact comparison among five bridge designs. *Int. J. Life Cycle Assess.* **2014**, *19*, 1948–1964. [CrossRef]

5. Ademovic, N. Life-cycle assessment of a masonry arch bridge. In Proceedings of the Quality Specifications For. Roadway Bridges, Standardization At. A European Level, Barcelona, Spain, 28 September 2018; pp. 1–16.

6. Long, A.E.; Basheer, P.A.M.; Taylor, S.E.; Rankin, B.G.I.; Kirkpatrick, J. Sustainable bridge construction through innovative advances. *Br. Eng.* **2008**, *161*, 183–188. [CrossRef]

7. Přikryl, R.; Török, A.; Theodoridou, M.; Gomez-Heras, M.; Miskovsky, K. Geomaterials in construction and their sustainability: Understanding their role in modern society. Sustainable Use of Traditional Geomaterials in Construction Practice. *Geol. Soc. Lond. Spec. Publ.* **2016**, *416*, 1–22. [CrossRef]

8. Du, G. Life Cycle Assessment of Bridges, Model Development and Case Studies. Ph.D. Thesis, Royal Institute of Technology, Estocolmo, Sweden, 2015.

9. Zhang, Y.-R.; Wu, W.-J.; Wang, Y.-F. Bridge life cycle assessment with data uncertainty. *Int. J. Life Cycle Assess.* **2016**, *21*, 569–576. [CrossRef]

10. Itoh, Y.; Hirano, T.; Nagata, H.; Hammad, A.; Nishido, T.; Kashima, A. Study on bridge type selection system considering environmental impact. *Doboku Gakkai Ronbunshu* **1996**, *553*, 187–199. [CrossRef]

11. Penadés-Plà, V.; García-Segura, T.; Martí, J.V.; Yepes, V. A review of multi-criteria decision-making methods applied to the sustainable bridge design. *Sustainability* **2016**, *8*, 1295. [CrossRef]

12. Milani, C.J.; Kripka, M. Evaluation of short span bridge projects with a focus on sustainability. *Struct. Infrastruct. Eng.* **2020**, *16*, 367–380. [CrossRef]

13. Santoro, J.F.; Kripka, M. Minimizing environmental impact from optimized sizing of reinforced concrete elements. *Comput. Concr.* **2020**, *25*, 111–118.

14. Tormen, A.F.; Pravia, Z.M.C.; Ramires, F.B.; Kripka, M. Optimization of steel-concrete composite beams considering cost and environmental impact. *Steel Compos. Struct.* **2020**, *34*, 409–421.

15. Penadés-Plà, V.; Yepes, V.; García-Segura, T. Robust decision-making design for sustainable pedestrian concrete bridges. *Eng. Struct.* **2020**, *209*, 109968. [CrossRef]

16. Zastrow, P.; Molina-Moreno, F.; García-Segura, T.; Martí, J.V.; Yepes, V. Life cycle assessment of cost-optimized buttress earth-retaining walls: A parametric study. *J. Clean Prod.* **2017**, *140*, 1037–1048. [CrossRef]

17. Goh, K.C.; Yang, J. Developing a life-cycle costing analysis model for sustainability enhancement in road infrastructure design. In Proceedings of the Second Infrastructure Theme Postgraduate Conference: Rethinking Sustainable Development—Planning, Infrastructure Engineering, Design and Managing Urban Infrastructure, Brisbane, Queensland, 26 March 2009.

18. Fauzi, R.T.; Lavoei, P.; Sorelli, L.; Heidari, M.D.; Amor, B. Exploring the current challenges and opportunities of life cycle sustainability assessment. *Sustainability* **2019**, *11*, 636. [CrossRef]

19. Swarr, T.E.; Hunkeler, D.; Klöpffer, W.; Pesonen, H.-L.; Ciroth, A.; Brent, A.C.; Pagan, R. Environmental life-cycle costing: A code of practice. *Int. J. Life Cycle Assess.* **2011**, *16*, 389–391. [CrossRef]

20. Janjua, S.Y.; Sarker, P.K.; Biswas, W.K. A review of residential buildings' sustainability performance using a life cycle assessment approach. *J. Sustain. Res.* **2019**, *1*, 1–29.

21. Almeida, J.M.M.R.M.O. Sistema de Gestão de Pontes com Base em Custos de Ciclo de Vida. Ph.D. Thesis, Universidade do Porto, Porto, Portugal, 2013.

22. Veganzones Muñoz, J.J.; Pettersson, L.; Sundquist, H.; Karoumi, R. Life-cycle cost analysis as a tool in the developing process for new bridge edge beam solutions. *Struct. Infrastruct. Eng.* **2016**, *49*, 1737–1746. [CrossRef]

23. Milani, C.J.; Kripka, M. Diagnosis of pathologies in bridges of the road system in Brazil. *Constructii* **2012**, *13*, 26–34.

24. Carbonell, A.; González-Vidosa, F.; Yepes, V. Design of reinforced concrete road vaults by heuristic optimization. *Adv. Eng. Softw.* **2011**, *42*, 151–159. [CrossRef]

25. García-Segura, T.; Yepes, V.; Frangopol, D.M. Multi-Objective Design of Post-Tensioned Concrete Road Bridges Using Artificial Neural Networks. *Struct. Multidiscipl. Optim.* **2017**, *56*, 139–150. [CrossRef]

26. Penadés-Plà, V.; García-Segura, T.; Yepes, V. Robust Design Optimization for Low-Cost Concrete Box-Girder Bridge. *Mathematics* **2020**, *8*, 398. [CrossRef]

27. Sierra, L.A.; Yepes, V.; García-Segura, T.; Pellicer, E. Bayesian network method for decision-making about the social sustainability of infrastructure projects. *J. Clean. Prod.* **2018**, *176*, 521–534. [CrossRef]

28. Navarro, I.J.; Yepes, V.; Martí, J.V. Social life cycle assessment of concrete bridge decks exposed to aggressive environments. *Environ. Impact Assess. Rev.* **2018**, *72*, 50–63. [CrossRef]

29. Molina-Moreno, F.; Martí, J.V.; Yepes, V. Carbon embodied optimization for buttressed earth-retaining walls: Implications for low-carbon conceptual designs. *J. Clean. Prod.* **2017**, *164*, 872–884. [CrossRef]

30. Yepes, V.; Martí, J.V.; García, J. Black Hole Algorithm for Sustainable Design of Counterfort Retaining Walls. *Sustainability* **2020**, *12*, 2767. [CrossRef]

31. Bansal, S.; Singh, A.; Singh, S.K. Sustainability evaluation of two iconic bridge corridors under construction using Fuzzy Vikor technique: A case study. *Alconpat* **2017**, *7*, 1–14. [CrossRef]

32. Steele, K.; Cole, G.; Parke, G.; Clarke, B.; Hardin, J. The application of life cycle assessment technique in the investigation of brick arch highway bridges. In Proceedings of the Conference for the Engineering Doctorate in Environmental Technology; 2002. Available online: https://www.mbhplc.co.uk/bda/Sustain-Life-Cycle.pdf (accessed on 8 April 2020).

33. Li, X.; Guo, L. Study on civil engineering sustainable development strategy. In Proceedings of the 3rd International Conference on Management, Education, Information and Control, Shenyang, China, June 2015; pp. 405–412.

34. Internaltional Organization for Standardization. *Environmental Management—Life Cycle Assessment-Principles and Framework*; ISO 14040:2006; ISO: Geneva, Switzerland, 2006.

35. Internaltional Organization for Standardization. *Environmental Management—Life Cycle Assessment-Requirements and Guidelines*; ISO 14040:2006; ISO: Geneva, Switzerland, 2006.

36. European Commission-Joint Research Centre-Institute for Environment and Sustainability. *International Reference Life Cycle Data System (ILCD) Handbook: Review Schemes for Life Cycle Assessment*, 1st ed.; Publications Office of the European Union: Luxembourg, 2010.

37. Frischknecht, R.; Jungbluth, N.; Althaus, H.-J.; Doka, G.; Dones, R.; Hirschier, R.; Hellweg, S.; Humbert, S.; Margni, M.; Nemecek, T.; et al. *Implementation of Life Cycle Impact Assessment Methods*; Ecoinvent Report No. 3; Swiss Centre for Life Cycle Inventories: Dübendorf, Switzerland, 2004.

38. Recipe. ReCiPe is the Most Recent and Harmonized Indicator Approach Available in Life Cycle Impact Assessment. 2017. Available online: https://www.pre-sustainability.com/recipe (accessed on 12 December 2019).

39. SIMAPRO. SimaPro Database Manual Methods Library. SIMAPRO: San Francisco, CA, USA, April 2016. Available online: https://www.pre-sustainability.com/download/DatabaseManualMethods.pdf (accessed on 28 June 2017).

40. Goedkoop, M.; Oele, M.; Leijting, J.; Ponsioen, T.; Meijer, E. *Introduction to LCA with SimaPro*; PRé Consultants: Amersfoort, The Netherlands, 2016.

41. Jolliet, O.; Margni, M.; Charles, R.; Humbert, S.; Payet, J.; Rebitzer, G.; Rosenbaum, R. IMPACT2002+: A New Life Cycle Impact Assessment Methodology. *Int. J. Life Cycle Assess.* **2003**, *8*, 324–330. [CrossRef]

42. European Commission—Joint Research Centre. *International Reference Life Cycle Data System (ILCD)Handbook Recommendations for Life Cycle Impact Assessment in the European Context*, 1st ed.; EUR 24571 EN; Publications Office of the European Union: Luxemburg, 2011.

43. Lippiatt, B.C. *BEES 4.0: Building for Environmental and Economic Sustainability: Technical Manual and User Guide*; NISTIR 7423; National Institute of Standards and Technology: Gaithersburg, MD, USA, 2007.

44. Brundtland, R. *Nosso Futuro Comum*; FGV: Rio de Janeiro, RJ, Brasil, 1991.

45. Siche, J.R.; Agostinho, F.; Ortega, E.; Romeiro, A. Sustainability of nations by indices: Comparative study between environmental sustainability index, ecological footprint and the emergy performance índices. *Ecol. Econ.* **2008**, *66*, 628–6378. [CrossRef]

46. Waas, T.; Hugé, J.; Verbruggen, A.; Wright, T. Sustainable development: A bird's eye view. *Sustainability* **2011**, *3*, 1637–1661. [CrossRef]

47. Boulenger, M. Life Cycle Assessment of Concrete Structures Using Public Databases: Comparison of a Fictitious Bridge and Tunnel. Master's Thesis, Royal Institute of Technology (KTH), Stockholm, Sweden, 2011.

48. Kuhlmann, U.; Beck, T.; Fischer, M.; Friedrich, H.; Kaschner, R.; Maier, P.; Mensinger, M.; Pfaffinger, M.; Sedlbauer, K.; Ummenhofer, T. Ganzheitliche bewertung von stahl-und verbundbrücken nach kriterien der nachhaltigkeit. *Stahlbau* **2011**, *80*, 703–710. [CrossRef]

49. Ahlborn, T. Sustainability: For the concrete bridge engineering community. *Perspective* **2008**, *2*, 16–18.

50. Bader, P. Sustentabilidade: Do modelo à implementação. *Revista Ecol.* **2012**, *21*, 184.

51. Furuta, H.; Nakatsu, K.; Ishibashi, K.; Miyoshi, N. Optimal bridge maintenance of large number of bridges using robust genetic algorithm. *Struct. Congr.* **2014**, *v.1*, 2282–2291.

52. Life Cycle Initiative. What is Life Cycle Thinking? June 2017. Available online: http://www.lifecycleinitiative.org/starting-life-cycle-thinking/what-is-life-cycle-thinking/ (accessed on 28 June 2017).

53. Penadés-Plà, V.; Martínez-Muñoz, D.; García-Segura, T.; Navarro, I.J.; Yepes, V. Environmental and social impact assessment of optimized post-tensioned concrete road bridges. *Sustainability* **2020**, *12*, 4265. [CrossRef]

54. Sagemo, A.; Storck, L. Comparative Study of Bridge Concepts Based on Life-Cycle Cost Analysis and Lifecycle Assessment. Master's Thesis, Chalmers University of Technology, Gothenburg, Sweden, 2013.

55. Maier, P.; Kuhlmann, U.; Gervásio, H.; Perdigão, V.; Martins, N.; Barros, P. Obras de Arte Mistas—AnáLise Holística Aplicada a Casos Europeus. 7° Congresso Rodoviário Português-Novos Desafios Para a Atividade Rodoviária. 2013. Available online: http://www.crp.pt/docs/A45S117-7CRP_prog_net.pdf (accessed on 8 April 2020).

56. Hammervold, J.; Reenaas, M.; Brattebo, H. Environmental life cycle assessment of bridges. *J. Bridge. Eng.* **2013**, *8*, 153–161. [CrossRef]

57. Nielsen, D.; Raman, D.; Chattopadhyay, G. Life cycle management for railway bridge assets. *Proc. Inst. Mech. Eng. Part F J. Rail Rapid Transit* **2013**, *227*, 570–581. [CrossRef]

58. García-Segura, T.; Yepes, V.; Frangopol, D.M.; Yang, D.Y. Lifetime reliability-based optimization of post-tensioned box-girder bridges. *Eng. Struct.* **2017**, *145*, 381–391. [CrossRef]

59. Navarro, I.J.; Martí, J.V.; Yepes, V. Reliability-based maintenance optimization of corrosion preventive designs under a life cycle perspective. *Environ. Impact Assess. Rev.* **2019**, *74*, 23–34. [CrossRef]

60. Mara, V.; Haghani, R.; Sagemo, A.; Storck, L.; Nilsson, D. Comparative study of different bridge concepts based on life-cycle cost analyses and life-cycle assessment. In Proceedings of the Asia-Pacific Conference on FRP in Structures—APFIS, Melbourne, Australia, 11–13 December 2013.

61. Kuhlmann, U.; Maier, P.; Zinke, T.; Ummenhofer, T.; Pfaffinger, M.; Mensinger, M.; Schneider, S.; Lenz, K.; Beck, T.; Friedrich, H. Nachhaltigkeitsanalysen von Stahlverbundbrücken—Ganzheitliche Bewertung von Autobahnbrücken mit Hilfe von quantitativen Methoden. *Stahlbau* **2014**, *83*, 476–486. [CrossRef]

62. Du, G.; Karoumi, R. Life cycle assessment of a railway bridge: Comparison of two superstructure designs. Structure and Infrastructure Engineering: Maintenance, Management. *Life-Cycle Des. Perform.* **2013**, *9*, 1149–1160.

63. Tapia, C. Pursuing Life-Cycle Sustainability for Bridges Subjected to Multiple Threats. Master's Thesis, Rice University, Houston, TX, USA, 2014.

64. Pang, B.; Yang, P.; Wang, Y.; Kendall, A.; Xie, H.; Zhang, Y. Life cycle environmental impact assessment of a bridge with different strengthening schemes. *Int. J. Life Cycle Assess.* **2015**, *20*, 1300–1311. [CrossRef]

65. Sabatino, S.; Frangopol, D.M.; Dong, Y. Sustainability-informed maintenance optimization of highway bridges considering multi-attribute utility and risk atitude. *Eng. Struct.* **2015**, *102*, 310–321. [CrossRef]

66. Almeida, J.O.; Teixeira, P.F.; Delgado, R.M. Life cycle cost optimisation in highway concrete bridges management. *Struct. Infrastruct. Eng.* **2015**, *11*, 1263–1276. [CrossRef]

67. Bizjak, K.F.; Lenart, S. Life cycle assessment of a geosynthetic-reinforced soil bridge system—A case study. *Geotext. Geomembr.* **2018**, *46*, 543–558. [CrossRef]

68. Huijbregts, M.A.J.; Steinmann, Z.J.N.; Elshout, P.M.F.; Verones, F.; Vieira, M.D.M.; Hollander, A.; Zijp, M.; van Zelm, R. ReCiPe2016: A harmonised life cycle impact assessment method at midpoint and endpoint level. *Int. J. Life Cycle Assess.* **2017**, *22*, 138–147. [CrossRef]

69. Kripka, M.; Yepes, V.; Milani, C.J. Selection of sustainable short-span bridge design in Brazil. *Sustainability* **2019**, *11*, 1307. [CrossRef]

70. Padgett, J.E.; Tapia, C. Sustainability of natural hazard risk mitigation: Life cycle analysis of environmental indicators for bridge infrastructure. *J. Infrastruct. Syst.* **2013**, *19*, 395–408. [CrossRef]

71. Arya, C.; Amiri, A.; Vassie, P. A new method for evaluating the sustainability of bridges. *Proc. Inst. Civ. Eng.-Struct. Build.* **2015**, *168*, 441–453. [CrossRef]

72. Hatami, A.; Morcous, G. Life-cycle cost assessment for bridge management: An application to Nnebraska bridges. *Comput. Civ. Build. Eng.* **2014**, *2014*, 1779–1787.

73. Marchetti, O. *Pontes de Concreto Armado*; Blucher: São Paulo, Brazil, 2008.

74. Instituto Brasileiro de Informação em Ciência e Tecnologia—IBICT. *Manual do Sistema ILCD: Sistema Internacional de Referência de Dados do Ciclo de vida de Produtos e Processos—Guia Geral para Avaliações do Ciclo de Vida—Orientações Detalhadas*; IBICT: Brasília, Brasil, 2014.
75. Prefeitura de Joinville, P.; Prefeitura de Joinville e Defesa Civil. Estadual Instalam kit de Transposição na Estrada dos Portugueses. 2019. Available online: https://wwwold.joinville.sc.gov.br/noticia/11591-prefeitura+de+Joinville+e+Defesa+Civil+Estadual+instalam+kit+de+transposi%C3%A7%C3%A3o+na+Estrada+dos+Portugueses.html (accessed on 31 October 2019).

International Journal of
Environmental Research and Public Health

Article

Customized ViNeRS Method for Video Neuro-Advertising of Green Housing

Arturas Kaklauskas [1,*], **Edmundas Kazimieras Zavadskas** [1,*], **Bjoern Schuller** [2], **Natalija Lepkova** [1], **Gintautas Dzemyda** [3], **Jurate Sliogeriene** [1] and **Olga Kurasova** [3]

[1]   Department of Construction Management and Real Estate, Faculty of Civil Engineering, Vilnius Gediminas Technical University, Sauletekio av. 11, LT-10223 Vilnius, Lithuania; natalija.lepkova@vgtu.lt (N.L.); jurate.sliogeriene@vgtu.lt (J.S.)
[2]   Department of Computing, Faculty of Engineering, Imperial College London, South Kensington Campus, London SW7 2BU, UK; bjoern.schuller@imperial.ac.uk
[3]   Cognitive Computing Group, Institute of Data Science and Digital Technologies, Vilnius University, Universiteto st. 3, LT-01513 Vilnius, Lithuania; gintautas.dzemyda@mii.vu.lt (G.D.); olga.kurasova@mii.vu.lt (O.K.)
*   Correspondence: arturas.kaklauskas@vgtu.lt (A.K.); edmundas.zavadskas@vgtu.lt (E.K.Z.); Tel.: +370-5-274 5234

Received: 2 March 2020; Accepted: 23 March 2020; Published: 27 March 2020

Abstract: The implementation of advertising for green housing usually involves consideration of individual differences among potential buyers, their desires for residential unit features as well as location impacts on a selected property. Much more rarely, there is consideration of the arousal and valence, affective behavior, emotional, and physiological states of possible buyers of green housing (AVABEPS) while they review the advertising. Yet, no integrated consideration of all these factors has been undertaken to date. The objective of this study was to consider, in an integrated manner, the AVABEPS, individual differences, and location impacts on property and desired residential unit features. During this research, the applications for the above data involved neuromarketing and multicriteria examination of video advertisements for diverse client segments by applying neuro decision tables. All of this can be performed by employing the method for planning and analyzing and by multiple criteria and customized video neuro-advertising green-housing variants (hereafter abbreviated as the ViNeRS Method), which the authors of this article have developed and present herein. The developed ViNeRS Method permits a compilation of as many as millions of alternative advertising variants. During the time of the ViNeRS project, we accumulated more than 350 million depersonalized AVABEPS data. The strong and average correlations determined in this research (over 35,000) and data examination by IBM SPSS tool support demonstrate the need to use AVABEPS in neuromarketing and neuro decision tables. The obtained dependencies constituted the basis for calculating and graphically submitting the ViNeRS circumplex model of affect, which the authors of this article developed. This model is similar to Russell's well-known earlier circumplex model of affect. Real case studies with their related contextual conditions presented in this manuscript show a practical application of the ViNeRS Method.

Keywords: green housing; neuro decision matrix; neuro correlation matrix; video neuro-advertising; COPRAS and ViNeRS Methods; multivariate design and multiple criteria analysis

---

1. Introduction

Soon it will be fifty years since researchers such as Fisk [1] and Henion and Kinnear [2] integrated ecological questions into the marketing approach and presented concepts such as ecological marketing. Notwithstanding the ubiquity of green/environmental research in marketing works, rarely did such experiential investigations lead companies to incorporate and operationalize green marketing in their normal professional activities [3].

Green marketing focuses on the arousal and valence, affective behavior, emotional, and physiological states of possible buyers of green housing (AVABEPS). Based on such AVABEPS (e.g., happiness), buyer shopping needs and priorities can be identified. A brief overview of several studies in this area follows. Using housing satisfaction as explanatory variables, Zhang et al. [4] believe that housing satisfaction and home ownership significantly contribute to overall happiness. Functional and emotional value, for many, is likely the dominant value dimension that, among personal health causes, consumers seek in social marketing. Emotional worth is linked to numerous affective states that can be either positive or negative [5]. When personal health causes are considered, this can mean either the suppression of negative affective states or the promotion of positive affective states [6]. The occupants of green buildings, in general, were more satisfied than those in conventional buildings [7]. As the geographical detector model shows, overall satisfaction with urban livability is significantly and positively affected by all six dimensions of urban livability, with convenient transportation, the natural environment, and environmental health having the biggest impact [8].

The definition of green marketing positioning is coined to address a firm's holistic positioning relative to the natural environment [9]. It indicates the tailoring of a circular economy positioning to keep the value of resources, materials, and products for as long as possible [10]. A life cycle assessment (LCA) can avoid a narrow viewpoint on sustainability concerns [11]. In the late 1980s, new instruments were invented such as LCA, with which ecological considerations could be introduced into marketing decisions [12].

Papadas et al. [9] propose that tactical actions (i.e., the use of resources, materials and products, and green pricing guidelines) suggest elasticity to executives for correcting their green marketing plan according to micro- and macro-environmental variations. It is possible to accentuate the advantages to residents of green buildings (material, water, and energy efficiency; reusable, recycled, and low-impact structure materials; waste lessening; low-carbon machineries; inhabitant health and inside environmental quality; and renewable energy) and their environments (air and water pollution reduction and human health, green architecture, green-built environment), depending on the micro-, meso- and macro-environment during the time of the advertising. It is also possible to point out the environmental reputation, values, culture, and behavior of the companies that constructed these buildings. The holistic advertising of the aforementioned green buildings and their environments in time and space would result in a more synergetic marketing effect. To achieve such a purpose, the effort was to establish the needs for green housing at the locales under analysis.

An empirical study on tourist segmentation that Bigné and Andreu [13] presented used the consumption emotions by dimensions of pleasure and arousal as its basis. The enjoyment of leisure and tourism services evokes such consumption emotions. That emotion is suitable as a segmentation variable was supported by these obtained results. Greater levels of pleasure and arousal indicated experiences of increased satisfaction. Furthermore, these also indicated increases in favorable behavioral intentions, which meant greater loyalty and willingness to pay more [13]. The emotional profiles and segments of tourists along with their post-consumption evaluations of satisfaction related to their intentions to recommend, as per Hosany and Prayag [14]. Del Chiappa et al. [15] have studied emotions as a potential variable for segmenting museum audiences. Positive emotions were reported for the audience segment that perceived the museum as being a unique attraction, and this segment also reported greater satisfaction with their museum experience [15].

Emotion can encourage good decision-making, according to Nobel Laureate Simon [16], who has analyzed the role of emotions in decision-making and concluded a lack of any intrinsic conflict between rationality and emotions. Furthermore, there is Pham [17], who proposes using fleeting feelings as credible sources of information when it comes to consumption decisions. The somatic marker hypothesis states that there are bodily signs relevant to experiencing a positive outcome that makes people feel happy and motivates them to continue seeking the same sort of behavior. Negative perception of a bodily sign will result in feelings of sadness, which initiates an internal warning to avoid acting in some certain way. The specific situation thus triggers a certain bodily sign and serves to guide behavior, because its basis relates to some past experience. This process reinforces choice as offering a more advantageous outcome, which could be considered adaptive [18]. The 80/20 rule for decision-making about purchases is popular worldwide. This rule proposes that purchasing decisions are based 80% on emotions and 20% on logic. The authors of this article have also engaged in studies indicating the strong role of emotions while analyzing decision-making alternatives.

It is usual to advertise green housing taking into account individual differences, location impacts on property variables, and the looked-for qualities of a dwelling [4,7]. But looking at variables such as the emotional, affective, and physiological reactions, and arousal and valence [19–21] that characterize a potential buyer is less common. Real estate and construction sector companies have simply ignored these factors in their advertising campaigns and never considered making an integrated multivariant design of video ad alternatives. This research is an attempt to fill this knowledge gap by examining all the factors mentioned above. For that purpose, the ViNeRS Method developed by the authors is used. This integrated study could highlight interrelated aspects that have never been analyzed before.

This article begins with the introduction followed by Section 2 on methodology, Section 3, which contains case studies for illustration, Section 4 on the correlational analysis between emotions and a built environmental state and closes with the conclusions.

2. Method

The initial discussion will be pertinent to the life process of a green and energy-efficient building and the quantitative and qualitative aspects of its marketing. A multisensory, green, and energy-efficient housing neuromarketing method was developed to integrate various aspects of the green marketing process, such as energy, the environment, health, economics, laws/regulations, innovation, the microclimate, and social, cultural, ethical, psychological, religious, ethnic, and other related matters with the life cycle of a built environment (see Figure 1). A number of researchers have analyzed various marketing variables. Jamal and Sharifuddin [22], for instance, analyzed marketing literature and concluded that the concept of perceived value is well established in that context and has been used to examine variables that play a role in future purchase decisions, as well as in the use of services and products.

Dangelico and Vocalelli [23] believe that the qualities that make a product green must be valuable and perceivable. When strength-related attributes (e.g., being long lasting) of a product are valued, sustainability could be an advantage, because, compared with other products, green products are often seen as healthier, safer, and gentler [24].

Functional and emotional values, for many, are likely the dominant value dimensions that consumers seek in social marketing for personal health causes. Emotional value relates to various affective states that can be either positive or negative [5]. When personal health causes are considered, this can mean either the suppression of negative affective states or the promotion of positive affective states [6].

Hedonic value and utilitarian value are important aspects of business and retail strategies. Other crucial concerns in management and marketing are customer satisfaction and brand loyalty [25]. The nature and theory of marketing experience, however, suggest that, specifically in a shopping context, the link between satisfaction and hedonic value should be stronger than that between satisfaction and utilitarian value [26].

Marketers promoting a product with mainly hedonic characteristics (enjoyment, identification, prestige, and general positive experiences) should focus on collaborating with marketers who can serve an appealing leadership function [27].

Figure 1 presents the general layout of the multisensory, green, and energy-efficient housing neuromarketing method, which the authors of this research developed.

Occupants in green buildings were generally more satisfied than those in conventional buildings [7]. As the geographical detector model shows, all six dimensions of urban livability, including convenient transportation, a natural environment, and good environmental health, significantly and positively have the greatest impact on overall satisfaction with urban livability [8].

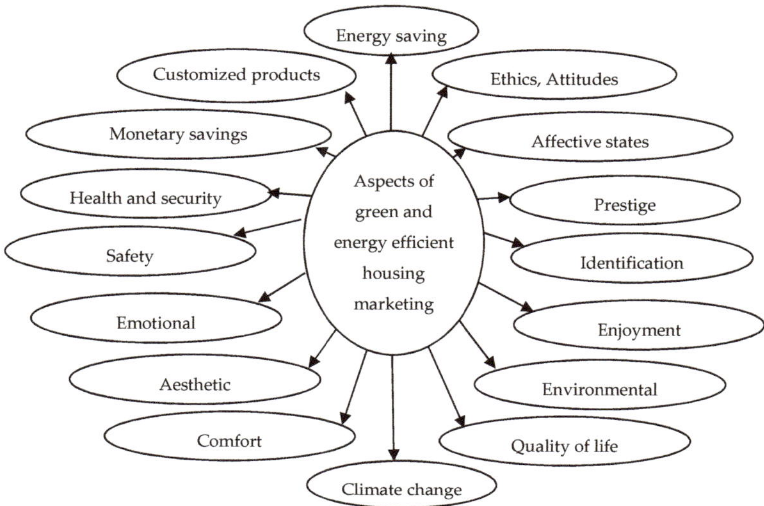

Figure 1. Quantitative and qualitative aspects of green and energy efficient marketing for analyzing the life cycle of a built environment.

ViNeRS is a customization method. Mass customization means that each individual customer gets what he or she needs, but mass production efficiency can still be ensured [28]. As defined by Pine [29], mass customization is the efficient production of individually customized offerings in high volumes and at a low cost.

The multivariant planning and the multiple criteria analysis of video neuro-advertising variants (hereafter referred to as the ViNeRS Method) goes through seven stages. The key phases of the ViNeRS Method appear in Figure 2.

Brief descriptions of these phases follow.

This article aims to showcase the offers to users made possible by the proposed ViNeRS Method.

2.1. The Research Problem and Hypotheses

Advertising for green housing units must integrate the looked-for qualities of a dwelling and the respective location impacts on the proposed property, with consideration of how the arousal and valence, affective behavior, emotional, and physiological states, affect the attitudes of possible buyers of green housing (AVABEPS), and individual differences of the potential buyer. Usually, advertising for green housing is executed in consideration of the individual differences, the looked-for qualities of a dwelling, and the location impacts on the property [4,7]. However, consideration of the emotional, affective and physiological reactions as well as arousal and valence [20,21] of a potential buyer occurs markedly less often. Despite this, the advertising campaigns of real estate companies simply do not consider these factors and the multivariant design of ad alternatives in an integrated manner at all. This research aims to fill this gap by examining all the above factors and the multivariant planning of variants in an integrated study. This research could support the highlighting of interrelated aspects, which have not been analyzed previously.

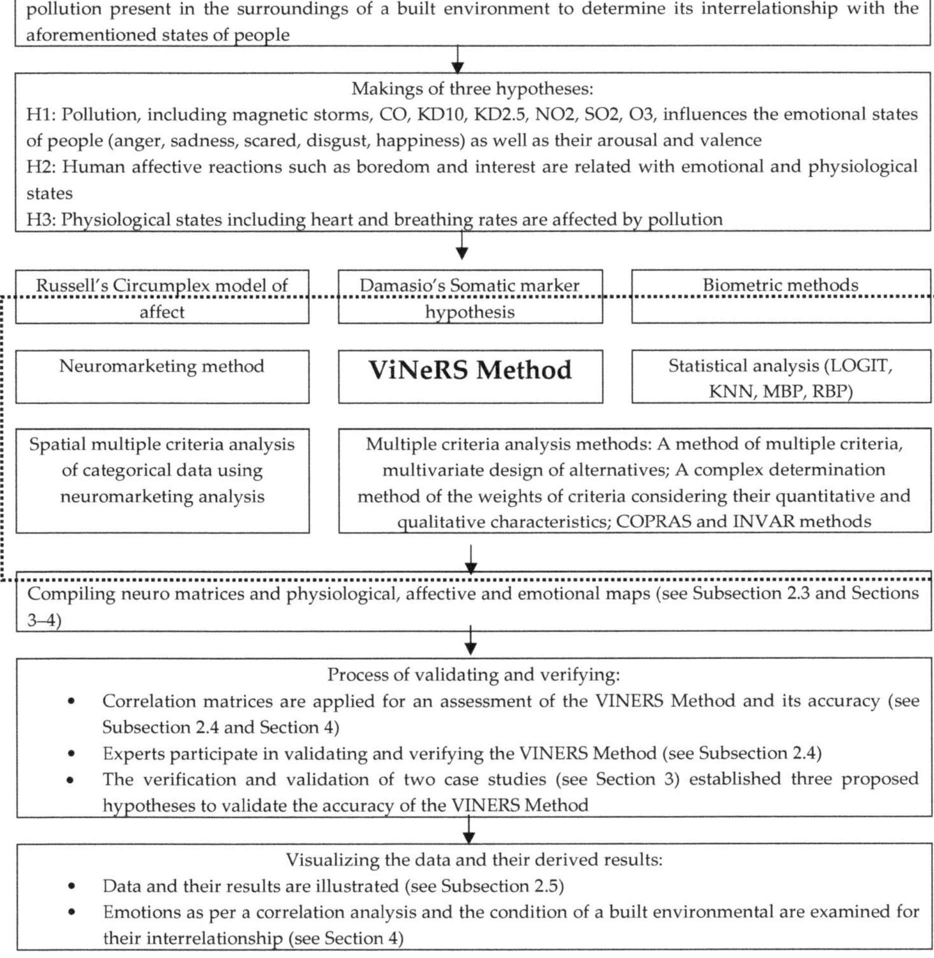

Figure 2. Main phases of the Video Neuroadvertising Recommender System (ViNeRS) Method.

Development of the ViNeRS Method was the objective in conducting this research. This method is meant to investigate the emotional, affective, and biometrical states of people, considered potential buyers, in real time, along with their arousal and valence and its interdependency. Additionally, an investigation is made of the pollution present in the surroundings of a built environment to determine its interrelationship with the aforementioned states of people.

The combined experiences of the authors of this article, together with intuition and the analysis of scholarly literature, contributed to formulating three hypotheses (see Figure 2) for this research.

The application of the ViNeRS Method, which the authors of this article developed, substantiated these hypotheses.

2.2. ViNeRS Method

The ViNeRS Method consists of a methodological integration of the Somatic Marker Hypothesis proposed by Damasio [18], Russell's circumplex model of affect [30], and other components, including biometric methods [31]; use of neuro-marketing analysis methods with categorical data for a spatial multiple criteria analysis employing, for example, AVABEPS maps [32]; LOGIT model, k-nearest neighbors algorithm (KNN), Marquardt backpropagation (MBP) algorithm, and recurrent backpropagation (RBP) statistical analyses, neuromarketing methods, application of multiple criteria, multivariant design of alternatives [33,34], and multiple criteria analysis methods for establishing criteria weights [33,34]; and the COPRAS [33–35] and the INVAR methods [36]. The multiple criteria, multivariant design and multiple criteria analysis methods itemized here include a detailed summary along with a number of examples of practices previously described [33–35]. There is also a detailed overview of the COPRAS technique, which also includes examples [33–35].

2.3. Compiling Neuro Matrices and Developing AVABEPS Maps

Four quantitative and qualitative strata of data were accumulated in this stage of the research. The work proceeded by a systematic study of these data.

The authors of this contribution believe that only a neuro decision matrix (which includes criteria, their values and weights) can ensure that the apartment for sale and its surroundings can be described in detail, and that the arousal and valence, affective behavior, emotional, and physiological states, affect the attitudes of possible buyers of green housing (AVABEPS) as well as how individual differences in people present in a public space can be tracked to achieve a more systematic analysis of the neuromarketing process related to an apartment. The neuro decision matrix is a tool that offers real-time mapping of the AVABEPS of people present in a public space. The matrix also offers integrated analysis of expert judgement results retrieved from various databases (location impacts on property, individual differences, features of an apartment).

The result of our research is unique, with integrated data related to the apartment for sale, its surroundings and potential buyers (arousal and valence, emotional, affective and physiological reactions, and individual differences). In the development of a neuro decision matrix, a key stage is determining what set of criteria will describe the alternative video advertisements that promote apartments, which units of measurement will be used, and what the weights and values will be. The quantitative and qualitative data of these alternative video advertisements are an important contributor to neuromarketing efficiency because they comprehensively describe the alternatives considered in the research.

The outcomes from the examination of ad options are delivered in the neuro decision matrix. The columns describe the ad variants under analysis, and the rows describe the data on the criteria that thoroughly define the video alternatives for the green housing unit under deliberation. A more detailed description of the neuro decision matrix appears in the case study write-up.

Equipment, which consisted of remote, biometric analysis devices (Respiration Sensor X4M200, the H.264 Indoor Mini Dome IP Camera and FaceReader 7.1) establish the affective attitudes (interest, boredom), emotional states (anger, sadness, scared, disgust, happiness), valence, arousal, and physiological states (breathing and heart rates). Measurements were taken every second. Examinations of passersby regarding their affective attitudes and emotional, affective, and physiological states took place at four intersections in Vilnius under analysis, from November 2017 until February 2020. Over 350 million pieces of data were collected during this period.

The accumulation of six layers of data validated these three hypotheses (see Figure 2) and aided the completion of this research:

- Potential buyers' individual differences: location impacts on property and desired residential unit features—data gained from the Lithuanian Department of Statistics (X_1–X_6).

- Green housing unit attributes (material, water, and energy efficiency; reusable, recycled, and low-impact structural materials; waste lessening; low-carbon machineries; inhabitants health and inside environmental quality; renewable energy)—data gained from the Lithuanian Department of Statistics (X_8, X_9), the Environmental Protection Agency (X_{11}), real estate brokers and experts in the field (X_7, X_{10}, X_{12}, X_{13}).
- Location impacts (urban quality and infrastructure (X_{14}) and green spaces (X_{15}))—data gained from the real estate brokers and experts in the field.
- AVABEPS data measured by the respiration sensor X4M200 and FaceReader 7.1 (X_{16}–X_{19}).

The fourth presented layer of data served as the basis for compiling maps locating the emotional, affective and physiological reactions of potential buyers of residential units.

This marketing segmentation involved the division of a wide-ranging customer base containing existing and potential consumers into subgroups of customer segments grounded on the above-named features. The general goal of this segmentation was to recognize the subgroups of customers that are the most likely purchasers of green housing units or the greatest money-making customers. Consequently, these possible buyers can be targeted according to their demographics, behavior, or any other added segments of significance. Marketing segmentation accepts that diverse marketing segments need various advertising alternatives with different dwelling variants (prices, qualities, location impacts, and other variables). The marketing segmentation goal here is to generate profiles of the main potential dwelling buyers.

Assuming, for example, that passers-by of a certain age group in a district under deliberation feel better (i.e., display more indicators of positive emotions such as happiness) and show fewer indicators of negative emotions (e.g., angry, sad) compared to other age groups with analogous indicators; then, the weight of importance for this age group is greater. The converse is also true. This corresponds with results gained by other researchers [37–39].

In the future, additional criteria may be added that will allow users to choose manually the location impacts, individual differences, or desired features of a dwelling along with other criteria for consideration when developing video ad alternatives. When the multiple criteria analysis of suitable video ad alternatives aimed at some specific segment is performed, potential buyers are shown video advertisements of dwellings that are best matched to their needs. Once a decision is made regarding the advertising aimed at a segment of specific, potential green housing buyers, the pool of video advertisements is reduced, ideally to include only those of interest to a potential buyer of a green housing unit. Nonetheless, in reality, only a fraction of the ads in that limited pool will actually appeal to a buyer.

Compilations of biometric, physiological and emotional maps based on the developed neuro decision tables [32] are accomplished.

2.4. Validating and Verifying for Applying the Method in Practice

Initially, the special, custom-made ViNeRS Method was verified for an evaluation of its accuracy. The application of correlation matrices is the means to prove that the results obtained by the ViNeRS Method are indeed relevant to a situation in reality. The principal goal of this phase was to define any prevailing associations between the AVABEPS emotional states (anger, sadness, scared, disgust, happiness), arousal and valence, affective reactions, and physiological states (heart and breathing rates) of the passers-by in question, and built-environment pollution (SO_2, NO_2, CO, $KD_{2.5}$, KD_{10}, O_3, and magnetic storms). The correlations determined between the variables (the neuro correlation matrix) support the necessity to use AVABEPS variables in neuro decision tables and neuromarketing.

This article presents realistic, in-depth and detailed descriptive case studies in a naturalized setting to serve as a practical demonstration of the ViNeRS Method. The overall objective of these case studies is to analyze a specific case by applying the ViNeRS Method for a better understanding of the developed technique.

Validation proved to be the means for establishing the accuracy of the customised ViNeRS Method. A pilot experiment on the practical application of the ViNeRS Method was performed in order to evaluate its efficiency and usability so as to improve the method prior to introducing it into real estate brokering practice. The testing of the ViNeRS Method employed the black-box testing method. The tester was provided with information about the results gained by the ViNeRS Method together with the corresponding data.

There were 18 experts (residents, real estate brokers, and potential real estate buyers) who checked the ViNeRS Method to make sure it was meeting stakeholder expectations. All their reviews, assessing how well stakeholder expectations were met, were presented as a report. This report also included recommended improvements, citations from existing best practices that analyze human emotions and biometric parameters in a public space, and statements on actions taken by the ViNeRS Method developers to improve the method since its last review. The results of this experiment suggest that the analyses of human emotions and the biometric parameters of a public space empower the ViNeRS Method to generate buying opportunities for green housing that meet the needs of different stakeholders more efficiently.

2.5. Illustrations of Data and Results

The illustrations of the data and results derived by the ViNeRS Method can be in quantitative forms (tables, graphs, charts, and circles presenting relationships) and in conceptual forms (a written description of the quantitative part). Therefore, it is possible to provide presentations of different neuromarketing alternatives from different perspectives. We present an example of illustrations of data and results in Chapter 4, "Relation of human emotions based on correlation analysis between emotions and built environmental state.".

3. Practical Application

The verification and validation of two case studies established three proposed hypotheses to validate the accuracy of the ViNeRS Method. These hypotheses are the following:

Hypothesis 1 (H1). *Pollution, including magnetic storms, CO, KD_{10}, $KD_{2.5}$, NO_2, SO_2, O_3, influences the emotional states of people (anger, sadness, scared, disgust, happiness) as well as their arousal and valence.*

Hypothesis 2 (H2). *Human affective reactions such as boredom and interest are related with emotional and physiological states.*

Hypothesis 3 (H3). *Physiological states including heart and breathing rates are affected by pollution.*

3.1. Case Study 1: Neuro Correlation Matrix

To create a neuro correlation matrix, we need two metrics (variables). Experiments were conducted to gather data for the neuro correlation matrix. For that purpose, we tracked AVABEPS of passers-by; all data was anonymized. The experiments (still ongoing) were launched on 6 November 2017 at seven intersections across Vilnius, Lithuania. Table 1 shows a database snippet with AVABEPS variables of passers-by recorded at four intersections in Vilnius. The AVABEPS layers of data were collected and then processed, integrated, and analyzed. Figure 3 gives several AVABEPS data relationships as an example. With the help of the neuro correlation matrix, over 35,000 average and strong correlations were determined.

Table 1. A database snippet with AVABEPS variables of passers-by recorded at four intersections in Vilnius.

Age Groups	Šventaragio and Pilies Sts. intersection		Šventaragio St. and Gedimino Pr. intersection		Kudirkos St. and Gedimino Pr. intersection		Santariškių and Baublio Sts. intersection	
	Female	Male	Female	Male	Female	Male	Female	Male
(a) Happiness								
0–20	0.128	0.139	0.128	0.103	0.138	0.163	0.146	0.074
20–30	0.131	0.138	0.125	0.135	0.123	0.154	0.133	0.255
30–40	0.117	0.115	0.116	0.114	0.114	0.101	0.114	0.123
40–50	0.103	0.098	0.081	0.085	0.080	0.089	0.129	0.130
50–60	0.123	0.085	0.089	0.061	0.074	0.066	0.162	0.217
(b) Heart rate								
0–20	79,071	76,537	81,791	56,000	69,628	63,667		
20–30	78,779	76,931	75,002	83,914	72,351	70,953		
30–40	75,923	80,393	80,669	72,113	76,228	72,756		
40–50	77,147	74,872	87,742	78,144	86,521	82,111		
50–60	69,539	70,819	97,927	84,312	82,000	83,600		
(c) Sadness								
0–20	0.236	0.214	0.213	0.187	0.221	0.170	0.293	0.246
20–30	0.197	0.183	0.195	0.188	0.213	0.168	0.223	0.098
30–40	0.158	0.155	0.161	0.158	0.153	0.129	0.185	0.175
40–50	0.149	0.142	0.176	0.137	0.126	0.116	0.159	0.204
50–60	0.153	0.144	0.108	0.116	0.119	0.098	0.234	0.086
(d) Anger								
0–20	0.092	0.116	0.086	0.118	0.087	0.116	0.132	0.116
20–30	0.086	0.122	0.079	0.109	0.103	0.127	0.135	0.074
30–40	0.091	0.119	0.093	0.111	0.089	0.109	0.134	0.157
40–50	0.094	0.123	0.086	0.119	0.094	0.106	0.084	0.126
50–60	0.092	0.129	0.070	0.135	0.071	0.104	0.195	0.106
(e) Valence								
0–20	−0.149	−0.134	−0.138	−0.155	−0.128	−0.075	−0.225	−0.220
20–30	−0.102	−0.105	−0.105	−0.109	−0.138	−0.086	−0.158	0.120
30–40	−0.089	−0.102	−0.092	−0.104	−0.093	−0.100	−0.141	−0.147
40–50	−0.107	−0.121	−0.147	−0.131	−0.109	−0.104	−0.225	−0.167
50–60	−0.076	−0.126	−0.081	−0.156	−0.094	−0.106	−0.249	0.001

The neuro correlation matrix was applied to examine the relationships linking multiple metrics (variables). AVABEPS data were analyzed using IBM SPSS software. Correlational analysis results appear in Table 1.

The accumulation and analysis of more than 350 million data comprised this research, which enabled determinations of more than 35,000 average and strong correlations. The results of our correlational analysis (see Table 2) and the significance of the variable interrelations (see Table 3) support the first hypothesis that "Built environment pollution, including magnetic storms, CO, KD_{10}, $KD_{2.5}$, NO_2, SO_2, and O_3, influences the emotional states of people (anger, sadness, scared, disgust, and happiness) as well as their arousal and valence".

Our findings also show that human affective reactions such as boredom and interest are linked to emotional and physiological states (see Table 4). Global findings also confirm these relationships [40–43]. The correlational relationships we have determined (see Table 4) and findings by other researchers support our second hypothesis "Human affective reactions such as boredom and interest are related with emotional and physiological states".

(**a**) Graph of dependencies (r = 0.551654) between heart rate (▬) and KD$_{10}$ (▬▬▬) of passers-by.

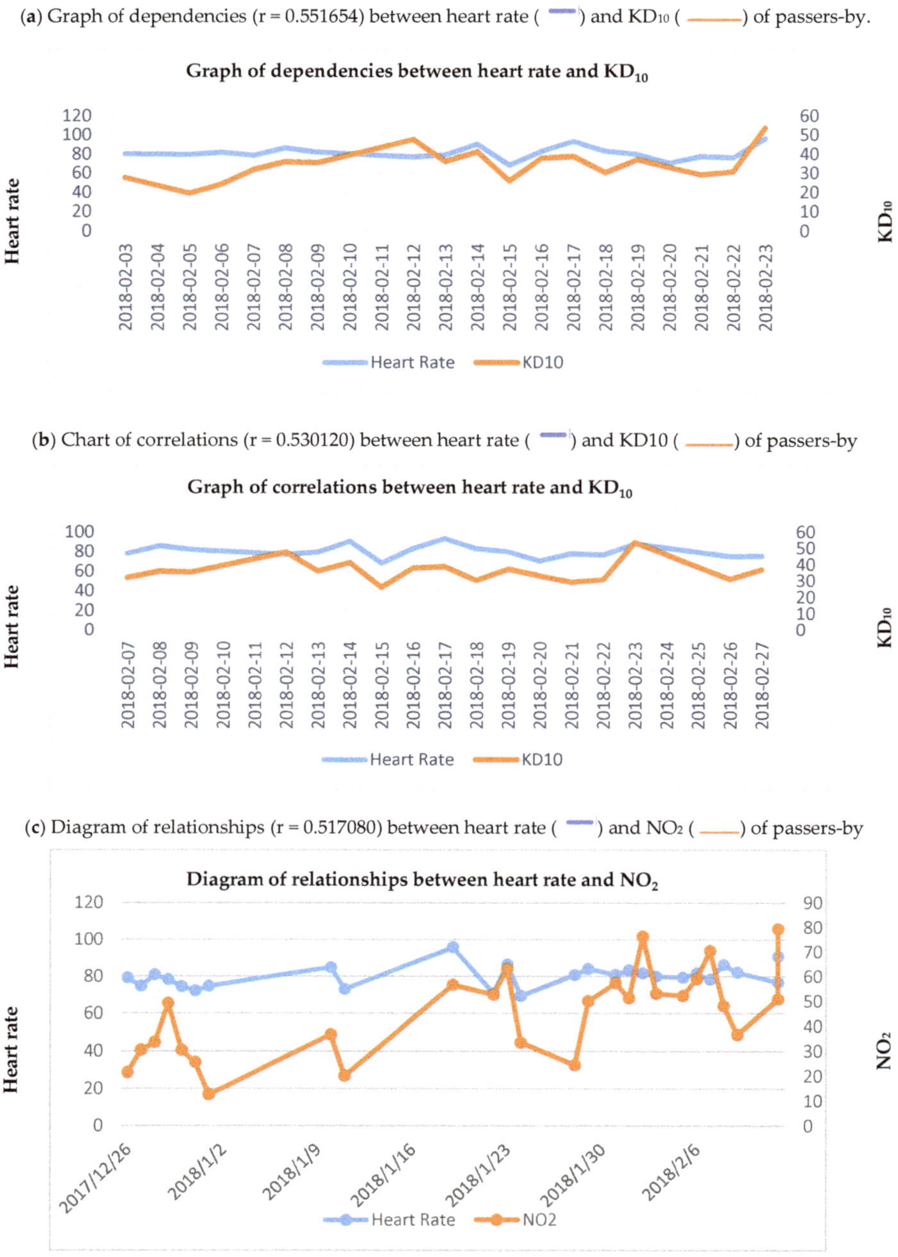

(**b**) Chart of correlations (r = 0.530120) between heart rate (▬) and KD10 (▬▬▬) of passers-by

(**c**) Diagram of relationships (r = 0.517080) between heart rate (▬) and NO$_2$ (▬▬▬) of passers-by

Figure 3. *Cont.*

(**d**) Graph of dependencies (r = 0.722840) between arousal ( ) and SO$_2$ ( ) of passers-by

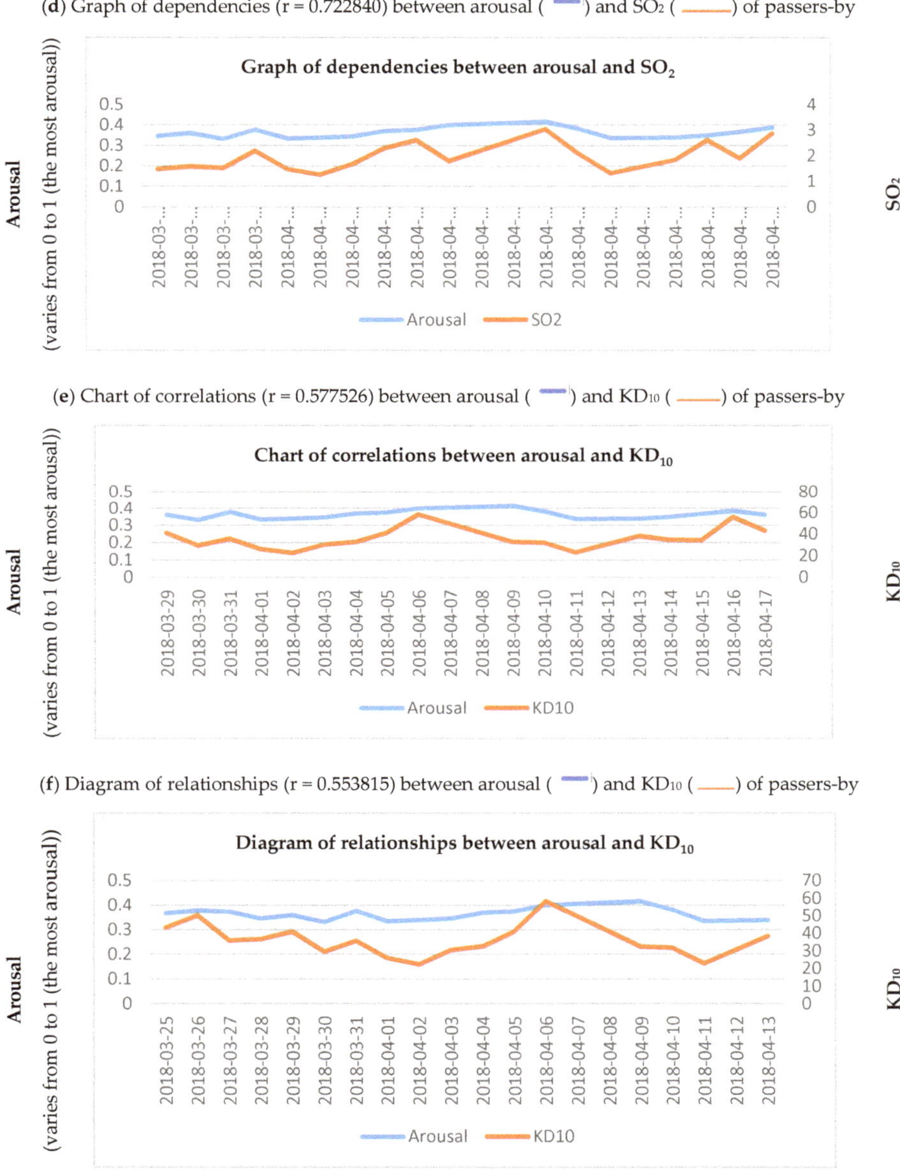

(**e**) Chart of correlations (r = 0.577526) between arousal ( ) and KD$_{10}$ ( ) of passers-by

(**f**) Diagram of relationships (r = 0.553815) between arousal ( ) and KD$_{10}$ ( ) of passers-by

Figure 3. *Cont.*

(**g**) Graph of dependencies (r = 0.528843) between arousal (▬) and CO (▬) of passers-by

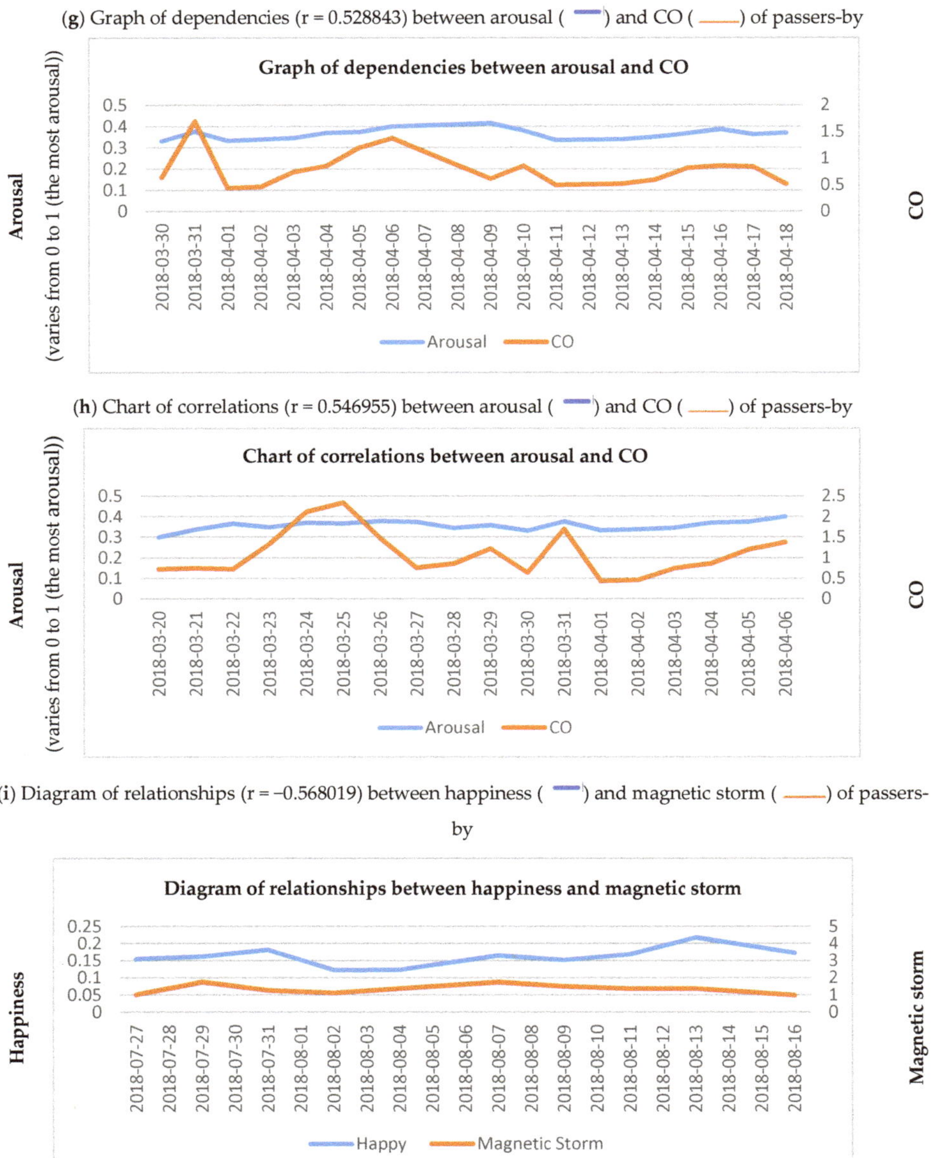

(**h**) Chart of correlations (r = 0.546955) between arousal (▬) and CO (▬) of passers-by

(**i**) Diagram of relationships (r = −0.568019) between happiness (▬) and magnetic storm (▬) of passers-by

Figure 3. *Cont.*

(**j**) Graph of dependencies (r = −0.517522) between happiness () and magnetic storm (————) of passers-by

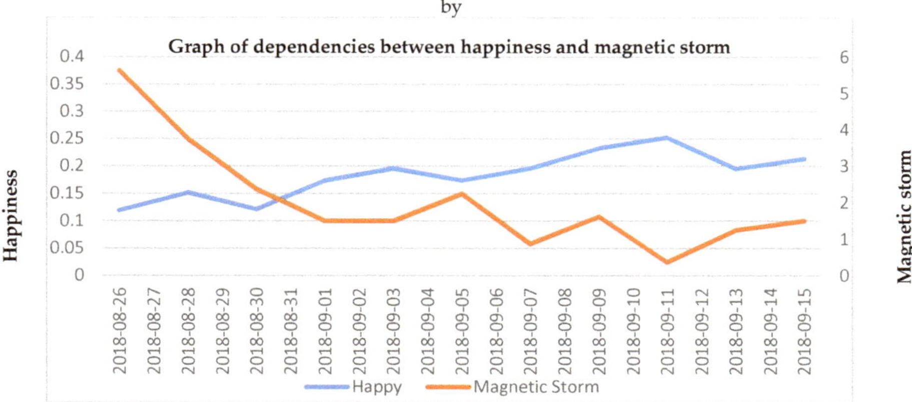

Figure 3. The dependency between average daily (**a**) heart rate and KD_{10}, (**b**) heart rate and KD_{10}, (**c**) heart rate and NO_2, (**d**) arousal and SO_2, (**e**) arousal and KD_{10}, (**f**) arousal and KD_{10}, (**g**) arousal and CO, (**h**) arousal and CO, (**i**) happiness and magnetic storm, (**j**) happiness and magnetic storm based on the values measured at six Vilnius intersections.

Table 2. AVABEPS and built environment pollution correlational analysis results.

	SO_2	$KD_{2.5}$	KD_{10}	NO_2	CO	O_3	Magnetic Storm
Anger	0.489 **	0.507 **	0.306	0.472 **	0.564 **	0.565 **	0.558 **
Valence	−0.613 *	−0.380 *	−0.417 *	−0.498 *	−0.298	−0.621 **	−0.572 **
Sadness	0.740 **	0.511 **	0.515	0.517 **	0.339	0.683 **	0.477 *
Arousal	0.698 *	0.614 **	0.566 **	0.086	0.635 **	0.719 **	−0.170
Scared	0.510	0.510 **	0.605 **	0.501 *	0.550 **	0.571 *	0.402 *
Disgust	0.286	0.181	0.576 **	0.624 **	0.418	0.351 *	0.513 **
Heart rate	0.399 *	0.772 **	0.526 **	−0.353 *	−0.077	0.412	0.590 **
Happiness	−0.788 **	0.695 **	−591 **	−0.217	−0.674 **	−0.673 **	−0.319 *

* The correlation is significant at $p < 0.05$, ** The correlation is significant at $p < 0.01$.

Table 3. Significance of links between AVABEPS and built environment pollution variables.

	SO_2	$KD_{2.5}$	KD_{10}	NO_2	CO	O_3	Magnetic Storm
Anger	+	+	−	+	+	+	+
Valence	+	+	+	+	−	+	+
Sadness	+	+	−	+	−	+	+
Arousal	+	+	+	−	+	+	−
Scared	−	+	+	+	+	+	+
Disgust	−	−	+	+	−	+	+
Heart rate	+	+	+	+	−	−	+
Happiness	+	+	+	−	+	+	+

Having examined the correlational relationships, we have determined that physiological states including the heart rate and breathing rate are affected by built environment pollution. Global findings show that environmental pollution affects the heart rate [44–47] and the breathing rate [48] of people. The correlational relationships we have determined (see Table 5) and findings by other researchers support the third hypothesis that "Physiological states including heart and breathing rates are affected by built environment pollution".

Table 4. Potential housing buyers' affective reactions such as boredom and interest and their relation to emotional and physiological states.

	Happy	Angry	Arousal	Sad	Scared	Disgusted	Surprised	Heart Rate	RPM
Boredom	−0.951(5)	−0.515(26) −0.530(20) −0.530(20)	−0.751(19) −0.526(19)	−0.565(19) −0.555(19)	−0.504(20) −0.557(6)	−0.522(19)	0.579(20) 0.616(19) 0.769(19) 0.684(19)	−0.658(7) −0.680(6)	0.510(17) 0.516(16)
Interest	0.507(20) 0.634(20) 0.724(20)	0.509(13) 0.512(12)	0.885(23) 0.697(19)	−0.564(6)	−0.751(7)	−0.516(15)	0.568(25)	0.505(13) 0.555(12)	0.587(20)

The quantities above are correlations, followed by the number of days examined for the precise correlation specified in parentheses.

Table 5. The effect of built environment pollution on physiological states such as the heart rate and breathing rate.

	SO$_2$	KD$_{2.5}$	KD$_{10}$	NO$_2$	CO	O$_3$
RPM	0.817(19) 0.780(19)	0.583(22) 0.521(21)	0.587(14) 0.605(13)	0.601(19) 0.559(19)	0.534(18) 0.650(15)	0.515(134) 0.561(133)
Heart Rate	0.502(12) 0.530(20)	0.782(8) 0.779(7)	0.552(18) 0.530(17)	0.517(26) 0.719(12)	0.591(9) 0.539(9)	0.521(11) 0.564(11)

The quantities above are correlations, followed by the number of days examined for the precise correlation specified in parentheses.

3.2. Case Study 2: Multiple Criteria Analysis of the Segmentation Matrices

Buyers' individual differences, dwelling characteristics, and location impacts are very significant factors during the selection of an apartment. Previous studies have analyzed individual buyer differences, housing characteristics, and location impacts [49].

Potential buyers were offered views of video advertisements arranged by Vilnius real estate developers featuring various projects offered to the market at the time, ranging from economic to luxury classes. A team was formed consisting of ten experts working as real estate agency brokers, developers, and analysts in Vilnius City. The experts evaluated the advertisement offered and assigned it to an appropriate group of district residents defined by the Department of Statistics under research. There were four age groups divided conditionally: Group I aged 20-30 years, Group II aged 31-40 years, Group III aged 41-60 years and Group IV aged over 60 years.

Group I and II. Ages 20-40 years, the first two groups, have the main purchasing power in Lithuania. The main reason is that ~70% of real estate (RE) acquisitions occur with the participation of a bank, i.e., buyers employ a bank loan. Meanwhile, banks finance specifically these two age groups the most favorably, on their own accord. The ages of these buyers permit them to take a loan with the longest repayment term, and it is likely the incomes of such buyers will only rise. Representatives of these age groups are the most active. Their children still attend kindergartens and schools and participate in various activity groups. Therefore, it is especially important to these buyers to have a location that is accessible as much by automobile as by public transport.

Group III, aged 41–60 years, are buyers who are already buying the second or third housing unit in their life. Their values and customs have already formed, so they are less vulnerable to fashion trends. Their children have grown up, so their priorities become matters such as comfort, a stable neighborhood, and nature. This category of buyers frequently look for real estate for investment purposes and to ensure stable incomes for themselves upon becoming pensioners.

Group IV, aged over 60 years, are buyers who frequently want to sell their large dwellings and move into smaller units, which are less expensive and easier to maintain. They are also the potential buyers for the smallest dwellings that they help their children or grandchildren to purchase.

A segmentation of potential green housing buyers in two phases was applied for this study. The first, geographical segmentation, involved the analysis of real estate subdivided into segments of the residential districts under deliberation. The second, demographic (main source of earnings, gender, age, marital status, education, families with children) and customer psychographic and behavioral (happy, angry, valence, sad, and heart rate) segmentation, involved compiling the sum matrices of neuro decision-making for the residential districts under deliberation. The evaluations of the psychographic and behavioral segments of potential housing unit buyers for this research consist of calculated weights of the criteria. The two above-described segmentation phases allow real estate brokers to optimize the available marketing resources by contacting a maximum number of relevant potential customers. Similarly, McGarigal et al. [50] and Strong and Jacobson [51] used a two-step segmentation technique to identify the optimal number of green housing submarkets by considering the characteristics of green housing and attributes of the neighborhoods.

3.3. The Sum Segmentation of Four Neighborhoods

In our approach, the first step is the pre-segmentation, wherein a large dataset is grouped into smaller subsegments. In our case, the urban district we were looking at was divided into four neighborhoods (Naujamiesčio, Verkių, Old Town, Žirmūnų). In selecting input variables for our analysis of the sum segmentation of four neighborhoods (Naujamiesčio, Verkių, the Old Town, Žirmūnų), we considered four types of green housing characteristics. Other researchers performed similar studies looking into green housing clusters. Bourassa et al. [52] and Poudyal et al. [53] believe that, in delineating green housing submarkets, four types of green housing characteristics (socioeconomic, structural, neighborhood, and locational) are commonly regarded as the most important factors. To the set of green housing attributes, Jun [54] also added input segment variables such as building age, sales price, floor size, and the number of apartment units in the block; socioeconomic variables such as the average household income and the household head's education level; and neighborhood and location-related variables such as urban parks, a nearby subway and highway interchanges, and others.

In selecting input variables for the sum segmentation of the four neighborhood (Naujamiesčio, Verkių, Old Town, Žirmūnų) analysis, we considered four types of green housing characteristics:

- individual differences of potential buyers (X_1–X_6): age (20–30, 31–40, 41–60 and over 60 years, X_1), gender (male and female, X_2), education (higher, high and special secondary, secondary, basic, elementary, incomplete elementary school, X_3), marital status (married, divorced, widowed, never married, X_4), seven main source of earnings (X_5) and families with children (X_6),
- apartment attributes (X_7–X_{13}): dwelling price (X_7), type of residential housing unit (X_8), ownership form of residential dwelling (X_9), building materials (X_{10}), air pollution and noise (X_{11}), energy usage (X_{12}), and aesthetic attributes (X_{13}),
- location impacts (X_{14}, X_{15}): urban quality (infrastructure) (X_{14}) and green spaces (X_{15}).
- AVABEPS data (happiness (X_{16}), interest (X_{17}), valence (X_{18}), and arousal (X_{19})) (see Table 6).

3.4. Reclassification of the Housing Subsegments from the First Stage Into a Relevant Number of Segments

The second stage involves reclassifying the subsegments from the first stage into a relevant number of segments (see Table 7).

Twenty video advertisements were provided to the group of experts taken from submitted real estate offerings assigned to the corresponding group of potential green housing buyers (see Appendix A Table A1). These data provided the basis for compiling the Old Town aggregated segmentation neuro decision table. All experts denoted the points in the table on their own accord, ranging from 1 to 10 regarding the acquisition/interest potential of a real estate project, considering the requirements along with trends, acquisitions, and stereotypes in the market. Analogous to sum segmentation, neuro decision tables were also compiled for the other Vilnius City districts under analysis (Naujamiestis, Verkiai and Žirmūnai).

Table 6. Sum segmentation, neuro housing decision-making matrix.

Indicators Defining Options	Sub-Indicators Defining Options	Measuring Units	Weight	*	A_1	...	A_j	...	A_n
Individual differences of buyers									
Age (X_1)	X_{11} (20–30 years)	Points	q_1	+	x_{111}	...	x_{11j}	...	x_{11n}
	X_{12} (31–40 years)	Points	q_2	+	x_{121}	...	x_{12j}	...	x_{12n}
	X_{13} (41–60 years)	Points	q_3	+	X_{131}	...	x_{13j}	...	x_{13n}
	X_{14} (over 60 years)	Points	q_4	+	x_{141}	...	x_{14j}	...	x_{14n}
Gender (X_2)	X_{21} (male)	Points	q_5	+	x_{211}	...	x_{21j}	...	x_{21n}
	X_{22} (female)	Points	q_6	+	x_{221}	...	x_{22j}	...	x_{22n}
Education (X_3)	X_{31} (higher)	Points	q_7	+	x_{311}	...	x_{31j}	...	x_{31n}
	X_{32} (high and special secondary)	Points	q_8	+	x_{321}	...	x_{32j}	...	x_{32n}
	X_{33} (secondary)	Points	q_9	+	x_{331}	...	X_{33j}	...	X_{33n}
	X_{34} (basic)	Points	q_{10}	+	X_{341}	...	x_{34j}	...	x_{34n}
	X_{35} (elementary)	Points	q_{11}	+	X_{351}	...	X_{35j}	...	x_{35n}
	X_{36} (incomplete elementary school)	Points	q_{12}	+	X_{361}	...	x_{36j}	...	x_{36n}
Marital status (X_4)	X_{41} (married)	Points	q_{13}	+	X_{411}	...	x_{41j}	...	x_{41n}
	X_{42} (divorced)	Points	q_{14}	+	X_{421}	...	x_{42j}	...	x_{42n}
	X_{43} (widowed)	Points	q_{15}	+	X_{431}	...	x_{43j}	...	x_{43n}
	X_{44} (never married)	Points	q_{16}	+	X_{441}	...	x_{44j}	...	x_{44n}
Main source of earnings (X_5)	X_{51} (salary/work compensation)	Points	q_{17}	+	X_{511}	...	x_{51j}	...	x_{51n}
	X_{52} (income from own or family business)	Points	q_{18}	+	X_{521}	...	x_{52j}	...	x_{52n}
	X_{53} (income from agricultural activities)	Points	q_{19}	+	X_{531}	...	x_{53j}	...	x_{53n}
	X_{54} (ownership or investment income)	Points	q_{20}	+	X_{541}	...	x_{54j}	...	x_{54n}
	X_{55} (pension)	Points	q_{21}	+	X_{551}	...	x_{55j}	...	x_{55n}
	X_{56} (governmental support)	Points	q_{22}	+	X_{561}	...	x_{56j}	...	x_{56n}
	X_{57} (support by family and/or other persons)	Points	q_{23}	+	X_{571}	...	x_{57j}	...	x_{57n}
Families with children (X_6)	X_{61} (families with children aged 0–17 yrs.)	Points	q_{24}	+	X_{611}	...	x_{61j}	...	x_{61n}
	X_{62} (families with no children aged 0–17 yrs.)	Points	q_{25}	+	x_{621}	...	x_{62j}	...	x_{62n}
Price (X_7)	X_{71}	Euro per sq. m.	q_{26}	−	x_{711}	...	x_{71j}	...	x_{71n}
Type of residential housing unit (X_8)	X_{81} (one unit house)	Points	q_{27}	+	x_{811}	...	x_{81j}	...	x_{81n}
	X_{82} (two-unit house)	Points	q_{28}	+	x_{821}	...	x_{82j}	...	x_{82n}
	X_{83} (multi-unit building dwelling)	Points	q_{29}	+	x_{831}	...	x_{83j}	...	x_{83n}
Ownership form of residential dwelling (X_9)	X_{91} (home owner resident)	Points	q_{30}	+	x_{911}	...	x_{91j}	...	x_{91n}
	X_{92} (resident in a rental unit)	Points	q_{31}	+	x_{921}	...	X_{92j}	...	x_{92n}
Building materials (X_{10})		Points	q_{32}	+	x_{1011}	...	x_{101j}	...	x_{101n}
Noise and air pollution (X_{11})		Points	q_{33}	+	x_{1101}	...	x_{110j}	...	x_{110n}
Energy consumption (floor heating, renewable energy sources, etc.) (X_{12})		Points	q_{34}	+	x_{1201}	...	x_{120j}	...	x_{120n}
Aesthetic features (X_{13})		Points	q_{35}	+	x_{1301}	...	x_{130j}	...	x_{130n}
Environmental influences									
Urban quality (infrastructure) (X_{14})		Points	q_{36}	+	X_{1401}	...	X_{140j}	...	X_{140n}
Green spaces (X_{15})		Points	q_{37}	+	X_{1501}	...	X_{150j}	...	X_{150n}
AVABEPS data									
Happiness (X_{16})		Points	q_{38}	+	X_{1601}	...	X_{160j}	...	X_{160n}
Interest (X_{17})		Points	q_{39}	+	X_{1701}	...	X_{170j}	...	X_{170n}
Valence (X_{18})		Points	q_{40}	+	X_{1801}	...	X_{180j}	...	X_{180n}
Arousal (X_{19})		Points	q_{41}	+	X_{1901}	...	X_{190j}	...	X_{190n}

Table 7. Sum neuro housing decision matrix compiled during the Old Town second-stage segmentation.

Indicators Defining Options	Sub-Indicators Defining Options	*	Weight	Measu-ring Units	Housing Unit Video ad Alternatives under Comparison						
					1	2	3	17	18	19	20
Individual differences of buyers											
Age	31–40 years	+	0.1826	Points	7	6	7	5	7	5	6
Gender	Male	+	0.1826	Points	7	9	8	6	8	6	9
Education	Higher	+	0.0925	Points	8	7	7	7	8	7	8
Marital status	Married	+	0.0989	Points	7	6	7	8	9	8	9
Main source of earnings	Salary/work compensation	+	0.1398	Points	7	8	7	4	7	5	7
Families with children	Families with no children	+	0.0560	Points	8	8	8	8	7	8	7
Apartment-style unit attributes											
Price (G1)	Average price (1 euro sq. m.)	–	0.8	€/m^2	1830	1450	1700	4050	1290	3710	1320
Type of residential housing unit	Multi-unit building dwelling	+	0.1662	Points	8	9	8	7	9	7	9
Ownership form of residential dwelling	Resident in a rental unit	+	0.0730	Points	7	8	8	5	8	5	7
Building materials		+	0.096	Points	7	5	7	9	6	9	7
Noise and air pollution		+	0.08	Points	8	8	7	8	8	8	7
Energy usage		+	0.184	Points	6	7	7	8	9	9	7
Aesthetic properties		+	0.04	Points	7	7	8	9	8	9	7
Environmental influences											
Urban quality (infrastructure)		+	0.144	Points	7	7	8	9	6	9	7
Green spaces		+	0.096	Points	7	7	7	6	7	7	6
AVABEPS data											
Happiness		+	0.1	Points	0.135	0.135	0.135	0.135	0.135	0.135	0.135
Interest		+	0.1	Points	0.013	0.013	0.013	0.013	0.013	0.013	0.013
Valence		+	0.1	Points	−0.131	−0.131	−0.131	−0.131	−0.131	−0.131	−0.131
Arousal		+	0.1	Points	0.330	0.330	0.330	0.330	0.330	0.330	0.330
Significance Q_j of housing unit video alternatives					0.0845	0.0887	0.0853	0.1057	0.0926	0.1005	0.0917
Priority of housing unit video alternatives					18	10	14	1	6	2	8
Utility degree N_j of housing unit video alternatives (%)					79.99	83.92	80.76	100	87.59	95.07	86.78

*—The + (−) specifies that either a greater or lower criterion value means greater (lower) significance for customers.

Having established the buyer segments in the Old Town, it is possible to perform more relevant marketing to achieve the greatest level of success. The individual differences among buyers, apartment attributes, and location impacts for each buyer segment assist in targeting ads more effectively.

There may be a great deal of possible information that real estate agents could research, use and define, however, the best place to start is with the information that brokers can use in practice. Most importantly, it also contains numerous targeting options that go hand in hand with the buyer segments of brokers. The purpose should always be to understand customers better for more effective communication, as well as to gain the ability to target ads more precisely [55].

As real estate agents learn new information, the buyer segments are likely to change, and, with growing business, brokers may even discover entirely new buyer segments. Defined buyer segments can ensure better ad targeting and communication by brokers. From an increased engagement in marketing, the time taken to define brokers' buyer segments can help businesses succeed, as this will enable them to know and understand their core customers better [55].

In determining the Old Town neighborhood purchaser segments, we establish with as-broad-as-possible alternatives and then personalized housing toward more concrete potential customers (see Table 7).

Based on the sum neuro decision matrix compiled for the Old Town district during the first-step segmentation, a markedly more personalized neuro decision table is compiled during the second segmentation step (see Table 7). Interested groups can continue to perform the process of dwelling segmentation based on such decision matrices. During such a segmentation process, dwellings may be assembled into similar submarkets relevant to the individual differences of potential buyers (X_1–X_6), the attributes of green housing units (X7–X13), location impacts (X14, X15), and AVABEPS data (X_{16}–X_{19}). By using the COPRAS [35] method and the data from Table 6, the effectiveness of the housing unit video ads in question has been determined. It is obvious that the seventeenth video ad (N_{17} = 100%) is the most effective. N_j can vary between 0% and 100%.

4. Relation of Human Emotions based on Correlation Analysis between Emotions and Built Environmental State

4.1. Built Environment Data for Analysis

Correlations were measured using seven features, describing human emotions (angry, sad, scared, disgusted, and happy), valence, and arousal, with seven features characterizing the built environment (SO_2, $KD_{2.5}$, KD_{10}, NO_2, CO, O_3, and Magnetic Storm). As a result, we obtained a matrix of correlations that is given in Table 8. In most cases, the absolute value of correlations exceeds 0.5. There are positive and negative correlations.

Table 8. Correlation matrix of potential housing buyers' emotions with features characterizing the built environment.

	SO_2	$KD_{2.5}$	KD_{10}	NO_2	CO	O_3	Magnetic Storm
Angry	0.489	0.507	0.306	0.472	0.564	0.565	0.558
Valence	−0.613	−0.380	−0.417	−0.498	−0.298	−0.621	−0.572
Sad	0.740	0.511	0.515	0.517	0.339	0.683	0.477
Arousal	0.698	0.614	0.566	0.086	0.635	0.719	−0.170
Scared	0.510	0.510	0.605	0.501	0.550	0.571	0.402
Disgusted	0.286	0.181	0.576	0.624	0.418	0.351	0.513
Happy	−0.788	0.695	−0.591	−0.217	−0.674	−0.673	−0.319

The peculiarities of the experiments were as follows:

1. In most cases, the correlations were measured independently of each other.
2. The number of experiments was different for each measurement of correlation.

4.2. Multidimensional Scaling for visual Analysis of Correlations

The option to analyze the correlation matrix of human emotions with features characterizing the built environment is Multidimensional Scaling (MDS). This method is the most popular method for a visual representation of multidimensional data in a low-dimensional manner [56–58]. It has a number of realizations using artificial neural networks and in combinations with neural networks, too.

In general, when solving real-world data analysis problems, the analyzed objects (items) X_1, X_2, ... , X_m are characterized by some features x_1, x_2, ... , x_n, where n is the number of features, and m is the number of objects. The features x_1, x_2, ... , x_n can achieve some numerical values. A set of these values characterizes a particular object $X_i = (x_{i1}, x_{i2}, ... , x_{in})$, $i \in \{1, ... , m\}$, where i is the order number of the object. If the objects are described by more than one feature, the data characterizing the objects are called multidimensional data. Visual analysis of such data helps gain a deeper insight into the data and to directly interact with the data. In our case, the objects are five human emotions (angry (X_1), sad (X_3), scared (X_5), disgusted (X_6), and happy (X_7)), valence (X_2), and arousal (X_4) ($m = 7$). There are seven features characterizing the built environment ($n = 7$): SO_2 (X_1), $KD_{2.5}$ (X_2), KD_{10} (X_3), NO_2 (X_4), CO (X_5), O_3 (X_6), and Magnetic Storm (X_7).

Low-dimensional visualization requires preserving proximities between objects X_1, X_2, ... , X_m as much as possible. MDS ensures such an objective, as it is cluster-preserving. MDS requires estimating the coordinates of new points $Y_i = (y_{i1}, y_{i2})$, $I = 1, \ldots, n$, in a lower-dimensional space Rd ($d = 2$) by minimizing some stress function depending on $Y_1, Y_2, \ldots, Y_m$. An example of the stress function may be as follows:

$$E_{MDS}(Y) = \sum_{i<j}\left(d\left(Y_i, Y_j\right) - d\left(X_i, X_j\right)\right)^2 \tag{1}$$

Here, $d(X_i, X_j)$ is the proximity between two human emotions, X_i, X_j; $Y = (Y_1, Y_2, \ldots, Y_m)$ is a set of points of lower dimensionality, and ($d<n$); d (Yi, Yj) is the Euclidean distance between the points, Y_i, Y_j, in our case. More stress functions are available in Medvedev et al. [57]. The optimization problem is quite complicated because of the number of variables, which is equal to $d{\times}n$, in the general case. Such a large number of variables is determined by the fact that we need to find d coordinates of n points. Moreover, the stress function is multimodal, i.e., it has many minima. Usually, gradient-based optimization methods are applied. One of the commonly used algorithms is stress minimization using majorization (SMACOF), which is based on iterative majorization guaranteeing a monotonic convergence. Here, the optimization process is started from initial values of $Y_1, Y_2, \ldots, Y_m$. A particular rule changes these. Finally, such values find that the value of the stress function is as minimal as possible. Two ways are usually applied to the selection of the initial values of $Y_1, Y_2, \ldots, Y_m$. In the simplest way, the values are random numbers from interval (0,1). However, the principal component approach is often used.

The 7×7 matrix of proximities d (Xi, Xj) between human emotions are presented in Table 9. Elements of the table correspond to the proximity of a particular pair of emotions. We observe high similarities between several pairs of emotions. The proximities of these pairs are tinted in yellow.

Table 9. Table of proximities of potential housing buyers' emotions, valence, and arousal for Multidimensional Scaling.

	Angry	**Valence**	**Sad**	**Arousal**	**Scared**	**Disgusted**	**Happy**
Angry	0.000	2.625	0.424	0.911	0.340	0.559	2.604
Valence	2.625	0.000	2.780	2.616	2.680	2.450	1.226
Sad	0.424	2.780	0.000	0.842	0.352	0.669	2.762
Arousal	0.911	2.616	0.842	0.000	0.759	1.138	2.705
Scared	0.340	2.680	0.352	0.759	0.000	0.503	2.689
Disgusted	0.559	2.450	0.669	1.138	0.503	0.000	2.534
Happy	2.604	1.226	2.762	2.705	2.689	2.534	0.000

4.3. Visualization Results and Their Interpretation

Results of the visualization of features are presented in Figure 4.

Dots in the figures correspond to the emotions. These are named by the corresponding dot. It is possible to evaluate their similarities visually. The more similar emotions appear closer on the plane. The visual presentation on the plane is invariant to the angle of rotation of all points and to the representation's mirror image. Therefore, the results in Figure 4 may have many interpretations. However, there is no tendency to draw an analogy with the well-known graphical representation of Russell's circumflex model of emotions for acoustic stimuli [30,59,60]. This is the model where the horizontal axis represents the valence dimension, and the vertical axis represents the arousal dimension.

The main unifying feature of these results is the derivation of formal estimates of similarities of human emotions depending on built environmental features using data science, including big data analysis, visual data analysis and artificial intelligence. The basis for these estimates consists of the influence of the built environment on emotions. Figure 4 displays the different influences of built environmental features on human emotions. There are four clusters of emotions, valence, and arousal:

- Happy
- Valence
- Aroused
- Sad, Scared, Angry, Disgusted

Figure 4. Visualization of a set of human emotions based on built environmental features.

Note that these clusters are obtained based on the proximities of emotions, where the proximities are obtained from the correlation of emotions with built environmental features. Therefore, the built environment influences emotions differently. A greater distance between the points in Figure 4 means a greater specificity of the influence.

The conclusions may be as follows. Russell's circumplex model of affect [30] proposes that emotions are spread in a two-level round plane, comprising arousal and valence magnitudes. Slight changes in the built environment may lead to emotional changes. The state of the arousal requires more built environmental changes, but not so much when compared with happy and positive valence. Presume the human emotional state is somewhere close to happy or positive valence. The built environment has to change drastically, if the desired emotion (happy) or positive valence has to change to sad, scared, angry, or disgusted. This means it is very difficult to control a positive human emotion when the built environmental features only vary slightly. However, this is possible for the emotions sad, scared, angry, and disgusted.

5. Conclusions

Many real estate brokers have noticed that modern buyers are becoming more and more selective. A tremendous amount of effort is required to analyze a tremendous number of variants to arrive at an actual purchase of an apartment.

The performance of a real estate advertisement takes into consideration the individual differences of potential buyers, the desired features of the property, and how the location impacts on the property. These constitute the similarities of this research to prior research studies. However, there was no integrated consideration at the time of the real estate advertising of the physiological, emotional and affective responses of clients and the aforementioned aspects employing a neuro decision matrix. Furthermore, there was no multivariant planning performed on customized, video neuro-advertising variants and multiple criteria analysis. Therefore, the area of the research was expanded. The ViNeRS technique was applied, which includes a combination of physiological, biometric and multiple criteria analysis and multivariant design methods, Damasio's Somatic Marker Hypothesis, AVABEPS maps, and the employment of a statistical analysis method. The developed ViNeRS Method permits this to be performed. Additionally, this research employed big data (consisting of over 350 million recordings) gained from performing analyses on the arousal and valence, affective behavior, emotional and physiological states of possible buyers of green housing (AVABEPS) in seven intersections of Vilnius City from 7 November 2017 to the point of writing the article in February 2020. Based on this data, over 35,000 strong and average correlations were determined, and they prove the need to use the AVABEPS variables analyzed in this research in neuromarketing. As part of our research, the AVABEPS data were applied in neuromarketing and multicriteria examinations of video advertisements for diverse client segments by applying the neuro decision tables.

The significance of the variable interrelations (see Table 3) and our correlational analysis (see Table 2) support the first hypothesis that "Built environment pollution, including magnetic storms, CO, KD_{10}, $KD_{2.5}$, NO_2, SO_2, O_3, influences the emotional states of people (anger, sadness, scared, disgust, and happiness), arousal and valence" (see Section 4). Another point of our findings is that boredom and interest, i.e., human affective reactions, are linked to emotions and physiological states (see Table 4). These relationships have been also confirmed by global findings. Findings by other researchers and the correlational relationships we have identified (see Table 4) support the second hypothesis that "Human affective reactions such as boredom and interest are related with (other) emotional and physiological states". We have examined the correlational relationships, and our findings show that built environment pollution affects physiological states such as the breathing rate and heart rate. The effect of built environmental pollution on the breathing rate and heart rate of individuals is also evident in global findings. Findings by other researchers and the correlational relationships we have identified (see Table 5) support the third hypothesis that "Physiological states including heart and breathing rates are affected by built environment pollution". Two case studies were verified and validated. Thereby, the ViNeRS Method was deemed accurate by three of the proposed hypotheses.

The obtained dependencies constituted the basis for calculating and graphically submitting the ViNeRS circumplex model of affect, which the authors of this article had developed. This model is similar to the earlier popular Russell's circumplex model of affect.

Future research should involve establishing high, medium or low importance for an advertising process by correlating AVABEPS metrics, climatic conditions (air temperature, humidity, average wind velocity, atmospheric pressure, and apparent temperature), and built environment pollution (magnetic storm, SO_2, $KD_{2.5}$, KD_{10}, NO_2, CO and O_3). A detailed analysis of the parameter measurements taken in Vilnius with observed strong correlations is needed, because these would greatly affect potential real estate buyers. Rapid decision-making would thus become possible, and problems could be circumvented. The situation at hand offers the best advantage for gain. Furthermore, in the future, we intend to apply various intelligent decision support systems to improve the ViNeRS Method.

Author Contributions: Data curation, A.K., N.L., G.D., J.S., O.K.; Methodology, A.K. and E.K.Z.; Supervision, A.K.; Writing—original draft, A.K., E.K.Z., B.S., N.L., G.D., J.S., O.K.; Writing—review and editing, A.K., E.K.Z., B.S., N.L., G.D., J.S., O.K. All authors have read and agreed to the published version of the manuscript.

Funding: This project has received funding from European Regional Development Fund (project No 01.2.2-LMT-K-718-01-0073) under grant agreement with the Research Council of Lithuania (LMTLT). The sponsor had no involvement in the study design, the collection, analysis and interpretation of data, or the preparation of the article.

Conflicts of Interest: The authors declare no conflict of interest.

Appendix A

Table A1. Video advertisements on housing units and their descriptions.

Line No.	Name of RE Project	Video, Tour, Gallery	RE Project Descriptions	Targeted Age Group *
1.	Ozo kvartetas (apartment-style dwelling units)	http://ozokvartetas.lt/virtualus-turas/	Small, economy class apartment-style units designated for a young, modern person without a family. New buildings surround this project. It has an average distance from the city center. The entire, required infrastructure is nearby.	I, IV
2.	Ozo parkas (apartment-style dwelling units)	http://ozoparkas.lt/galerija/ozo-parkas/	This project is economy to middle class. Buyers are offered apartment-style, 1 to 3-room dwelling units. This is an entire residential micro-region with new buildings. A recreational zone is alongside the park. The target audience consists of young families; however, persons of various ages are accommodated.	II, IV
3.	Miesto ritmu (apartment-style dwelling units in the Šnipiškės district)	https://www.miestoritmu.lt/lt/galerija	Average class project located in the city's central district. Offers to buyers are for apartment-style, 1 to 4-room dwelling units. A center of numerous city businesses neighbors the project. The target group of buyers consists of employed persons. Excellent accessibility with the city's historical center as well as with the other districts is available.	II, III
4.	Žirmūnų skveras (apartment-style dwelling units in the Žirmūnai district)	http://zirmunuskveras.lt/galerija/	Average class project is in a central district of the city. Offers to buyers are for apartment-style, 1 to 4-room dwelling units. Excellent accessibility with the city's historical center as well as with the other districts is available. A river runs along the project. Pedestrian walkways including a recreational zone are available.	II, III
5.	Raitininkų sodai (apartment-style dwelling units in the Žirmūnai district)	https://www.yit.lt/bustas/nauji-butai-vilniuje/raitininku-sodas	Upper average class project in the city's center accommodates families with small children along with older persons. The construction quality is high. Excellent accessibility with the city's historical center as well as with the other districts is available. A river runs along the project. Pedestrian walkways including a recreational zone are available.	II, III

Table A1. *Cont.*

Line No.	Name of RE Project	Video, Tour, Gallery	RE Project Descriptions	Targeted Age Group *
6.	Namų pynės (apartment-style dwelling units in the Žirmūnai district)	http://www.namupynes.lt/galerija	Average class project is in a central district of the city. Offers to buyers are for apartment-style, 1 to 3-room dwelling units. Excellent accessibility with the city's historical center as well as with the other districts is available. A river runs along the project. Pedestrian walkways including a recreational zone are available.	II, III
7.	Rinktinės namai (apartment-style dwelling units in the Verkiai district)	https://www.youtube.com/watch?v=eFpwwkm9Zd4(video-filmas)	Average class project is in a central district of the city. Offers to buyers are for apartment-style, 2 to 4-room dwelling units. Excellent accessibility with the city's historical center as well as with the other districts is available. Many business centers of the city neighbor the project. The target buyer group consists of employed persons.	II, III
8.	Veikmės parko namai (apartment-style dwelling units in the Baltupiai district)	http://www.veikme.lt/uploads/files/dir248/dir12/8_0.php	A modern average class project in one of the bedroom districts. Nature surrounds the project. It displays exceptional architectural and engineering decisions that employ passive home technologies.	II, III
9.	Levandų namai (apartment-style dwelling units in the Pašilaičiai district)	https://www.youtube.com/watch?v=sfQMUE6A5F4(video-filmas)	Economy class project is in one of the more distant districts of the city. Offers of small apartment-style, dwelling units are for young families or older single people.	I, II
10.	Elguvos deimantai (apartment-style dwelling units in the Karoliniškės-Ozas district)	https://www.youtube.com/watch?v=UpcKkYncjVo(video-filmas)	Elguvos Deimantai is a low-rise, construction project located in the Karoliniškės district. It merges with its neighboring woodlands. Fencing encompasses the entire project. Its construction materials are high quality containing heat insulation. It contains underground and ground parking and storage sheds. Potential buyers include older people, generally businesspeople and high-level specialists.	II, III
11.	Kalvarijų St. multi-unit building dwelling	https://www.youtube.com/watch?v=AI_scq37sFA(video-filmas)	Apartment-style dwelling units in a residential micro-region are located near the city's center. The units are part of a monolithic building of an old construction that has been remodeled. Potential buyers are single persons or families of average and lower incomes.	I, IV

Table A1. *Cont.*

Line No.	Name of RE Project	Video, Tour, Gallery	RE Project Descriptions	Targeted Age Group *
12.	Pavasaris (apartment-style dwelling units in the Lazdynai district)	https://www.youtube.com/watch?v=7FxqyynUjDo(video-filmas)	Economy class project in one of the city's bedroom residential districts is at a distance from the city's center. Still, it assures the entire, needed infrastructure and full services. It neighbors a recreational zone. Potential buyers include families satisfied to live at a distance from the city's center along with existing residents of the district.	II, IV
13.	Žirgų 1 (houses, cottages)	http://www.zirgu1.lt/360/; http://www.zirgu1.lt/gyvenviete-2/	This is an economy class square of cottages located in the Vilnius suburbs. It is a price alternative dwelling, the same as many other cottages in this class. Its location is at a distance from the city's center. Potential buyers include families desiring greater spaciousness for whom the distance from the city's center has no relevance.	II, III
14.	Valakampių krantas (cottages)	http://valakampiukrantas.lt/?gclid=EAIaIQobChMI3PXHlOmA3gIVBhUYCh1vJg45EAMYASAAEgLpWPDBwE	Luxury class project built with high quality buildings and modern engineering decisions. The neighborhood is solid and stable. Nature surrounds the project. It is for those who desire a comfortable life style but have no need to be near the city's center. It suits persons in higher income and older groups.	III
15.	Baltupio krantas (cottages)	https://www.youtube.com/watch?v=cVIhfHqvfSc(video-filmas); http://baltupiokrantas.lt/	Luxury class project is located in a bedroom district at about an average distance from the city's center. It suits buyers who desire greater privacy, spaciousness and nature but without sacrificing ease of accessibility with the city's center. It is suitable for older people.	II, III
16.	Brick house in the Kalnėnai district	https://www.youtube.com/watch?v=NE3aMYbmS24(video-filmas)	Brick house constructed in 2004 in the Kalnėnai district, which is under intensive development in Vilnius. This is a fully equipped, spacious house with a garage and other subsidiary facilities. It has a convenient driveway. It suits families; though, the social infrastructure is still not fully developed (it has no nurseries, shops or the like).	II, III

Table A1. *Cont.*

Line No.	Name of RE Project	Video, Tour, Gallery	RE Project Descriptions	Targeted Age Group *
17.	Live square	http://livesquare.lt/galerija-vizualizacijos/	Luxury class project located in the city's center. It is equally as suitable as an investment or for a modern, comfortable residence. The highest standards in construction quality are assured. Excellent accessibility in all directions combines with an excellent infrastructure and solid neighborhood. Potential buyers consist of older people, generally businesspeople or high-level specialists.	II, III
18.	Karaliaučiaus slėnis	https://www.karaliauciausslenis.lt/galerija/181?c=v	Economy class cottages harmoniously constructed in a square, located in a bedroom district. This is an alternative to an apartment-style unit by price, the same as are many other cottages of this class. However, it is at a distance from the city's center. This project was awarded for harmonious development in 2016. Potential buyers include families requiring greater space for whom the city's center is not a relevant aspect.	II, III
19.	CNTRL	http://www.cntrl.lt/apie-projekta/48	Luxury class project located in the city's center is as suitable for an investment as for a modern, convenient residence. The highest standard in construction quality is assured. Accessibility is excellent in all directions. The infrastructure is excellent and the neighborhood, solid. Potential buyers include older people, generally businesspeople or high-level specialists.	III
20.	Vilniaus vakarai	http://vilniausvakarai.lt/butai-kotedzai/	Economy class apartment-style units and cottages in a bedroom district are available in an area with an undeveloped infrastructure. Thus potential buyers must be mobile. It suits those who search from a natural setting and who want relaxation from the city's hubbub.	II, III

* Target age groups: I—(20–30) II—(31–40) III—(41–60) IV—(60+).

References

1. Fisk, G. *Marketing and the Ecological Crisis*; Harper & Row: New York, NY, USA, 1974.
2. Henion, K.E.; Kinnear, T.C. A guide to ecological marketing. In *Ecological Marketing Columbus*; American Marketing Association: Chicago, IL, USA, 1976.
3. Fuentes, C. How green marketing works: Practices, materialities and images. *Scand. J. Manag.* **2015**, *31*, 192–205. [CrossRef]
4. Zhang, F.; Zhang, C.; Hudson, J. Housing conditions and life satisfaction in urban China. *Cities* **2018**, *81*, 35–44. [CrossRef]
5. Sánchez-Fernández, R.; Iniesta-Bonillo, M.Á. Consumer perception of value: Literature review and a new conceptual framework. *J. Consum. Satisf. Dissatisfaction Complain. Behav.* **2006**, *19*, 40.
6. Zainuddin, N. Value creation in social marketing for the continued use of wellness services. In Proceedings of the Australian and New Zealand Marketing Academy Conference 2011, Perth, Australia, 28–30 November 2011; pp. 1–8.
7. Geng, Y.; Ji, W.; Wang, X.; Lin, B.; Zhu, Y. A review of operating performance in green buildings: Energy use, indoor environmental quality and occupant satisfaction. *Energy Build.* **2019**, *183*, 500–514. [CrossRef]
8. Zhan, D.; Kwan, M.P.; Zhang, W.; Fan, J.; Dang, Y. Assessment and determinants of satisfaction with urban livability in China. *Cities* **2018**, *79*, 92–101. [CrossRef]
9. Papadas, K.K.; Avlonitis, G.J.; Carrigan, M. Green marketing orientation: Conceptualization, scale development and validation. *J. Bus. Res.* **2017**, *80*, 236–246. [CrossRef]
10. MacArthur, E. *Towards the Circular Economy: Accelerating the Scale-Up Across Global Supply Chains*; Technical Report; World Economic Forum: Colony, Switzerland, 2014.
11. Cooper, J.S.; Fava, J.A. Life-Cycle assessment practitioner survey: Summary of results. *J. Ind. Ecol.* **2006**, *10*, 12–14. [CrossRef]
12. Belz, F.; Peattie, K. *Sustainability Marketing: A Global Perspective*, 2nd ed.; Wiley: Hoboken, NJ, USA, 2009; p. 352.
13. Bigné, J.E.; Andreu, L. Emotions in segmentation: An empirical study. *Ann. Tour. Res.* **2004**, *31*, 682–696. [CrossRef]
14. Hosany, S.; Prayag, G. Patterns of tourists' emotional responses, satisfaction, and intention to recommend. *J. Bus. Res.* **2013**, *66*, 730–737. [CrossRef]
15. Del Chiappa, G.; Andreu, L.; Gallarza, M. Emotions and visitors' satisfaction at a museum. *Int. J. Cult. Tour. Hosp. Res.* **2014**, *8*, 420–431. [CrossRef]
16. Simon, H. *Administrative Behavior*, 4th ed.; The Free Press: New York, NY, USA, 1997.
17. Pham, M.T. The logic of feeling. *J. Consum. Psychol.* **2004**, *14*, 360–369. [CrossRef]
18. Damasio, A.R. *Descartes' Error: Emotion, Reason, and the Human Brain*; Grosset/Putnam: New York, NY, USA, 1994.
19. Daimi, S.N.; Saha, G. Classification of emotions induced by music videos and correlation with participants' rating. *Expert Syst. Appl.* **2014**, *41*, 6057–6065. [CrossRef]
20. Lee, W.; Norman, M.D. Affective computing as complex systems science. *Procedia Comput. Sci.* **2016**, *95*, 18–23. [CrossRef]
21. Gauba, H.; Kumar, P.; Roy, P.P.; Singh, P.; Dogra, D.P.; Raman, B. Prediction of advertisement preference by fusing EEG response and sentiment analysis. *Neural Netw.* **2017**, *92*, 77–88. [CrossRef] [PubMed]
22. Jamal, A.; Sharifuddin, J. Perceived value and perceived usefulness of halal labeling: The role of religion and culture. *J. Bus. Res.* **2015**, *68*, 933–941. [CrossRef]
23. Dangelico, R.M.; Vocalelli, D. "Green Marketing": An analysis of definitions, strategy steps, and tools through a systematic review of the literature. *J. Clean. Prod.* **2017**, *165*, 1263–1279. [CrossRef]
24. Luchs, M.G.; Naylor, R.W.; Irwin, J.R.; Raghunathan, R. The sustainability liability: Potential negative effects of ethicality on product preference. *J. Mark.* **2010**, *74*, 18–31. [CrossRef]
25. Mehmood, K.; Hanaysha, J. The Strategic Role of Hedonic Value and Utilitarian Value in Building Brand Loyalty: Mediating Effect of Customer Satisfaction. *Pak. J. Soc. Sci.* **2015**, *35*, 1025–1036.
26. Jones, M.A.; Reynolds, K.E.; Arnold, M.J. Hedonic and utilitarian shopping value: Investigating differential effects on retail outcomes. *J. Bus. Res.* **2006**, *59*, 974–981. [CrossRef]

27. Klein, K.; Melnyk, V. Speaking to the mind or the heart: Effects of matching hedonic versus utilitarian arguments and products. *Mark. Lett.* **2016**, *27*, 131–142. [CrossRef]
28. Piller, F.; Reichwald, R.; Tseng, M. Competitive advantage through customer centric enterprises. *Int. J. Mass Cust.* **2006**, *1*, 157–165.
29. Pine, B.J. The state of mass customization and why authenticity in business is the next big issue, B. Joseph Pine II in an interview with Frank Piller. *Mass Cust. Open Innov. News*, 2007.
30. Russell, J.A. A circumplex model of affect. *J. Personal. Soc. Psychol.* **1980**, *39*, 1161–1178. [CrossRef]
31. Kaklauskas, A. Biometric and intelligent decision making support. In *Intelligent Systems Reference Library, 81*; Springer International Publishing: Berlin/Heidelberg, Germany, 2015; 220p.
32. Kaklauskas, A.; Jokubauskas, D.; Čerkauskas, J.; Dzemyda, G.; Ubartė, I.; Skirmantas, D.; Podviezko, A.; Šimkutė, I. Affective analytics of demonstration sites. *Eng. Appl. Artif. Intell.* **2019**, *81*, 346–372. [CrossRef]
33. Kaklauskas, A.; Zavadskas, E.K.; Raslanas, S. Multivariant design and multiple criteria analysis of building refurbishments. *Energy Build.* **2005**, *37*, 361–372. [CrossRef]
34. Kaklauskas, A. *Multiple Criteria Decision Support of Building Life Cycle*; Technika: Vilnius, Lithuania, 1999.
35. Zavadskas, E.K.; Peldschus, F.; Kaklauskas, A. *Multiple Criteria Evaluation of Projects in Construction*; Technika: Technical University, Institute of Technological and Economic Development: Vilnius, Lithuania, 1994; p. 226.
36. Kaklauskas, A. Degree of project utility and investment value assessments. *Int. J. Comput. Commun. Control (IJCCC)* **2016**, *11*, 666–683. [CrossRef]
37. Ballas, D. What makes a "happy city"? *Cities* **2013**, *32*, 39–50. [CrossRef]
38. Paralkar, S.; Cloutier, S.; Nautiyal, S.; Mitra, R. The sustainable neighborhoods for happiness (SNfH) decision tool: Assessing neighborhood level sustainability and happiness. *Ecol. Indic.* **2017**, *74*, 10–18. [CrossRef]
39. Austen, K. Cities of dreams. *New Sci.* **2013**, *220*, 50. [CrossRef]
40. Dahlen, E.R.; Martin, R.C.; Ragan, K.; Kuhlman, M.M. Boredom proneness in anger and aggression: Effects of impulsiveness and sensation seeking. *Personal. Individ. Differ.* **2004**, *37*, 1615–1627. [CrossRef]
41. Vittersø, J.; Overwien, P.; Martinsen, E. Pleasure and interest are differentially affected by replaying versus analyzing a happy life moment. *J. Posit. Psychol.* **2009**, *4*, 14–20. [CrossRef]
42. Nederkoorn, C.; Vancleef, L.; Wilkenhöner, A.; Claes, L.; Havermans, R.C. Self-inflicted pain out of boredom. *Psychiatry Res.* **2016**, *237*, 127–132. [CrossRef] [PubMed]
43. Zhang, Q.; Chen, X.; Zhan, Q.; Yang, T.; Xia, S. Respiration-based emotion recognition with deep learning. *Comput. Ind.* **2017**, *92*, 84–90. [CrossRef]
44. Brook, R.D. Cardiovascular effects of air pollution. *Clin. Sci.* **2008**, *115*, 175–187. [CrossRef] [PubMed]
45. Mills, N.L.; Donaldson, K.; Hadoke, P.W.; Boon, N.A.; MacNee, W.; Cassee, F.R.; Sandström, T.; Blomberg, A.; Newby, D.E. Adverse cardiovascular effects of air pollution. *Nat. Clin. Pract. Cardiovasc. Med.* **2009**, *6*, 36–44. [CrossRef] [PubMed]
46. Simkhovich, B.Z.; Kleinman, M.T.; Kloner, R.A. Air pollution and cardiovascular injury epidemiology, toxicology, and mechanisms. *J. Am. Coll. Cardiol.* **2008**, *52*, 719–726. [CrossRef] [PubMed]
47. Gold, D.R.; Litonjua, A.; Schwartz, J.; Lovett, E.; Larson, A.; Nearing, B.; Allen, G.; Verrier, M.; Cherry, R.; Verrier, R. Ambient pollution and heart rate variability. *Circulation* **2000**, *101*, 1267–1273. [CrossRef] [PubMed]
48. Engström, E.; Forsberg, B. Health impacts of active commuters' exposure to traffic-related air pollution in Stockholm, Sweden. *J. Transp. Health* **2019**, *14*, 100601. [CrossRef]
49. Kaklauskas, A.; Zavadskas, E.K.; Banaitis, A.; Meidute-Kavaliauskiene, I.; Liberman, A.; Dzitac, S.; Ubarte, I.; Binkyte, A.; Cerkauskas, J.; Kaminske, A.; et al. A neuro-advertising property video recommendation system. *Technol. Forecast. Soc. Chang.* **2018**, *131*, 78–93. [CrossRef]
50. McGarigal, K.; Cushman, S.; Stafford, S. *Multivariate Statistics for Wildlife and Ecology Research*; Springer: Berlin/Heidelberg, Germany, 2000; p. 283.
51. Strong, N.A.; Jacobson, M.G. Assessing agro-forestry adoption potential utilizing market segmentation: A case study in Pennsylvania. Small-Scale Forest, Economics. *Manag. Policy* **2005**, *4*, 215–228.
52. Bourassa, S.; Hoesli, M.; Peng, V.S. Do green housing submarkets really matters? *J. Green Hous. Econ.* **2003**, *12*, 2–28.
53. Poudyal, N.C.; Hodgesa, D.G.; Merrett, C.D. A hedonic analysis of the demand for and benefits of urban recreation parks. *Land Use Policy* **2009**, *26*, 975–983. [CrossRef]
54. Jun, M.-J. Quantifying welfare loss due to longer commute times in Seoul: A two-stage hedonic price approach. *Cities* **2009**, *84*, 75–82.

55. Lazazzera, A. Better Brand: Ecommerce Branding Guide. 2015. Available online: https://www.abetterlemonadestand.com/branding-guide/ (accessed on 10 January 2020).
56. Groenen, P.; Mathar, R.; Trejos, J. Global optimization methods for multidimensional scaling applied to mobile communication. In *Data Analysis: Scientific Modeling and Practical Applications*; Gaul, W., Opitz, O., Schander, M., Eds.; Springer: Berlin/Heidelberg, Germany, 2000; pp. 459–475.
57. Medvedev, V.; Dzemyda, G.; Kurasova, O.; Marcinkevičius, V. Efficient data projection for visual analysis of large data sets using neural networks. *Informatica* **2011**, *22*, 507–520.
58. Dzemyda, G.; Kurasova, O.; Žilinskas, J. *Multidimensional Data Visualization: Methods and Applications*; Springer: Berlin/Heidelberg, Germany, 2013; p. 75.
59. Posner, J.; Russell, J.A.; Peterson, B.S. The circumplex model of affect: An integrative approach to affective neuroscience, cognitive development, and psychopathology. *Dev. Psychopathol.* **2005**, *17*, 715–734. [CrossRef] [PubMed]
60. Nardelli, M.; Valenza, G.; Greco, A.; Lanata, A.; Scilingo, E.P. Recognizing emotions induced by affective sounds through heart rate variability. *IEEE Trans. Affect. Comput.* **2015**, *6*, 385–394. [CrossRef]

International Journal of
*Environmental Research
and Public Health*

Article

Enhancing Sustainability and Resilience through Multi-Level Infrastructure Planning

Jorge Salas [1] and Víctor Yepes [2,*]

[1] School of Civil Engineering, Universitat Politècnica de València, 46022 Valencia, Spain; jorsaher@doctor.upv.es
[2] ICITECH, Universitat Politècnica de València, 46022 Valencia, Spain
* Correspondence: vyepesp@cst.upv.es; Tel.: +34-963-879-563; Fax: +34-963-877-569

Received: 12 January 2020; Accepted: 3 February 2020; Published: 4 February 2020

Abstract: Resilient planning demands not only resilient actions, but also resilient implementation, which promotes adaptive capacity for the attainment of the planned objectives. This requires, in the case of multi-level infrastructure systems, the simultaneous pursuit of bottom-up infrastructure planning for the promotion of adaptive capacity, and of top-down approaches for the achievement of global objectives and the reduction of structural vulnerabilities and imbalances. Though several authors have pointed out the need to balance bottom-up flexibility with top-down hierarchical control for better plan implementation, very few methods have yet been developed with this aim, least of all with a multi-objective perspective. This work addressed this lack by including, for the first time, the mitigation of urban vulnerability, the improvement of road network condition, and the minimization of the economic cost as objectives in a resilient planning process in which both actions and their implementation are planned for a controlled, sustainable development. Building on Urban planning support system (UPSS), a previously developed planning tool, the improved planning support system affords a planning alternative over the Spanish road network, with the best multi-objective balance between optimization, risk, and opportunity. The planning process then formalizes local adaptive capacity as the capacity to vary the selected planning alternative within certain limits, and global risk control as the duties that should be achieved in exchange. Finally, by means of multi-objective optimization, the method reveals the multi-objective trade-offs between local opportunity, global risk, and rights and duties at local scale, thus providing deeper understanding for better informed decision-making.

Keywords: multi-scale assessment; hierarchical relational modeling; cascading impacts; adaptive capacity; infrastructure integrated planning; road network; decentralization optimization

---

1. Introduction

1.1. Implementation Planning as a Part of Resilient Planning

The concept of resilience was first introduced into ecological theory by Holling [1] as a measure of the capacity of a system to absorb change and external disturbance while maintaining key functions, and it is rapidly gaining ground in the urban sustainability literature [2]. In the field of urban infrastructure planning, resilient planning studies can refer to "planning" for a more resilient city, or to the "resiliency" of an urban planning, and, together, both approaches provide a constructive option for a controlled sustainable development of social–ecological systems [3]. While the first aspect focuses on the planning of actions leading to the improvement of a city's resilience, the second has to do with the implementation of these actions within an urban framework. Resilient infrastructure planning, in this context, refers to a more flexible, adaptable approach for dealing with dynamic problems arising from the implementation of an infrastructure plan. This means that resilient infrastructure planning requires not only the design of measures for the improvement of infrastructure resiliency, but also measures ensuring the best implementation of these actions [3]. However, there are currently few methods incorporating the design of an implementation strategy as a part of the planning process across multi-level governmental environments [4,5].

1.2. Implementation Planning and Decentralization

Several authors have pointed out the role of decentralization in providing urban systems with their required adaptive capacity. Sharifi and Yamagata [6] pointed out that decentralization is essential for enhancing local adaptive capacity, and that a shift towards bottom-up planning approaches must be made in order to improve the adaptability and flexibility of urban systems, and therefore contribute to achieving sustainable urban development. Gonzales and Ajami [7] proposed a methodology for improving resilience by adding flexibility at a local scale in urban water systems, while Leigh and Lee [8] showed how decentralization leads to greater adaptability of water systems for specific local contexts and operational changes. Additionally, Rogers [9] demanded that national policies and actions should be framed to facilitate local adaptation.

While recognizing the importance of flexibility at the local scale, resilient planning argues for the need of a regional and national perspective [2,9–11]. In the planning of road networks, this integrated outlook makes it possible to pursue overall objectives such as overall condition improvement [12–16] or safety performance [17], as well to contribute to the mitigation of the system's structural imbalances [18] and vulnerabilities [4,19]. Given the link between road networks and other essential facilities, such as hospital or schools, and their role in induced community vulnerability [20], reducing these networks' vulnerabilities and structural imbalances should be a primary objective for infrastructure planning. These pursuits, however, can be jeopardized by ignoring the negative cascading, cross-scale effects that actions taken at the local scale can bring to bear on global objectives [9,21–24], such as coordination problems in decentralized systems [25,26]. In other words, while decentralization contributes to the adaptive capacity demanded at the local scale, it also poses risks [5] and barriers [27] to the achievement of objectives at larger scales, which demands a proper balance of decentralization [11] that facilitates integrated planning formulation and its implementation [28].

1.3. Decentralized Systems: Balancing Adaptive Capacity and Hierarchical Control

The search for this balance between flexibility at the local scale and hierarchical control from central government has caused a debate among practitioners [27] whose ultimate purpose is to improve, in multi-level systems, the coordination between scales that is required [1,4,25,29,30]. This coordination, which is critical for implementing adaptation strategies in the transport sector [5] as well as in urban planning [27], highly depends on the system's decentralization level [25,26]. Consequently, determining the proper decentralization level in multi-scale infrastructure networks is a key issue for a system's design, implementation, and operation [2,31], and therefore for its resilient planning. However, there are currently very few studies affording implementation strategies that offer this balance between local adaptive capacity and the comprehensive perspective demanded for resilient planning. Ganzle et al. [32] pointed out the need for research specifically aimed at providing strategies for addressing the coordination problem arising from the implementation of integrated planning within multi-level governmental frameworks. Newman et al. [31] explored the effect of different decentralization levels in water systems, finding that a system's performance may be sensitive to the level of decentralization adopted, while Roozbahani et al. [33] evaluated the risks of urban water supply systems from bottom to top by means of hierarchical structure analysis. Gupta et al. [34] pointed out the tension between top-down (centralized) and bottom-up (decentralized) planning approaches, and they stated the need to balance them for improved adaptive capacity. Regmi et al. [24] remarked on the convenience of integrating both planning approaches, the lack of methods addressing this objective, and the need to bridge the gap between global policies and local strategies. Finally, Salas and Yepes [4] presented Multi-scale relational risk and opportunity (Ms-ReRO), a methodology which respectively represents adaptive capacity at the local scale and hierarchical control at the global scale as "right" and "duty" rules between hierarchically linked entities [25,26]. This method combines optimization for the design of plans of action (Figure 1, planning module) with quantitative risk assessment and multi-objective optimization (Figure 1, Ms-ReRO module) in order to afford decentralization configurations to minimize overall risks and maximize local adaptive capacity. These decentralization configurations are defined via the "rights" and "duties" embodied in the relational contracts linking entities of interdependent scales [25]. Through these contracts, top entities (i.e., countries) transfer some of their "right" to take decisions to entities below (i.e., regions) which, in exchange, must achieve a given "duty", or level of performance. By allowing these "rights" and "duties" to be regulated, the proposed framework enables the optimization algorithm to identify the trade-offs between risks and opportunities (Figure 1, Dynamic risk and opportunity simultaneous evaluation (D-ROSE) module), which makes it possible for the decision-maker to balance adaptive capacity and hierarchical control (Figure 1, Ms-ReRO module).

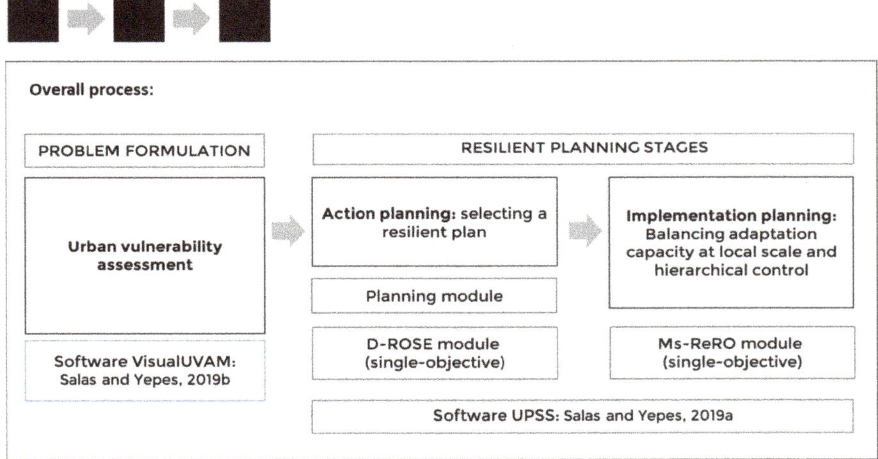

Figure 1. Overall process: vulnerability assessment and resilient planning.

However, in their article, Salas and Yepes [4] studied risks only from the economic cost perspective, pointing out the need to introduce additional objectives in future work. They also remarked on the limitations of their methodology in providing criteria for choosing among pareto-optimal decentralization alternatives, which requires a deeper analysis of the trade-offs between rights, duties, and global and local risks and provides opportunities for enhanced decision-making.

The aim of this paper was to contribute to the field of resilient planning by enabling, for the first time, a multi-objective balance of local adaptive capacity and global risk control in net infrastructure planning, as well as to provide a deep analysis of the trade-offs between decentralization configurations and the risks and opportunities they bear for multiple objectives. By means of the proposed resilient planning process, both actions and their implementation were planned, in a decentralized system case study, in order to mitigate the system's urban vulnerability, to improve the road network's current condition, and to minimize the economic cost.

The remainder of this paper is organized as follows. In the Methods section, each stage of the three-step process (Figure 1), namely urban vulnerability assessment, action planning, and implementation planning, is described. In the Case Study section, the whole process is illustrated through an actual case, the results of which are presented in the Results section. These results are then analyzed in the Discussion section to show whether the applied method contributed to resilient planning or not, and, finally, general conclusions are drawn in the closing section.

2. Methods

2.1. Step 1: Urban Vulnerability Assessment

Broadly, vulnerability can be understood as the susceptibility to suffer from, or the difficulty in coping with the negative effects of an event, and it has become a major concern for sustainable urban development [35–38]. In a prior work, Salas and Yepes [39] presented VisualUVAM, a software that affords the urban vulnerability assessment of cities, provinces, and regions of Spain. This software extended the scope of possible variables for the characterization of urban vulnerability (UV) from the three basic criteria adopted by the Spanish Observatory of Urban Vulnerability (OVU) to a wider set of 36 possible indicators, among which the method selected those most suitable according to several criteria. Based on this set of the most suitable indicators, the method yields a quantitative assessment of both the state of vulnerability at the end of a given time period and the risk of becoming more vulnerable during the next period.

In VisualUVAM, the selection among the 36 possible indicators for characterizing UV is addressed via a multi-objective optimization (MOO) problem in which expert judgment, statistical consistency, and robustness against data uncertainties are used as the criteria for the choosing of indicators (Figure 2).

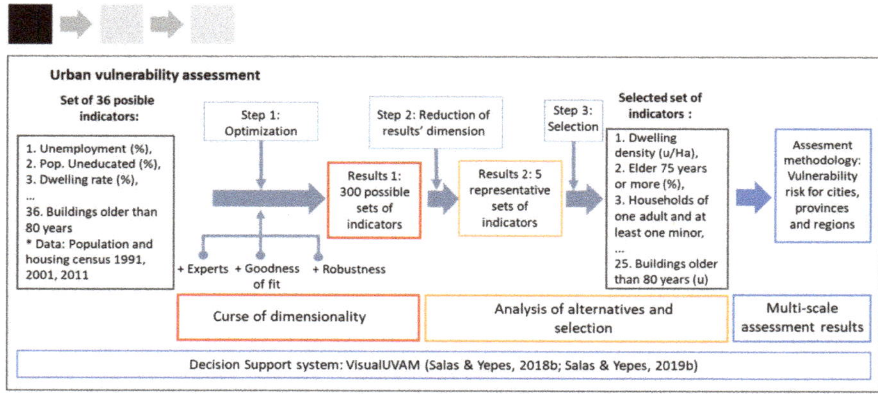

Figure 2. Step 1, urban vulnerability assessment: selection of indicators.

Since MOO usually yields large sets of solutions, giving rise to the so called "curse of dimensionality" problem [40], the assessment process also implements a cluster-analysis-based methodology that synthesizes the initial space of 300 solutions into a smaller, manageable one of 5 representative solutions (Figure 2) [39]. This enables the decision-maker to focus the analysis on the most promising alternatives and to select the most suitable, which affords a multi-scale evaluation of the risk of urban vulnerability of entities at city, province, region, and country scales [39]. Once the set of indicators has been selected, the method yields, for each of the cities, provinces, and regions being assessed, both the state of vulnerability (SV) at a given time and the risk of increasing vulnerability in the future.

2.2. Step 2: Resilient Planning I—Action Planning

The Urban Planning Support System (UPSS) [4] is a piece of software, programmed in Matlab, affording both the action planning and the implementation planning demanded by resilient planning. This software, however, still suffers from the lack of multi-objective capacity that this paper attempted to overcome. As to the action planning, UPSS includes planning and D-ROSE modules for the generation of planning alternatives and for evaluating the alternatives' risks and opportunities, which enables an informed selection of the most adequate planning alternative.

2.2.1. Planning Module: Generation of Planning Alternatives

Based on the infrastructure inventory (Figure 3), the planning module sought that combination of possible maintenance and construction actions [41] that would maximize the performance of the investment strategy according to three objectives, namely the mitigation of urban vulnerability, road condition improvement, and economic cost.

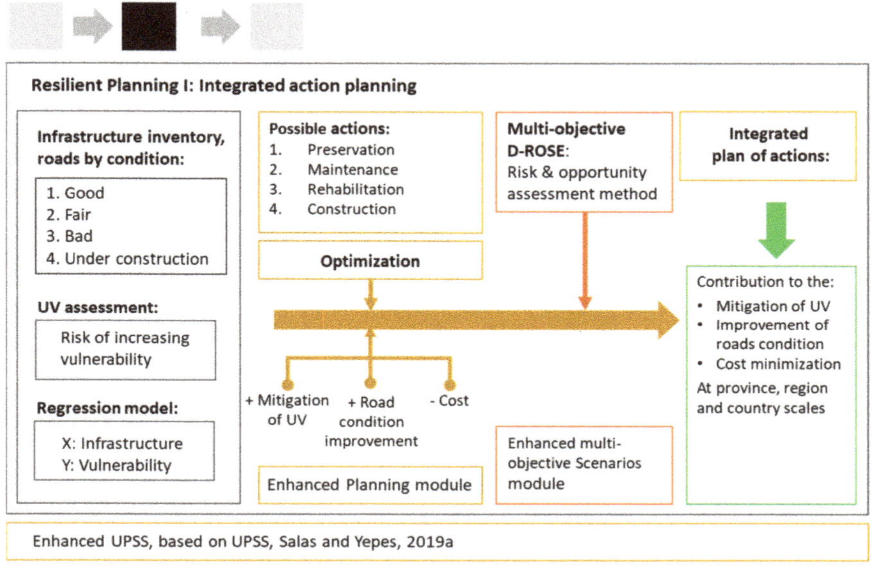

Figure 3. Step 2: planning of actions and action risk analysis.

Objective 1 was the mitigation of urban vulnerability. To evaluate the contribution of infrastructure to urban vulnerability mitigation, we first built a regression model based on the results of Step 1 and the road network's condition as described by the infrastructure inventory [4], which estimated the evolution of the risk of urban vulnerability in terms of the evolution of the road network's condition (Figure 3). This allowed the formulation of the urban vulnerability mitigation objective as follows:

$$\text{UVMNet} = \sum i,j \, \text{UVM}(\text{RCV}(\text{Plan}_j, \text{Inv}_j), \text{Mod}_i) \qquad (1)$$

where UVMNet is the urban vulnerability mitigation impact of the road network, i is each of the infrastructure system's hierarchical scales, j is each of the entities in the i scale, k is each of the actions planned for the j entity, and UVM is the evaluation of the RCV(Plan$_j$, Inv$_j$) road condition variable's evolution of the entity under the Mod$_i$ regression model.

Objective 2 was condition improvement. Building on prior work [4], we linked possible actions with condition improvement, which enabled us to estimate the condition improvement that a given set of actions would produce on the infrastructure inventory at the end of the analyzed period (Table 1). As we were planning for a 10 year period, in the case of actions with a shorter service life increase (Table 1, SLI) we assumed their repetition until the completion of the planning period [15]. For example, in the case of preservation, a treatment with a service life of 2.5 years, this action was considered to be applied four times over the 10 years of the analysis period (Table 1, column "Treatment/Period").

Table 1. Infrastructure condition variables and planning actions.

Infrastructure/Explanatory Variables:				Possible Action Variables:				
Description	Id	Unit	Type	Treatment Cost ($€/m^2$)	SLI (*)	PCI-CS (**)	Treatment/Period (***)	Period Cost
Net Infrastructures:								
Road condition variables								
Road condition: Good	1	m^2	Preservation	1.02	3	85	4	4
Road condition: Fair	2	m^2	Maintenance	23.24	10	60	1	23
Road condition: Poor	3	m^2	Rehabilitation	66.74	25	25	1	67
Road condition: Total	4	m^2	Construction	496	25	95	1	496

(*) Service life increase, based on Torres-Machí et al. (2017); (**) Pavement condition index condition score, based on Matin et al. (2017) and France-Mensah and O'Brien (2019); (***) Number of treatments required for a 10 year period.

Finally, we formulated the road network's condition improvement objective as the sum of the pavement condition index condition score (PCI-CS) improvements of all the entities of the road network being analyzed (Matin et al., 2017 [12]):

$$\text{RCI}_{Net} = \left(\sum\nolimits_j \Delta RC \, (\text{Plan}_j, \text{Inv}_j) \times CS \right) / \sum\nolimits_j R \, (\text{Plan}_j, \text{Inv}_j) \tag{2}$$

where RCINet is the road network's condition improvement of the j entities of the network, ΔRC is the transference function that transforms, based on Table 1, the actions of the Planj carried out over its Inv_j inventory into the evolution of the road condition variables, PCI-CS is the condition score attached to the road condition variables (Table 2), and R(Plan_j, Inv_j) is the quantity of roads in all conditions after carrying out the infrastructure plan.

Table 2. Actions included in the selected planning alternative for the region of Comunidad Valenciana.

	Initial Road Network Inventory		Actions Planned		Final Road Network Inventory (*)	
	Condition	Quantity (*)	Type	Quantity (*)	Variation	Total
Region: Comunidad Valenciana	Good	101.43	Preservation	90.34	14.34	115.78
	Fair	14.70	Maintenance	12.77	−3.61	11.09
	Poor	2.58	Rehabilitation	2.46	10.56	13.14
			Construction	10.21		
Province 1: Alicante	Good	40.26	Preservation	36.93	3.49	43.75
	Fair	2.83	Maintenance	2.68	0.50	3.33
	Poor	0.32	Rehabilitation	0.30	3.18	3.50
			Construction	3.83		0.00
Province 2: Castellón	Good	20.97	Preservation	16.05	−2.85	18.12
	Fair	0.29	Maintenance	0.25	4.63	4.92
	Poor	0.09	Rehabilitation	0.07	4.89	4.98
			Construction	1.75		0.00
Province 3: Valencia	Good	40.20	Preservation	37.35	13.71	53.90
	Fair	11.58	Maintenance	9.84	−8.73	2.84
	Poor	2.17	Rehabilitation	2.09	2.49	4.67
			Construction	4.62		0.00

(*) Surface in km^2.

Objective 3 was economic cost. As to the economic cost objective, the cost of each road network planning alternative was formulated as the product of the actions included and their unitary costs:

$$EC_{Net} = \sum I_{i,j,k}\, Action_{(i,j,k)} \times ICost_{(i,j,k)} \times IcostAsymm_{(i,j)} \tag{3}$$

where ECNet is the plan's cost, and $Action_{(I,j,k)}$ and $ICost_{(I,j,k)}$ are, respectively, the quantification of the actions included in the plan and the unitary costs of each of the k planned actions. IcostAsymm is a normalized asymmetry index that reflects different investment costs by entities of a given context, e.g., counties of a given province, provinces of a given region, or regions of a given country [4].

Objective 4 was the performance of the most vulnerable entities group of interest. Finally, in order to incorporate equity into the planning process and to provide proper visibility to the most vulnerable [6,42], we introduced as an additional objective the ratio between the most vulnerable group's performance [4] and the overall performance in the "Condition improvement" objective:

$$RCI_{Vul} = RCI_{Net}/RCI_{Hv} \tag{4}$$

where RCIVul is the road condition improvement ratio of the most vulnerable entities, while RCINet and RCIHv are, respectively, the net and the highly vulnerable entities group's condition improvement scores.

2.2.2. Scenario Module: Evaluation of Risk and Opportunities

The planning process implemented D-ROSE (Figure 3), an uncertainty analysis method capable of identifying a set of relevant scenarios and evaluating the risks and opportunities that these scenarios entail for each of the possible planning alternatives [4]. This method, however, lacks the multi-objective capacity required for analyzing planning alternatives against multiple risks [4], as was the case here. This multi-objective capacity implies that, for a proper selection of the most adequate planning alternative, the decision-maker should be enabled to simultaneously visualize the risks and opportunities borne by the set of relevant scenarios from all points of view, i.e., regarding all objectives. To address this, interactive visual analytics use different data visualization techniques, offering multiple, linked views of relevant information. Therefore, we implemented in the planning tool the capacity to simultaneously display risks and opportunities for all the objectives and planning alternatives to understand the trade-offs between the different risks, opportunities, and possible decisions [43].

2.3. Step 3: Resilient Planning II—Implementation Planning

As a final step, the process of resilient planning required the design of an implementation mechanism [44] that affords a proper balance between hierarchical control and adaptive capacity at the local scale across the road network's decentralization structure (Figure 4).

Ms-ReRO [4] is an uncertainty analysis method specifically designed for this purpose, based on hierarchical probabilistic relational modeling (HPRM) [45] and MOO, which affords an assessment of the global risks and opportunities at the central government (top) scale triggered by a plan's implementation at the municipal (local) scale. In this methodology, integrated planning implementation is represented as a hierarchical system of systems that are connected by relational contracts, and risks and opportunities are derived as the bottom-up cascading impacts produced by the actions performed at a local scale. In decentralized infrastructure systems, contracts between parties are key elements in the implementation scheme [23,25]. Contractual arrangements prescribing very precise actions work well at a tactical scale but not at a strategic, long-term scale [23]. Instead, Ms-ReRO includes a more flexible contractual framework based on the concept of relational contract [25], which defines "right" and "duty" [25,26] rules across contracting parties. By means of this, the proposed framework allows top entities to transfer the "right" to vary the initial plan to the entity below, which, in exchange, is obliged to achieve a given outcome, i.e., to perform a "duty".

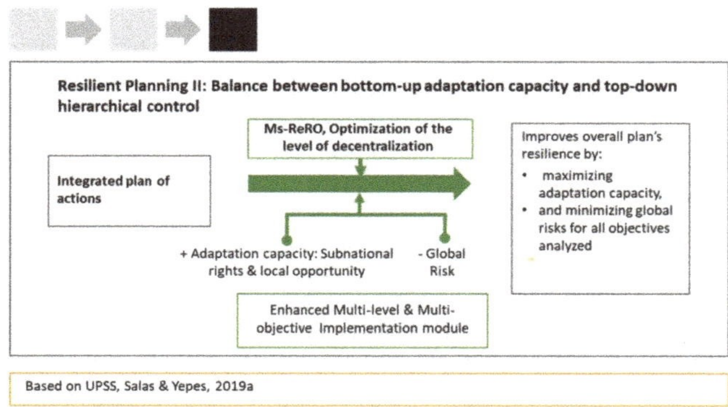

Figure 4. Step 3: implementation planning and risk analysis.

In HPRM, actions are performed at the bottom scale and their consequences are then bottom-up propagated, therefore impacting at the top scale; the aim is to regulate these actions and impacts by means of the relational contract's rules. In consequence, we first allowed variation the quantities specified in the baseline plan within certain limits (rights), which were modeled as the lower and the higher bounds for each action. We then imposed, as a restriction of the choice between the right's bounds for each action, that their joint effect had to fall within a given performance range, which we called "duty" and which represented the maximum possible deviation from the performance expected to be accomplished by each entity in the integrated plan. In consequence, in terms of the generation of simulations at the bottom scale, which will subsequently be bottom-up propagated through the relational system, we only admitted those meeting the conditions specified by the relational contracts at the bottom scale, i.e., we eliminated failing simulations from the set of possible realizations generated by the Monte Carlo simulation method. Finally, the choosing of actions between the rights' bounds by local entities is affected by their behavioral preferences, which we formalized by means of triangular probability distribution functions (PDF) functions. For this purpose, we employed a stochastic approach since it allowed integration into a single object of the rights' upper and lower bounds and the local preferences as the lower, central, and upper points of a triangular PDF function [4].

This relational framework enabled, through MOO, the balancing of hierarchical control and adaptive capacity by simultaneously minimizing the risk of failure at the top scale (global) while maximizing the opportunity to achieve better performance at the local scale.

Decentralization Objective 1 was global risk minimization:

$$\text{RGlobal(AP,IP)} = \text{P(FAP,IP,T)} \times \text{I(FAP,IP,T)} \tag{5}$$

where RGlobal (AP,IP) is the risk, for the implementation plan IP of the action plan AP, of achieving a result worse than that of the failure condition F; P(FAP,IP,T) is the probability of achieving anF failure condition at the system's T top scale; and I(FAP,IP,T) is the impact of this failure. In Ms-ReRO, the F failure condition consists of a performance worse than the previously set up pessimistic threshold.

The probability of failure, in turn, was defined as

$$\text{P}(F_{AP,IP,T}) = \text{N}(\text{Sims}^{F}{}_{AP,IP,T})/\text{N}(\text{Sims}_{AP,IP,T}) \tag{6}$$

where $\text{N}(\text{Sims}^{F}{}_{AP,IP,T})$ is the number of simulations achieving failure, while $\text{N}(\text{Sims}_{AP,IP,T})$ is the total number of simulations performed following the method described by Salas and Yepes (2019a).

Finally, failure's impact was formulated as

$$\text{I}(F_{AP,IP}) = \text{mean}(f(\text{Sims}^{F}{}_{AP,IP})) - f(\text{BL}_{AP,IP}) \tag{7}$$

where mean(f(Sims$^F_{AP,IP}$)) is the mean of the performances achieved by failing simulations, and f(BL$_{AP,IP}$) is the value achieved in the realization of the baseline plan of actions.

Decentralization Objective 2 was local opportunity maximization. Conversely to the risk, we modeled opportunity based on the simulations improving a given level of performance. Therefore, opportunity at local scale was formulated as

$$O_{Local}(AP,IP) = P(W_{AP,IP,B}) \times I(W_{AP,IP,B}) \tag{8}$$

where P($_{WAP,IP,B}$) is the probability of achieving a "W" windfall condition at the system's "B" bottom scale, I($_{WAP,IP,B}$) is the impact of the windfall condition, and B is each of the entities at bottom scale.

Decentralization Objectives 3 and 4 were related to the relational framework's flexibility maximization. As to the improvement of entities' capacity of varying the plan, we implemented this by maximizing the sum of the means of the rights bestowed by scale across the whole relational system. This flexibility was also improved via maximization of the range within which each entity was allowed to deviate from their duties, i.e., from their intended result.

In sum, decentralization Objectives 2–4 represented the maximization of the system's adaptive capacity at local scale, while decentralization Objective 1 accounted for the minimization of the system's risk of failing in the attainment of the required global performance. In seeking to achieve these goals, the MOO problem operated over the "rights" and "duties", which therefore became the MOO's decision variables, and were formulated as the percentage in which entities are allowed to deviate from the baseline plan, in the case of rights, or from the expected performance in the case of duties [4].

3. Case Study: Resilient Road Network Planning in Provinces of Spain

3.1. Information Collection Process

3.1.1. Information Required for Urban Vulnerability Assessment

Following prior work, the compilation of the quantitative information was downloaded from the website of the National Institute of Statistics, comprising 36 indicators for each of the 403 cities (264 of which are from the province of Valencia), 52 provinces (including Ceuta and Melilla), and 19 regions (including the autonomous cities of Ceuta and Melilla as regions) that composed the elaborated database [39]. This information was collected for the years 1991, 2001, and 2011, allowing analysis of the evolution of urban vulnerability in the periods 1991–2001 and 2001–2011.

Along with the quantitative information, we also gathered the qualitative information regarding experts' preferences for the indicators best representing urban vulnerability, required by the assessment process [39]. Based on the analytic hierarchy process (AHP) multi-criteria technique [46], we asked the experts to pairwise compare the 36 indicators of the quantitative database, which were structured in three levels so that only in one case was the number of indicators to be compared greater than five. Basically, this structuring of indicators was a transposition of the conceptual framework adopted by the Spanish OUV, to which some indicators were added.

Further, to avoid the problem of inconsistent judgment elicitation [47], we developed a software application, programmed in Matlab, that provided experts with real-time feedback on their judgments' consistency, enabling them to interactively revise their judgements until they became acceptable [39]. As an outcome, we obtained the experts' relative preferences for indicators as weights, which were incorporated into the experts' preferences objective in the optimization process (Figure 2).

3.1.2. Information Required for Urban Infrastructure Planning

As to the gathering of quantitative information on road conditions, we resorted to the data available from the Local Infrastructure and Equipment Survey (EIEL), which included a wide range of infrastructures present in municipalities of 50,000 habitants or fewer in all Spanish regions, with the exception, due to their specific organizational regimes, of the Basque Country and Navarra [4].

Since the planning process required a regression model correlating the evolution of urban vulnerability and that of the condition of urban infrastructure (Section 2.2.1), we retrieved from the EIEL the data corresponding to those employed for the assessment of UV in Step 1 (Section 2.1), i.e., between the years 2000 and 2010, and structured it based on the city, province, and region (autonomous communities) scales. However, since in Spain, road network planning is under the jurisdiction of the state, regions, and provinces, but not of cities, we excluded the latter scale from our database and settled on provinces as the bottom scale. We then sought to achieve objectives at the national (top) scale by building planning alternatives from a provincial scale, which is an approach more akin to actual road network decision-making than doing it from a municipal perspective.

3.2. Running of the Process

3.2.1. Step 1: Urban Vulnerability Assessment

The assessment of urban vulnerability was performed via VisualUVAM, a software that covered all the steps of the urban vulnerability assessment process described in the methodology (Section 2.1). Following the guidance afforded by the software, we first generated a set of 300 pareto-optimal combinations of indicators which, by means of the visual analytics and cluster analysis techniques implemented in the tool, were synthesized into a more manageable set of nine possible combinations. We then undertook a process of analysis that culminated in the selection of the combination of indicators deemed most appropriate [39].

3.2.2. Step 2: Action Planning

Based on the results of the urban vulnerability assessment carried out in the previous step, and on the gathered information of the road network's condition, the UPSS planning module (Section 2.2.1) provided an initial set of 300 pareto-optimal planning alternatives (Figure 3). The planning alternatives were then filtered by means of the implemented cluster analysis method [39], reducing the initial set of 300 possible solutions to a set of 11 representative, relevant alternatives, which were further analyzed by the scenario module (Section 2.2.2). By means of D-ROSE, we generated random scenarios and evaluated the risks and opportunities that these scenarios bore for each relevant planning alternative. In this case, we employed the scenario module to swap the range of possible decentralization combinations and therefore represent the impacts that different levels of decentralization had on each possible plan. Subsequently, trade-offs between risks, opportunities, and planning alternatives were evaluated and, after the analysis of these results, the most adequate plan was chosen for implementation.

3.2.3. Step 3: Implementation Planning

Based on the planning alternative selected in Step 2, the UPSS implementation module (Section 2.3) simultaneously sought, through the optimization of the system's level of decentralization, the minimization of global risks and the maximization of local adaptation (Figure 4). This afforded a set of optimal configurations of the relational contract's rights and opportunities, from which it was possible to draw out the trade-offs between global risk and local adaptive capacity for each objective. These trade-offs were then analyzed from a multi-objective perspective, which enabled us to balance different risks, opportunities, and possible decisions and accordingly choose the most adequate implementation plan.

4. Results

4.1. Step 1: Urban Vulnerability Assessment

Figure 5 shows the results of the state of UV, the evolution of UV state, and the risk of increasing UV for provinces of Spain, which revealed how urban vulnerability is, in general, more present in coastal and highly populated provinces [39]. Based on this information, the 30% most vulnerable entities were identified and grouped for the incorporation of their specific interest in the search, in Step 2, for optimal infrastructure plans.

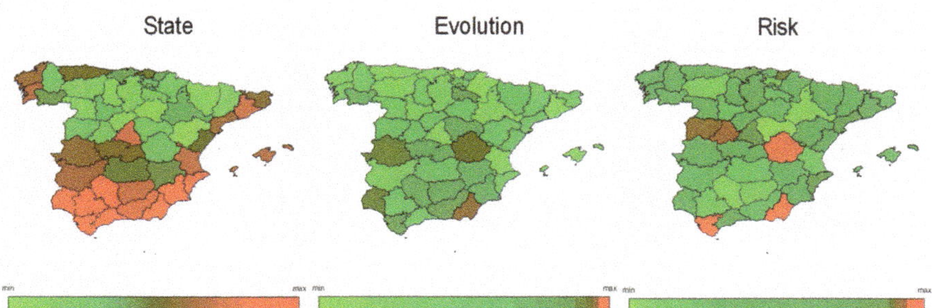

Figure 5. Results by province of the urban vulnerability assessment process: UV State in 2011, evolution of UV state between 2001 and 2011, and risk of increasing UV from 2011 onwards.

4.2. Step 2: Planning of Actions

The trade-offs between planning objectives (Figure 6) showed that the overall urban vulnerability mitigation (UVM (Net)) and road network condition improvement (RCI (Net)) objectives were aligned, which was consistent with the idea of the contribution of net infrastructures to the mitigation of urban vulnerability [4,19]. These objectives were also aligned with the maximization of the RCI(Vuln), which expressed the ratio of road condition improvement of the most vulnerable entities within the total, showing how, in some cases, it was possible to reconcile particular interests with general interests. As expected, all these objectives were in conflict with the network economic cost (EC (Net)) minimization which, since the results were pareto-optimal, could be used as an ex-post budgetary restriction by setting up the maximum economic cost allowed in the implemented selection controls.

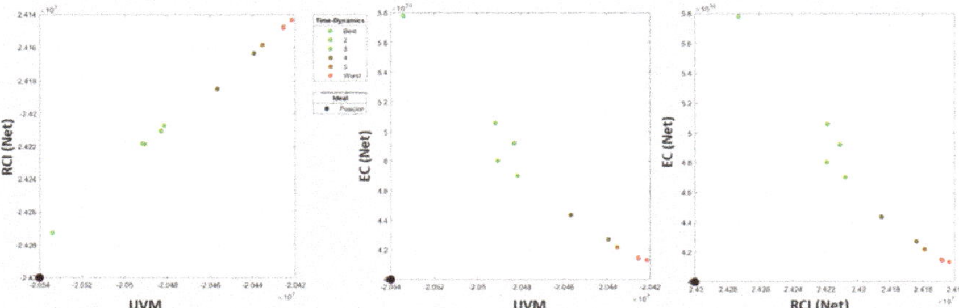

Figure 6. Trade-offs between objectives.

Figure 7 presents, for each objective, the results of the Monte Carlo simulation carried out over each of the planning alternatives, and also reflects the direct correlation between closeness to ideal and worst risk and opportunity performance. Relevant solutions performing well at the UVM objective did so at the RCI, while they performed badly at the EC objective. In effect, as we moved toward the right (ideal) in planning alternatives for the UVM and RCI objectives, the simulations' results passed from above the optimistic threshold to below the pessimistic threshold, which indicated movement from opportunity to risk. Conversely, at the EC objective, which was opposite to UVM and RCI, better (cheaper) solutions were placed at the left and worse (expensive) at the right, and, consequently, simulations improving the expected performance are on the right, while those worsening it are on the left side.

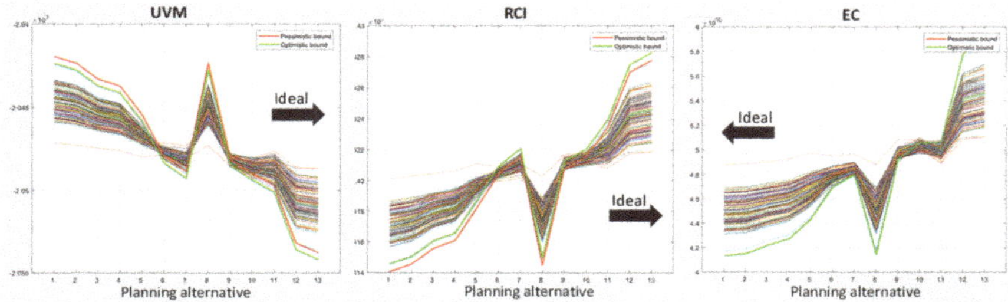

Figure 7. Results of the Monte Carlo simulation for each planning alternative. Simulation results exceeding pessimistic or optimistic bounds bear, respectively, risks or opportunities. The order of the planning alternatives in the horizontal axis indicates lower to higher UVM performance.

As to the multi-objective analysis of the results, Figure 8 portrays the risks and opportunities of each planning alternatives for the set of 100 scenarios generated. The analysis of these results showed that planning alternatives 7, 8, and 9 were the most relevant for our decision, since in all objectives they represented the turning point from opportunity to risk or vice versa. Finally, we selected alternative 9 due to our bias toward solutions improving especially the condition index of the most vulnerable entities, which in Figure 8 were to the right.

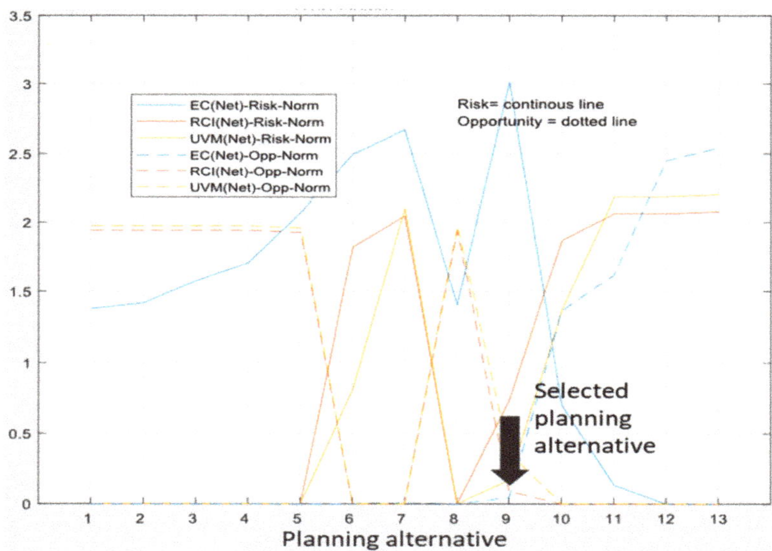

Figure 8. Normalized risks (Risk-Norm) and opportunities (Opp-Norm) of planning alternatives by objectives in terms of distance, in standard deviations, from the lowest risk/opportunity.

Each of the generated planning alternatives is a baseline plan specifying the quantity of each possible action that should be carried out at the bottom (provincial) scale to bring about the planned performance at the top (country) scale. Table 2 shows the specific results of the selected plan for the region of Comunidad Valenciana and its provinces.

4.3. Step 3 Implementation Planning

The implementation planning module (Section 3.2.3) afforded a set of pareto-optimal solutions for the configuration of the rights and duties which made up the system of relational contracts. As shown in Figure 9, economic risk reduction at the top scale was inversely correlated with opportunity increase at the local scale, which, on the other hand, had a clear inverse correlation with increase in the relaxing of duties. Finally, Figure 9 also shows a strong inverse correlation between increasing rights at subnational scales and reducing economic risks at the national scale, which, on the other hand, was directly correlated with increasing flexibility in duties. Altogether, the set of solutions showed that increasing local opportunity was in opposition to global risk reduction, and that increasing rights led to increased global economic risks but not to increased economic opportunity, thus producing an asymmetry in the share of risks and opportunities. By increasing rights, we increase global risk, but we do not necessarily increase local opportunity. As to the relaxing of duties at the bottom scale, its increase was slightly associated with reductions of both risks and opportunities. However, when this increase came together with that of the rights, it played against risk reduction at the top (national) scale (Figure 9).

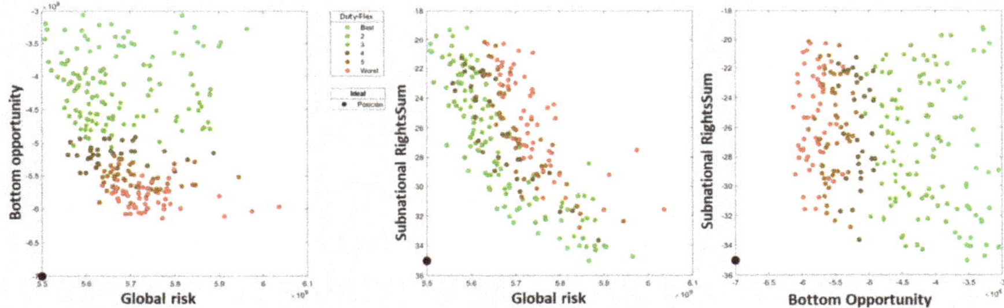

Figure 9. Results of the generation of decentralization solutions, trade-offs between global risk, opportunity at local scale, summation of sub-national rights, and flexibility in duties for the economic cost objective.

For a better understanding of how the decentralization model works, we resorted to global sensitivity analysis to evaluate the decision model in terms of output uncertainty and factor importance in order to gain a better understanding of how the model parameters affected the final outputs [48]. Regression-based and variance-based methods are two of the most commonly used approaches for global sensitivity analysis, and they perform almost equally well for quantifying output variance and contribution to variance of the input parameters, especially in the case of relatively small input uncertainties [49]. By incorporating the Matlab code [50] developed by Groen et al., [49] into our own Matlab software, we performed a global sensitivity analysis based on the squared standardized regression coefficients method. The results of the global sensitivity analysis showed that rights and duties at the province (bottom) scale were the driving factors in all objectives, but they were unequally distributed along objectives (Table 3). While duties at the province scale was the factor with the highest impact on global risk in the economic cost and mitigation of UV objectives, it had little impact on the road condition improvement global objective. Conversely, rights at the province (bottom) scale was the driving factor for opportunity at the bottom scale for all objectives, but also posed global risks for the road condition improvement objective.

Table 3. Results of the global sensitivity analysis of the decentralization model parameters.

Parameter	PDF (*)	Sensitivity Index								
		Economic Cost			Road Condition Improv.			Mitigation of UV		
		1 (**)	2 (**)	3 (**)	1 (**)	2 (**)	3 (**)	1 (**)	2 (**)	3 (**)
Rights:										
Regions (***)	5, 5, 21	8.50%	0.58%	13.67%	5.71%	0.12%	13.67%	10.88%	2.03%	13.67%
Provinces	4, 4, 14	1.34%	87.61%	86.33%	74.64%	99.34%	86.33%	1.73%	96.82%	86.33%
Duties:										
National	1, 1, 3	0.51%	0.01%	0.00%	0.11%	0.00%	0.00%	0.12%	0.00%	0.00%
Region	1, 1, 3	1.80%	0.01%	0.00%	0.47%	0.01%	0.00%	0.01%	0.00%	0.00%
Provinces	1, 2, 6	87.85%	11.79%	0.00%	19.07%	0.53%	0.00%	87.26%	1.15%	0.00%

(*) All parameters' uncertainties defined by triangular PDF (Min, Peak, Max) points; (**) 1: global risk; 2: bottom opportunity; 3: sum of subnational rights; (***) Summation of the sensitivy index of all spanish regions.

As to the selection of the proper decentralization configuration of the relational model, we resorted to cluster analysis to synthesis the initial set of solutions into another more manageable set, which we analyzed from the perspectives of all the objectives involved in the planning process (Figure 10).

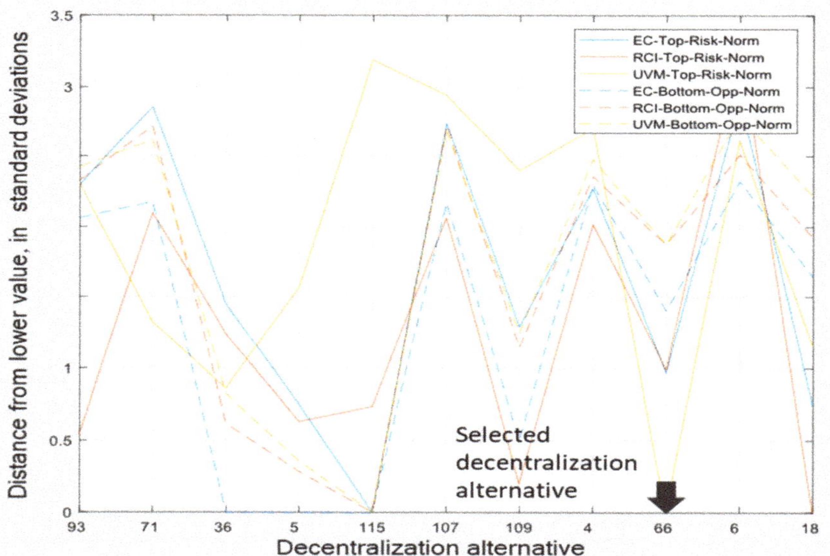

Figure 10. Implementation risks and opportunities borne by the representative decentralization solutions.

Decentralization alternatives 115, 66, and 18 achieved the lowest risks in terms of the economic cost, urban vulnerability mitigation, and road condition improvement, respectively, showing that there was not a unique best solution for all objectives and that some qualitative analysis is required to perform such selections. Alternative 115 also had the fewest opportunities of all, while alternatives 66 and 18 were the only alternatives with greater opportunity than risks in all objectives. Of these, alternative 18 presented a slightly better balance between risk and opportunity.

On the other hand, the performed global uncertainty analysis allowed some management implications to be drawn out [51]. On one hand, the focus should be put on the rights and duties bestowed at the bottom rather than at the top scale. On the other hand, there is no formula for balancing rights and duties that can be indiscriminately applied to all objectives, since relaxing duties, for example, would be risky at the global scale in terms of the economic cost and the mitigation of urban vulnerability points of view, but not much from a road condition improvement outlook, which would be highly affected, instead, by increased rights. Therefore, it is necessary to balance not only risk and opportunity, but also the objectives, selecting planning alternatives with better balance in those objectives prioritized by the decision-maker. In consequence, despite alternative 18 having an overall risk and opportunity balance slightly better than alternative 66, we selected the latter due to its lowest risk in the urban vulnerability mitigation objective, which we prioritized over the road network condition improvement. We therefore selected alternative 66, which enabled us to set up the relational contracts required for the implementation of the planning alternative (Table 4).

Table 4. Guidelines for decentralization alternative 66 for the issuing of relational contracts between central government and the region of Comunidad Valenciana, and between this region and its provinces.

	Rights:	**Region**	**Provinces**	**Duties:**	**Region**	**Provinces**
	Range	14%	5%	Range	3%	5%
	Actions	Lb (*)	Ub (*)	Objectives (**)	Lb	Ub
Country and Region: Comunidad Valenciana	Preservation	77.69	101.43	UVM(Net) (−)	-8.77×10^3	-9.31×10^3
	Maintenance	10.98	14.49	RCI(Net) (+)	9.73×10^3	1.03×10^4
	Rehabilitation	2.12	2.56	EC(Net) (−)	5.59×10^9	5.94×10^9
	Construction	8.78	8.25			
Region and Province 1: Alicante	Preservation	35.09	40.26	UVM(Net) (−)	-2.49×10^3	-2.75×10^3
	Maintenance	2.55	2.63	RCI(Net) (+)	3.10×10^3	3.42×10^3
	Rehabilitation	0.29	0.32	EC(Net) (−)	1.98×10^9	2.19×10^9
	Construction	3.64	1.41			
Region and Province 2: Castellón	Preservation	15.25	20.97	UVM(Net) (−)	3.13	3.46
	Maintenance	0.24	0.29	RCI(Net) (+)	3.38×10^2	3.74×10^2
	Rehabilitation	0.06	0.07	EC(Net) (−)	8.79×10^8	9.71×10^8
	Construction	1.67	2.07			
Region and Province 3: Valencia	Preservation	35.48	40.20	UVM(Net) (−)	-6.10×10^3	-6.74×10^3
	Maintenance	9.35	11.58	RCI(Net) (+)	6.09×10^3	6.73×10^3
	Rehabilitation	1.98	2.17	EC(Net) (−)	2.62×10^9	2.89×10^9
	Construction	4.39	4.77			

(*) Surface in km^2; (**) Negative and positive signs respectively indicate minimization and maximization.

5. Discussion

5.1. Action Planning

The MOO approach yielded trade-offs between the objectives involved in the planning process, i.e., maximization of urban vulnerability mitigation and road network condition improvement and minimization of economic cost. Additionally, it provided valuable information on the specific effects of the planning alternatives over the most vulnerable entities which, together with the analysis of the results of the risk and opportunity assessment, enabled us to select the most suitable planning alternative for its further implementation.

From a strategic point of view, each planning alternative represented a baseline plan containing the basic determinations required for the road network's maintenance and construction integrated planning, comprising regions and provinces of Spain. The solutions provided, based on the road network's current condition, the quantity of each action that should be performed for each entity at the bottom scale to attain a given performance at global scale, enabling their further development at the tactical level via relational contracts.

5.2. Implementation Planning

The results yielded by Step 3 showed how different decentralization solutions led to different risks and opportunities in the implementation of the selected planning alternative, and how the proposed methodology can be employed to find the most convenient decentralization solution. By means of this, the method afforded proper balance between risks at the top (national) level, opportunities at the bottom (provincial) scale, and rights bestowed through relational contracts. These rights represent the capacity to select actions other than those of the baseline plan, and, in consequence, are a way in which local entities can adapt the integrated planning to their circumstances and specific needs. Opportunity at the local scale, on the other hand, represents the potential positive effect that rights might have on local entities' performance, which is strongly correlated with subnational rights. Together, rights and local opportunity account for the demanded planning system's local adaptive capacity [7,9]. This plan's flexibility, as shown in Section 4.3, was in conflict with the reduction of risks at top scale, which reinforced the idea that, in infrastructure hierarchical systems, resilience at local scale does not necessarily lead to resilience at the global scale [9,23], and that some balance between global objectives and local adaptation is required [11,28,29]. Ms-ReRO addresses this issue by means of multi-objective optimization, which in our case afforded a set of optimal decentralization solutions from which it was possible to select an implementation plan achieving the demanded balance between global risk and local adaptive capacity.

Based on the trade-off between top-risk minimization and local adaptive capacity maximization, it was possible to select the proper action implementation plan, which included the guidelines required for issuing a system of relational contracts (Table 4). Contracts play a key role regarding the level of resilience level in a fragmented or decentralized infrastructure system [23], and should afford the means for dealing with the uncertainty always present in any infrastructure system's integrated plan's implementation and operation [25]. Relational contracts are a kind of contract specifically designed to alleviate relational problems between hierarchically dependent entities of decentralized systems [25], allowing the incorporation of both rights and duties, and they therefore provide the best framework for materializing the method's results. This approach, also allows multiple objectives to be taken into account by specifying in the relational contracts multiple duties to be carried out, which in our case were the expected performance of each entity in urban vulnerability mitigation, road network condition improvement, and economic cost.

As to the relationship between planning alternatives and implementation risks and opportunities, the results showed that, for each objective, planning solutions close to the ideal were prone to risk. In our case, this was due to the fact that in planning alternatives already close to the maximum or minimum possible quantity of a given action, transferring rights beyond this limit will be ineffective, thus producing an asymmetry in the PDF describing each entity's possible actions. For example, in a planning alternative preserving 97% of the roads in a good state, i.e., close to the maximum possible preservation quantity, bestowing rights of 15% means that the theoretical upper bound will exceed the real one, rendering ineffective 80% of the theoretical potential increase of actions. On the other hand, for the same example, its lower bound will fall from 95% down to 82.45%, thus producing an asymmetry that will be reflected in the behavior of the simulations generated (Figure 11), and therefore in the risks and opportunities attached to this decentralization configuration. This phenomenon has important implications for the issue of relational contracts, since their actions' upper and lower bounds will not necessarily match the range expressed by the rights embodied in the contract. In consequence, it is necessary to explicitly define, for every relational contract, the rights as the action's lower and upper bounds instead of only as a range (Table 4).

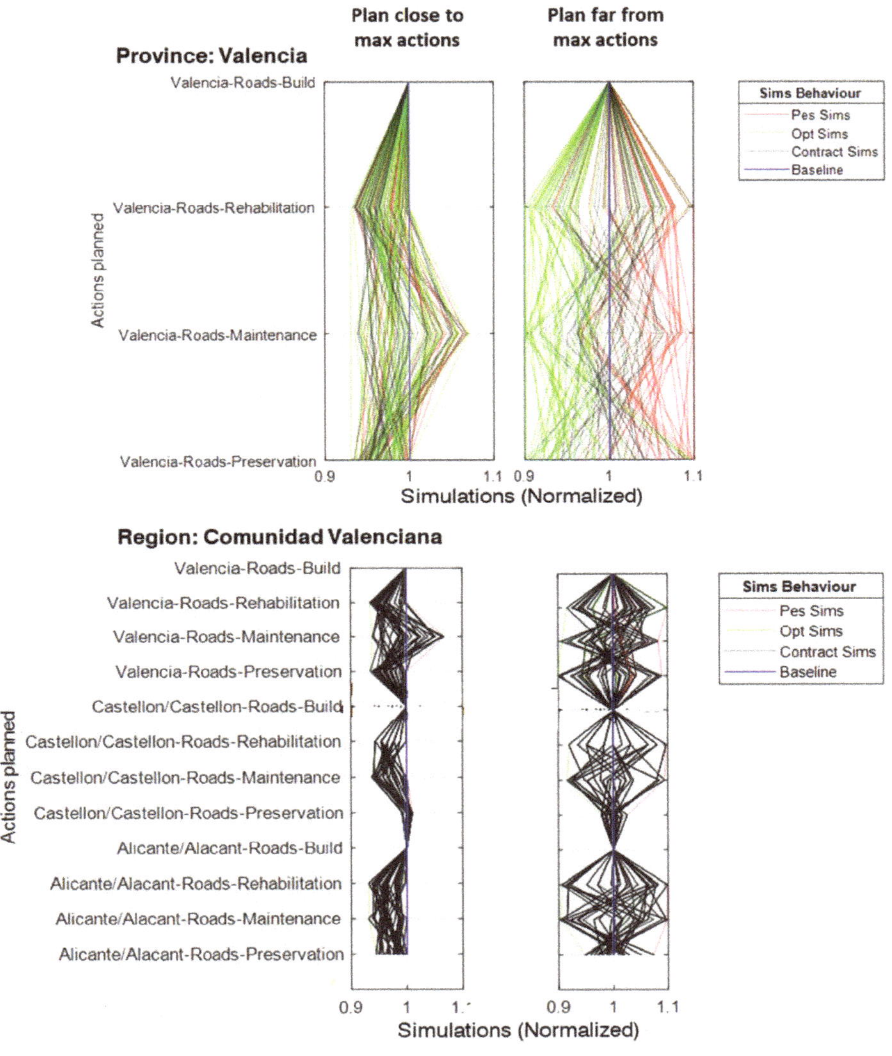

Figure 11. Example of simulations generated at provincial (bottom) and regional scales for different planning alternatives. Contractual simulations at the bottom scale were propagated to top scales. The decentralization configuration in both cases was the same.

As to how the plan's adaptive capacity, which is an abstract concept, can be materialized by local entities, Figure 11 plots, labeled as "contract sims", examples of possible variations over the baseline plan that could be carried out by local entities (provinces) without violating the relational contract, i.e., fulfilling the assigned "duties". These variations at the local scale will then combine with those of other provinces of the same region to determine the joint effect on the region's duties and so on, thus enabling the evaluation of the cascading impact of variations at a local scale over the objectives at the global scale. However, the presence of multiple duties–objectives embodied in the relational contract rules requires an additional control mechanism to simultaneously achieve them. In effect, this multi-objective dimension in the system of relational contracts means that variations at a local scale being acceptable from an economic point of view would not be acceptable from a road condition improvement outlook (planning alternative 114, Figure 10), and would therefore be rejected. Since there is not any one best solution for all objectives, a second-order analysis balancing not only risks and opportunities, but also the objectives themselves is necessary. In our case, we prioritized the minimization of risk for urban vulnerability mitigation and the balance between risk and opportunity, and therefore selected alternative 66; had we preferred risk minimization for road condition improvement, the best alternative would have been alternative 18.

The system of relational contracts helps in dealing with some kinds of uncertainty arising from the implementation of integrated plans across a territory, such as uncertainties related to the financial capacity of entities along the planning period. The proposed contractual framework enables local entities to adapt the baseline plan to their specific financial contexts by, for example, moving quantities from actions demanding heavy initial outlays, such as rehabilitation, to those requiring payments distributed over time, such as preservation. In contrast, it would be possible at some point that local entities have enough financial resources to undertake more demanding actions and are therefore willing to move quantities from preservation to rehabilitation, which they can do without trespassing upon the economic duty for the whole period. Another source of uncertainty would be the entities' capacity to bring about the baseline plan, which may contain actions that are difficult for them to perform due to, for example, human resource limitations. In this case, entities can ask the upper scale to partially assume the implementation of the baseline plan or to adapt it by increasing those activities for which they have enough resources. Local entities may also have a better knowledge on which roads have strategic importance for them that, within the system of relational contracts, can be used to improve the baseline plan. In the hypothetical case of a province with roads in a good state that are not completely preserved but are more important than any of the roads in bad states that are planned to be rehabilitated, local entities can automatically move economic resources from rehabilitation to preservation according to their aim, provided they still achieve their duties.

Finally, in governmental contexts, there are always institutional disputes surrounding any integrated, long-term planning that can prevent its implementation. By changing the triangular PDF modeling the behavior of the actions affected by the dispute [52], the method allows assessment of the impact on the local and the global objectives resulting from this change, which could be of help in promoting agreement between parties.

6. Conclusions

Resilient planning demands not only resilient actions but also resilient implementation [53]. Despite the vast amount of research devoted to developing methods for the planning of resilient actions, there have been very few studies investigating plans' implementation [4], which, in the case of net infrastructure planning, requires a proper balance between global risk minimization and local adaptive capacity maximization [9,11,28,29]. This paper contributes to resilient planning by, on one hand, extending the initial capacities of UPSS [4] to the search for road network investment plans and decentralization alternatives that are optimal from the perspectives of the network's condition improvement [54], contribution to urban vulnerability mitigation, and minimization of the economic cost. By integrating social sustainability aspects as a relevant criterion for the decision-making process, the method facilitates the adoption of a resilient plan of action, contributing to more sustainable development. On the other hand, this paper provides planners with a novel way of materializing a plan's adaptive capacity at the local (bottom) scale and risk control at the global (top) scale. By means of the rights and duties included in the provided decentralization solution, it is possible to set up a system of relational contracts in which the integrated plan is transferred from national to provincial entities, where it is finally executed according to the relational contract specifications.

Along the process, the improved planning support system afforded a plan of action for the Spanish road network with the best balance between closeness to ideal and risks entailed from a multi-objective perspective (Section 4.2). Additionally, the planning process provided a decentralization solution for the best implementation of the plan of actions across the Spanish governmental structure, consisting of the country, regional, and provincial levels (Section 4.3). This decentralization solution was then used in a novel way to shape a system of relational contracts between hierarchically dependent entities in which local adaptive capacity was formalized as the right to vary, within certain limits, the plan of action being implemented, and global risk control was materialized by means of the duties that should be achieved in exchange of the rights conferred. Overall, the presented method integrates, for the first time, the planning of resilient actions with the planning of their resilient implementation from a multi-objective point of view, thus contributing to the field of resilient planning.

In the selection of the most adequate planning alternative, the multi-objective capacity allowed the identification of key planning alternatives from the risk and opportunity points of view and, based on the alternatives' impact on the most vulnerable entities, the selection of the most appropriate one, contributing to the incorporation of equity into the planning process [42]. The results showed, on the other hand, that there was a clear relationship between closeness of the planning alternatives to the ideal and increased risks and, vice versa, alternatives farther from the ideal point were prone to opportunity, i.e., they had more chances of improving their expected performance. Regarding the selection of the proper implementation plan, the method's multi-objective capacity revealed that there were no clear trade-offs between the objectives' global risks and local opportunities. Instead, it was necessary to separately evaluate decentralization alternatives and select the most adequate according to the balance between risks and opportunities and the decision-maker's preferences for objectives. This evaluation prevents alternatives being chosen that perform well in a less important objective and badly in those more relevant to the decision-maker, as it affords improved global risk control in which adaptive capacity at the local scale is bound to the simultaneous accomplishment of a given level of performance for each objective. This paper also presents a novel approach for materializing a plan's adaptive capacity into actions. The use of relational contracts allows the contractual formulation of adaptive capacity as the rights bestowed to local entities, which enables them to vary the baseline plan to adapt it to their local circumstances and needs, in exchange for carrying out the duties assigned.

Additionally, the paper provides valuable insights into the relationships between planning objectives, planning alternatives, and their implementation's global risks and local adaptive capacity. As to the planning objectives, the results showed that the mitigation of urban vulnerability and road condition were aligned objectives, which was consistent with the idea of the net infrastructure contributing to the mitigation of urban vulnerability [4,11]. These objectives were also aligned with improving the road condition of the most vulnerable entities in particular, showing that, in some cases, it is possible to reconcile particular with general interest. On the other hand, the risk assessment of the planning alternatives revealed that the closer the alternatives were to the ideal in each objective, the riskier they were. Additionally, the comparison between closeness to ideal and risks showed the existence of turning points in the change of the trend from risk to opportunity that were especially relevant for multi-objective decision-making in the case of conflicting objectives. Regarding the implementation planning, it was possible to find a solution with the best balance between global risk and local opportunity for each objective. However, there was no decentralization solution that performed best for all objectives, which made it necessary to prioritize between objectives and choose accordingly.

Despite the remarkable outcomes, there were still limitations to this study. On one hand, there is still a need for a more systematic approach in the joint analysis of risks of different nature, as is the case. Multi-criteria methods such as AHP [55], Delphi [56], or Bayesian networks [57] can be used to build, based on experts' or decision-makers' preferences, a composite implementation risk index that would be of help in the selection of infrastructure planning alternatives. On the other hand, the proposed system of relational contracts may produce legal difficulties requiring specific research to overcome. For example, in Spanish legislation, maintenance and construction activities have different nature and are allocated in separated budget chapters, which requires specific contracts. Breaking down rights and opportunities by provinces in the decentralization framework, and conducting specific research on the interactions between these factors, would also be of use for territorial decision-making. This capacity, which is lacking in the proposed software, could be addressed by programming and incorporating variance-based global sensitivity methods such as Global sensitivity and uncertainty analysis GSUA [49] into the UPSS planning tool code. Finally, this paper studied the relationship between local adaptive capacity and global risks when actions were implemented at a local scale, which in our case was the provincial scale. However, the implementation of actions at this scale is still fragmented, since in provinces there are infrastructures of national, regional, and provincial ownership which are separately operated. In consequence, the framework of relational contracts is directly applicable only to road networks belonging to the same type of ownership, i.e., the networks of the national, regional, or provincial roads. This suggests the need for additional research supporting the development of a system of relational contracts in which the transference of actions between infrastructures of different owners could be regulated in order to achieve duties at the local scale.

Author Contributions: This paper represents a result of teamwork. J.S. and V.Y. jointly designed the research. J.S. developed the methodology and the software, carried out the investigation and drafted the original manuscript, and V.Y. revised and improved the manuscript. All authors have read and agreed to the published version of the manuscript.

Funding: This research was funded by the Spanish Ministry of Economy and Competitiveness, along with FEDER, grant number Project: BIA2017-85098-R.

Acknowledgments: The authors acknowledge the financial support of the Spanish Ministry of Economy and Competitiveness, along with FEDER funding (Project: BIA2017-85098-R). The authors also want to thank Julio Gómez-Pomar, former Spanish Secretary of Infrastructures, for his valuable insights on the uncertainties surrounding contracts between Public Administrations.

Conflicts of Interest: The authors declare no conflict of interest.

References

1. Holling, C.S. From complex regions to complex worlds. *Ecol. Soc.* **2004**, *9*, 11. [CrossRef]
2. Sharifi, A.; Yamagata, Y. Resilient urban planning: Major principles and criteria. *Energy Procedia* **2014**, *61*, 1491–1495. [CrossRef]
3. Chen, Z.; Qiu, B. Resilient Planning Frame for Building Resilient Cities. In *Building Resilient Cities in China: The Nexus between Planning and Science*; Chen, X., Pan, Q., Eds.; Springer: Dordrecht, The Netherlands, 2015; Volume 113, pp. 33–41. [CrossRef]
4. Salas, J.; Yepes, V. MS-ReRO and D-ROSE methods: Assessing relational uncertainty and evaluating scenarios' risks and opportunities on multi-scale infrastructure systems. *J. Clean. Prod.* **2019**, *216*, 607–623. [CrossRef]
5. Schulz, A.; Zia, A.; Koliba, C. Adapting bridge infrastructure to climate change: institutionalizing resilience in intergovernmental transportation planning processes in the Northeastern USA. *Mitig. Adapt. Strateg. Glob. Chang.* **2017**, *22*, 175–198. [CrossRef]
6. Sharifi, A.; Yamagata, Y. Resilience-Oriented Urban Planning. *Lect. Notes Energy* **2018**, *65*, 3–27. [CrossRef]
7. Gonzales, P.; Ajami, N.K. An integrative regional resilience framework for the changing urban water paradigm. *Sustain. Cities Soc.* **2017**, *30*, 128–138. [CrossRef]
8. Leigh, N.G.; Lee, H. Sustainable and Resilient Urban Water Systems: The Role of Decentralization and Planning. *Sustainability* **2019**, *11*, 918. [CrossRef]
9. Rogers, C.D.F. Engineering future liveable, resilient, sustainable cities using foresight. *Proc. Inst. Civ. Eng. Civ. Eng.* **2018**, *171*, 3–9. [CrossRef]
10. Wagenaar, H.; Wilkinson, C. Enacting Resilience: A Performative Account of Governing for Urban Resilience. *Urban Stud.* **2015**, *52*, 1265–1284. [CrossRef]
11. Wei, Y.D.; Li, H.; Yue, W. Urban land expansion and regional inequality in transitional China. *Landsc. Urban Plan.* **2017**, *163*, 17–31. [CrossRef]
12. France-Mensah, J.; O'Brien, W.J. Developing a Sustainable Pavement Management Plan: Tradeoffs in Road Condition, User Costs, and Greenhouse Gas Emissions. *J. Manag. Eng.* **2019**, *35*. [CrossRef]
13. Mao, X.; Wang, J.; Yuan, C.; Yu, W.; Gan, J. A Dynamic Traffic Assignment Model for the Sustainability of Pavement Performance. *Sustainability* **2019**, *11*, 170. [CrossRef]

14. Torres-Machi, C.; Pellicer, E.; Yepes, V.; Chamorro, A. Towards a sustainable optimization of pavement maintenance programs under budgetary restrictions. *J. Clean. Prod.* **2017**, *148*, 90–102. [CrossRef]

15. Torres-Machi, C.; Osorio, A.; Godoy, P.; Chamorro, A.; Mourgues, C.; Videla, C. Sustainable Management Framework for Transportation Assets: Application to Urban Pavement Networks. *Ksce J. Civ. Eng.* **2018**, *22*, 4095–4106. [CrossRef]

16. Ouma, Y.O.; Opudo, J.; Nyambenya, S. Comparison of Fuzzy AHP and Fuzzy TOPSIS for Road Pavement Maintenance Prioritization: Methodological Exposition and Case Study. *Adv. Civ. Eng.* **2015**. [CrossRef]

17. Gomes, S.V.; Cardoso, J.L.; Azevedo, C.L. Portuguese mainland road network safety performance indicator. *Case Stud. Transp. Policy* **2018**, *6*, 416–422. [CrossRef]

18. Heinitz, F.M. Consistency of state road network master plan development steps. *Case Stud. Transp. Policy* **2018**, *6*, 400–415. [CrossRef]

19. Rezaei, A.; Tahsili, S. Urban Vulnerability Assessment Using AHP. *Adv. Civ. Eng.* **2018**. [CrossRef]

20. Masi, A.; Santarsiero, G.; Chiauzzi, L. Development of a seismic risk mitigation methodology for public buildings applied to the hospitals of Basilicata region (Southern Italy). *Soil Dyn. Earthq. Eng.* **2014**, *65*, 30–42. [CrossRef]

21. Beilin, R.; Wilkinson, C. Introduction: Governing for urban resilience. *Urban Stud.* **2015**, *52*, 1205–1217. [CrossRef]

22. Cedergren, A.; Johansson, J.; Svegrup, L.; Hassel, H. Local success, global failure: Challenges facing the recovery operations of critical infrastructure breakdowns. In Proceedings of the 25th European Safety and Reliability Conference, Zurich, Switzerland, 7–10 September 2015; pp. 4343–4348. Available online: https://www.scopus.com/inward/record.uri?eid=2-s2.0-84959017927&partnerID=40&md5=e6ef4cacb9260cde65a111ae0d36a90a. (accessed on 27 July 2019).

23. Cedergren, A.; Johansson, J.; Hassel, H. Challenges to critical infrastructure resilience in an institutionally fragmented setting. *Saf. Sci.* **2018**, *110*, 51–58. [CrossRef]

24. Regmi, B.R.; Star, C.; Leal Filho, W. Effectiveness of the Local Adaptation Plan of Action to support climate change adaptation in Nepal. *Mitig. Adapt. Strateg. Glob. Chang.* **2016**, *21*, 461–478. [CrossRef]

25. Frank, J.; Martínez-Vázquez, J. Decentralization and Infrastructure: From Gaps to Solutions. In *Decentralization and Infrastructure in the Global Economy: From Gaps to Solutions*; Routledge: London, UK, 2015. [CrossRef]

26. Charbit, C.; Gamper, C. Coordination of infrastructure investment across levels of government. In *Decentralization and Infrastructure in the Global Economy: From Gaps to Solutions*; Routledge: London, UK, 2015. [CrossRef]

27. Lehmann, P.; Brenck, M.; Gebhardt, O.; Schaller, S.; Süßbauer, E. Barriers and opportunities for urban adaptation planning: analytical framework and evidence from cities in Latin America and Germany. *Mitig. Adapt. Strateg. Glob. Chang.* **2015**, *20*, 75–97. [CrossRef]

28. Jain, M.; Korzhenevych, A. Spatial Disparities, Transport Infrastructure, and Decentralization Policy in the Delhi Region. *J. Urban Plan. Dev.* **2017**, *143*. [CrossRef]

29. Wilkinson, C.; Porter, L.; Colding, J. Metropolitan planning and resilience thinking: A practitioner's perspective. *Crit. Plan.* **2010**, *17*, 25–44.

30. Hurtado, S.D.G. Is EU urban policy transforming urban regeneration in Spain? Answers from an analysis of the Iniciativa Urbana (2007–2013). *Cities* **2017**, *60*, 402–414. [CrossRef]

31. Newman, J.P.; Dandy, G.C.; Maier, H.R. Multiobjective optimization of cluster-scale urban water systems investigating alternative water sources and level of decentralization. *Water Resour. Res.* **2014**, *50*, 7915–7938. [CrossRef]

32. Ganzle, S.; Stead, D.; Sielker, F.; Chilla, T. Macro-regional Strategies, Cohesion Policy and Regional Cooperation in the European Union: Towards a Research Agenda. *Political Stud. Rev.* **2019**, *17*, 161–174. [CrossRef]

33. Roozbahani, A.; Zahraie, B.; Tabesh, M. Integrated risk assessment of urban water supply systems from source to tap. *Stoch. Environ. Res. Risk Assess.* **2013**, *27*, 923–944. [CrossRef]

34. Gupta, J.; Bergsma, E.; Termeer, C.J.A.M.; Biesbroek, G.R.; van den Brink, M.; Jong, P.; Klostermann, J.E.M.; Meijerink, S.; Nooteboom, S. The adaptive capacity of institutions in the spatial planning, water, agriculture and nature sectors in the Netherlands. *Mitig. Adapt. Strateg. Glob. Chang.* **2016**, *21*, 883–903. [CrossRef]

35. Rigillo, M.; Cervelli, E. Mapping Urban Vulnerability: the Case Study of Gran Santo Domingo, Dominican Republic. *Adv. Eng. Forum* **2014**, *11*, 142–148. [CrossRef]

36. Salas, J.; Yepes, V. Urban vulnerability assessment: Advances from the strategic planning outlook. *J. Clean. Prod.* **2018**, *179*, 544–558. [CrossRef]
37. Salas, J.; Yepes, V. A discursive, many-objective approach for selecting more-evolved urban vulnerability assessment models. *J. Clean. Prod.* **2018**, *176*, 1231–1244. [CrossRef]
38. Zhao, P.; Chapman, R.; Randal, E.; Howden-Chapman, P. Understanding resilient urban futures: A systemic modelling approach. *Sustainability* **2013**, *5*, 3202–3223. [CrossRef]
39. Salas, J.; Yepes, V. VisualUVAM: A Decision Support System Addressing the Curse of Dimensionality for the Multi-Scale Assessment of Urban Vulnerability in Spain. *Sustainability* **2019**, *11*, 2191. [CrossRef]
40. Kukkonen, S.; Lampinen, J. Ranking-dominance and many-objective optimization. In Proceedings of the 2007 IEEE Congress on Evolutionary Computation, Singapore, 25–28 September 2007; pp. 3983–3990. [CrossRef]
41. Navarro, I.J.; Martí, J.V.; Yepes, V. Reliability-based maintenance optimization of corrosion preventive designs under a life cycle perspective. *Environ. Impact Assess. Rev.* **2019**, *74*, 23–34. [CrossRef]
42. Adger, W.N. Vulnerability. *Glob. Environ. Chang.* **2006**, *16*, 268–281. [CrossRef]
43. Santos, J.; Ferreira, A.; Flintsch, G. A multi-objective optimization-based pavement management decision-support system for enhancing pavement sustainability. *J. Clean. Prod.* **2017**, *164*, 1380–1393. [CrossRef]
44. Zhang, M.; Chen, W.; Cai, K.; Gao, X.; Zhang, X.; Liu, J.; Wang, Z.; Li, D. Analysis of the Spatial Distribution Characteristics of Urban Resilience and Its Influencing Factors: A Case Study of 56 Cities in China. *Int. J. Environ. Res. Public Health* **2019**, *16*, 4442. [CrossRef]
45. Baudrit, C.; Taillandier, F.; Tran, T.T.P.; Breysse, D. Uncertainty Processing and Risk Monitoring in Construction Projects Using Hierarchical Probabilistic Relational Models. *Comput. Aided Civ. Infrastruct. Eng.* **2019**, *34*, 97–115. [CrossRef]
46. Saaty, T.L. How to make a decision-the analytic hierarchy process. *Eur. J. Oper. Res.* **1990**, *48*, 9–26. [CrossRef]
47. Singh, R.P.; Nachtnebel, H.P. Analytical hierarchy process (AHP) application for reinforcement of hydropower strategy in Nepal. *Renew. Sustain. Energy Rev.* **2016**, *55*, 43–58. [CrossRef]
48. Convertino, M.; Muñoz-Carpena, R.; Chu-Agor, M.L.; Kiker, G.A.; Linkov, I. Untangling drivers of species distributions: Global sensitivity and uncertainty analyses of MaxEnt. *Environ. Model. Softw.* **2014**, *51*, 296–309. [CrossRef]
49. Groen, E.A.; Bokkers, E.A.M.; Heijungs, R.; de Boer, I.J.M. Methods for global sensitivity analysis in life cycle assessment. *Int. J. Life Cycle Assess.* **2017**, *22*, 1125–1137. [CrossRef]
50. Evelyne Groen, Global Sensitivity Analysis. Available online: https://evelynegroen.github.io/Code/globalsensitivity.html (accessed on 1 February 2020).
51. Convertino, M.; Valverde, L.J., Jr. Portfolio Decision Analysis Framework for Value-Focused Ecosystem Management. *PLoS ONE* **2013**, *8*, e65056. [CrossRef]
52. García-Segura, T.; Penadés-Plà, V.; Yepes, V. Sustainable bridge design by metamodel-assisted multi-objective optimization and decision-making under uncertainty. *J. Clean. Prod.* **2018**, *202*, 904–915. [CrossRef]
53. McGlashan, A.; Verrinder, G.; Verhagen, E. Working towards More Effective Implementation, Dissemination and Scale-Up of Lower-Limb Injury-Prevention Programs: Insights from Community Australian Football Coaches. *Int. J. Environ. Res. Public Health* **2018**, *15*, 351. [CrossRef]
54. Yepes, V.; Torres-Machi, C.; Chamorro, A.; Pellicer, E. Optimal pavement maintenance programs based on a hybrid Greedy Randomized Adaptive Search Procedure Algorithm. *J. Civ. Eng. Manag.* **2016**, *22*, 540–550. [CrossRef]
55. Sierra, L.A.; Yepes, V.; Pellicer, E. A review of multi-criteria assessment of the social sustainability of infrastructures. *J. Clean. Prod.* **2018**, *187*, 496–513. [CrossRef]
56. Sierra, l.A.; Pellicer, E.; Yepes, V. Social sustainability in the life cycle of Chilean public infrastructure. *J. Constr. Eng. Manag.* **2016**, *142*, 05015020. [CrossRef]
57. Sierra, L.A.; Yepes, V.; García-Segura, T.; Pellicer, E. Bayesian network method for decision-making about the social sustainability of infrastructure projects. *J. Clean. Prod.* **2018**, *176*, 521–534. [CrossRef]

MDPI
St. Alban-Anlage 66
4052 Basel
Switzerland
Tel. +41 61 683 77 34
Fax +41 61 302 89 18
www.mdpi.com

International Journal of Environmental Research and Public Health Editorial Office
E-mail: ijerph@mdpi.com
www.mdpi.com/journal/ijerph